DATE DUE

Demco, Inc. 38-293

Advances in Genetics, Volume 75

Serial Editors

Theodore Friedmann
University of California at San Diego, School of Medicine, USA

Jay C. Dunlap
Dartmouth Medical School, Hanover, NH, USA

Stephen F. Goodwin
University of Oxford, Oxford, UK

Volume 75

Aggression

Edited by

Robert Huber

JP Scott Center for Neuroscience
Mind & Behavior, Biological Sciences
Bowling Green State University, Bowling Green, OH, USA

Danika L. Bannasch

Department of Population Health and Reproduction
School of Veterinary Medicine, University of California
Davis, CA, USA

Patricia Brennan

Department of Psychology
Emory University, Atlanta, GA, USA

Editorial Assistant

Kate Frishman

Bowling Green, OH, USA

AMSTERDAM • BOSTON • HEIDELBERG • LONDON
NEW YORK • OXFORD • PARIS • SAN DIEGO
SAN FRANCISCO • SINGAPORE • SYDNEY • TOKYO
Academic Press is an imprint of Elsevier

Academic Press is an imprint of Elsevier

525 B Street, Suite 1900, San Diego, CA 92101-4495, USA
225 Wyman Street, Waltham, MA 02451, USA
32 Jamestown Road, London, NW1 7BY, UK
Radarweg 29, POBox 211, 1000 AE Amsterdam, The Netherlands

First edition 2011

ISBN: 978-0-12-380858-5
ISSN: 0065-2660

For information on all Academic Press publications
visit our website at elsevierdirect.com

Printed and bound in USA

11   12   13   10 9 8   7 6 5 4 3 2 1

Contents

Contributors

Numbers in parentheses indicate the pages on which the authors' contributions begin.

Laura A. Baker (171) University of Southern California, Los Angeles, California, USA

Patricia A. Brennan (1, 215) Department of Psychology, Emory University, Atlanta, Georgia, USA

Ivan D. Chase (51) Department of Sociology, Stony Brook University, Stony Brook, New York, USA

Emil F. Coccaro (151) Clinical Neuroscience Research Unit, Department of Psychiatry, The University of Chicago Pritzker School of Medicine, Chicago, Illinois, USA

Yu Gao (255) Department of Psychology, Brooklyn College, New York, USA

Andrea L. Glenn (255) Department of Child and Adolescent Psychiatry, Institute of Mental Health, Singapore, Singapore

Kyle L. Gobrogge (121) Department of Psychology and Program in Neuroscience, Florida State University, Tallahassee, Florida, USA

James L. Goodson (83) Department of Biology, Indiana University, Bloomington, Indiana, USA

Roger T. Hanlon (23) Marine Resources Center, Marine Biological Laboratory, Woods Hole, Massachusetts, USA

Robert Huber (1) JP Scott Center for Neuroscience, Mind & Behavior, Biological Sciences, Bowling Green State University, Bowling Green, Ohio, USA

Jamie L. LaPrairie (215) Department of Psychology, Emory University, Atlanta, Georgia, USA

Patrik Lindenfors (7) Department of Zoology, and Centre for the Study of Cultural Evolution, Stockholm University, Stockholm, Sweden

Donna L. Maney (83) Department of Psychology, Emory University, Atlanta, Georgia, USA

Benjamin R. Nordstrom (255) Department of Psychiatry, University of Pennsylvania, Philadelphia, USA

Melissa Peskin (255) Department of Psychology, University of Pennsylvania, Philadelphia, USA

Adrian Raine (255) Departments of Psychiatry, Psychology and Criminology, University of Pennsylvania, Philadelphia, USA

Brittany A. Robinson (215) Department of Psychology, Emory University, Atlanta, Georgia, USA

Anna S. Rudo-Hutt (255) Department of Psychology, University of Pennsylvania, Philadelphia, USA

Birgitta S. Tullberg (7) Department of Zoology, Stockholm University, Stockholm, Sweden

Julia C. Schechter (215) Department of Psychology, Emory University, Atlanta, Georgia, USA

Robert A. Schug (255) Department of Criminal Justice, California State University, Long Branch, USA

William A. Searcy (23) Department of Biology, University of Miami, Coral Gables, Florida, USA

Kristine Seitz (51) Department of Biology, Stony Brook University, Stony Brook, New York, USA

Catherine Tuvblad (171) University of Southern California, Los Angeles, California, USA

Moira J. van Staaden (23) Department of Biological Sciences and JP Scott Center for Neuroscience, Mind & Behavior, Bowling Green State University, Bowling Green, Ohio, USA

Zuoxin W. Wang (121) Department of Psychology and Program in Neuroscience, Florida State University, Tallahassee, Florida, USA

Yaling Yang (255) Laboratory of Neuro Imaging, University of California, Los Angeles, USA

Rachel Yanowitch (151) Clinical Neuroscience Research Unit, Department of Psychiatry, The University of Chicago Pritzker School of Medicine, Chicago, Illinois, USA

1

Aggression

Robert Huber* and Patricia A. Brennan†

*JP Scott Center for Neuroscience, Mind & Behavior, Biological Sciences, Bowling Green State University, Bowling Green, Ohio, USA
†Department of Psychology, Emory University, Atlanta, Georgia, USA

Aggression ranks among the most misunderstood concepts in all the behavioral sciences. It is commonly viewed by the general public as an aberrant form of behavior, with situations of conflict pictured as unfavorable and stressful circumstances, brought about by amoral urges, in critical need of our cognitive control, and with negative consequences for all involved. Such a view fundamentally misunderstands the biological significance of the behaviors that occur during conflict. Deeply rooted in the demands of the natural world, an individual must fulfill its demands for self-preservation, defend its interests, or compete for limited vital resources. Basic tendencies for aggression are virtually ubiquitous throughout the animal kingdom, regardless of its bearer's neural or cognitive faculties, phylogenetic origins, or sociobiological circumstances. Just as widespread, however, are fundamental rules that govern physical conflict, such that cases of unbridled hostility are surprisingly rare. In most species, visual and elaborately ritualized displays effectively channel aggression, structure how individuals interact, and govern the conflict's resolution.

As we witness animals engaged in situations of conflict, we cannot help but be drawn in by the behavior's inherent relevance to our own biological roots. The knowledge that human aggression arises from our genetic heritage makes it all the more likely that it is of an adaptive nature. As we study the individuals and environments where aggression is most commonly displayed, we gain a better understanding of when aggression and violence may serve an adaptive function and when it may not. Current research points to the importance of delineating subtypes of aggression, focusing on such concepts as proactive and reactive, direct and indirect, and adolescent limited versus life course persistent. Each of these types of aggression has a distinct etiology and

Advances in Genetics, Vol. 75
0065-2660/11 $35.00
DOI: 10.1016/B978-0-12-380858-5.00016-2

utility, depending upon the social environment in which the individual must function. For example, reactive aggression is aggression that occurs in response to a threat in the external environment. It is easy to see how this type of aggression may have its basis in our inherent survival instincts. Nevertheless, a propensity for reactive aggression may be truly dysfunctional in environments that pose low levels of threat, like many of those in which children in Western societies are now raised. A more comprehensive understanding of how biology, behavior, and environment intersect is paramount in the study of human violence and aggression.

Studies of aggression, its motives, and its causes are of central interest to a wide range of academic disciplines—behavioral genetics, evolution, neuroscience, psychology, sociology, and criminology, to name just a few. Despite the wealth of empirical and theoretical attention, it is remarkable that a comprehensive synthesis of aggression has stubbornly remained elusive. This partly stems from the fact that the term "aggression" neither maps cleanly onto a monolithic behavioral phenomenon nor lends itself to representation by a simple explanatory concept. Another explanation for this paradox must reside in the absence of a unified, operational definition of aggression across disciplines, or even of a general agreement on what the term actually includes. For instance, most psychologists define aggression as "all behavior that is intended to cause bodily harm." Other widely adopted classifications of aggression recognize subtypes, ranging from competition between males, a mother's efforts to protect her offspring, or fighting as a learned response to cope with a particular situation. Biologists regard a definition focused solely on injury as insufficient as this excludes a wide range of threat behaviors directed at rivals, for example, birds that challenge their adversaries with song, an impala's exaggerated strutting as a signal of strength, or a wolf's territorial claims via scent markings. Moreover, there is little agreement on whether a predator's hunting behavior should be included. A lion chasing and killing a gazelle undoubtedly inflicts injury, but is this more akin to a cow cropping the top off a clump of grass, or to an elephant bull inflicting serious injury to a rival in battle? Moreover, behavior in aggressive encounters always balances contrasting impulses for approach and attack with a tendency to flee—rarely is either present entirely alone. To acknowledge the difficulty of disentangling these components, the term "agonistic behavior" has been introduced. The term specifically addresses the balance of forces for both attacking and fleeing and it accommodates all instances of attack and threat (i.e., offensive agonistic behavior) as well as escape and submission (i.e., defensive agonistic behavior).

As with any other characteristic, natural selection is assumed to enhance aggression's overall effectiveness. High ranking individuals are likely to display a favorable combination of strength, along with an ability to titer their levels of aggression, to pick fights that are winnable, and to only compete in those that are worth it. Hyperaggressive phenotypes exist in most systems.

These exhibit fighting that greatly exceeds the most effective norm, they readily launch the initial attack even in situations where they ought not to, are overly eager to escalate or retaliate, show a willingness to follow an excessively physical trajectory even when an opponent has already withdrawn, or fail to back down in situations where there is little prospect of winning. Such behaviors rarely make for an effective strategy, as they coincide with greater risk of injury or death, or in the best-case scenario, attaining a low rank.

A thorough analysis of aggression minimally demands that we (1) capture the essence of an inherently multifaceted phenotype, (2) address underlying elements of motivation that are not always readily observed or elicited, (3) understand the various scenarios and contexts that influence its expression, (4) decipher the neural, hormonal, and genetic causes that are at work, and (5) explore how its components are shaped by evolution. The chapters in this volume aim to provide a comprehensive overview of these topics as their authors unravel the individual behavioral and neural strands constituting situations of conflict.

Initial chapters of the volume characterize the elemental building blocks of aggression; they assess, precisely delineate, and account for aggression's different and unique components, and explain how intricate behavioral constructs often emerge from much simpler roots. The initial chapters review aggression from a predominantly evolutionary perspective. Conflicts are energetically costly and carry inherent risks. Natural selection offers a powerful conceptual tool as it focuses on an individual's behavioral strategies and decision making in ways that maximize its fitness. Evolution can only exert its influence on characters that depend, at least partially, on genetic underpinnings. Lindenfors and Tullberg (Chapter 2) discuss the significance of sexual selection as a key evolutionary structuring force in aggression. In most scenarios, ritualized displays take the place of unchecked, aggressive interactions. Game theory offers a powerful framework for why animals only tend to fight with great ferocity when a resource of exceptional value is at stake. Resources are rarely worth the risk of sustaining injury, and competing individuals will do best by resolving conflicts with ritualized displays only. Skill in assessing the relative strength of an opponent is key for navigating the demands, risks, and opportunities of social living. The review by van Staaden and colleagues (Chapter 3) discusses a prominent role for signaling aggressive behaviors, which permit individuals to obtain valid estimates of an opponent's true strength. Once an animal is bested by an opponent, it is always better to adopt submissive behavior and accept subordinate status, rather than risk something far worse. A wide range of attributes decides between victory and defeat. With prominent asymmetries in the size of weapons, strength, or agility, fights are often quickly resolved. In many instances, though, social success will depend also on an ability to form successful alliances, to harness cognitive skills, or to inherit status from high-ranking kin. A paired dominance relationship is established when prior encounters produce a lasting polarity in the outcome of future bouts. In its most common form, the past loser will be less likely to initiate further bouts against the winner or will retreat quickly if confronted. As individuals repeatedly meet and interact with others, higher order social organization emerges through a series of sequential dyadic interactions. Individuals of many species, including humans, tend to arrange themselves in largely linear social hierarchies. Although individual characteristics such as size, strength, or agility are relatively fixed and may indeed influence rank, Chase and Seitz (Chapter 4) illustrate that these qualities are more often overshadowed by contextual factors and chance events.

The search for proximate mechanisms underlying aggression requires us to view aggression's natural building blocks, to recognize the various factors that control them, and to effectively label their behavioral expression in the form of consistent and reliable phenotypes. Our understanding of the biological basis of aggression in all vertebrates, including humans, has been built largely upon discoveries first made in birds. An extensive literature indicates that hormonal mechanisms are shared between humans and many avian species. This recent development of hormonal, neuroendocrine, and genetic tools has established songbirds as powerful models for understanding the neural basis and evolution of vertebrate aggression. Maney and Goodson (Chapter 5) discuss the contributions of field endocrinology toward a theoretical framework linking aggression with sex steroids, explore evidence that the neural substrates of aggression are conserved across vertebrate species, and describe a promising new songbird model for studying the molecular genetic mechanisms underlying aggression. Voles have recently emerged as a key model for a genetic dissection of social behavior and its underlying neural mechanisms. Gobrogge and Wang (Chapter 6) discuss its utility for the study of aggression and review recent findings that illustrate the neurochemical mechanisms underlying pair bonding-induced aggression. Endogenous brain chemicals play a key role in the control of aggression in many taxa including humans. Neurotransmitters effectively pattern and modulate the expression of basic behavioral components. Genetic abnormalities in a number of neurotransmitter pathways have been implicated in aggression-related disorders. Yanowitch and Coccaro (Chapter 7) review work that demonstrates that neurotransmitter function is intricately linked to aggressive state.

Subsequent chapters focus on human aggression, with an emphasis on genetic and other biological factors. Human ingenuity for inflicting intentional harm is without equal, although warring tendencies may already be rooted in a deep, prehuman past. Instances of violence have been documented for a range of nonhuman apes and may have arguably wired into our genes when our more aggressive ancestors won against our less aggressive ancestors in terms of survival and reproduction. Aside from an unprecedented potential for carnage and destruction, humans are at the same time also capable of the most remarkable instances of compassion, understanding, and peaceful negotiation. The direction depends on each individual's ethical codes and moral norms driven by the societal expectations, good parenting, or social contexts. A clear vision has emerged where "natural" tendencies for aggression appear to be ubiquitous, but so too are a plethora of sophisticated mechanisms that keep conflicts in check, channel aggression, negotiate fighting signals, resolve conflicts, and ultimately govern social group structure.

Tuvblad and Baker (Chapter 8) demonstrate that genetic influences serve as powerful predictors of human aggression and violence. The relative influence of genetics depends upon developmental age, type of aggression, and the environmental context faced by the individual. LaPrairie and colleagues (Chapter 9) review research linking perinatal factors and aggression and conclude in a similar fashion that perinatal and neurodevelopmental factors influence the expression of aggressive behaviors. Nordstrom and colleagues (Chapter 10) further detail the importance of recognizing the role of brain functioning deficits in the risk for aggressive outcomes. Recent advances in imaging technology have enabled a far greater understanding of these influences on human behavior and the risks of criminal outcomes. Importantly, the chapters on human aggression also emphasize the fact that biology is not destiny and that, in the case of human aggression and violence, there is much that can and should be done in terms of early intervention and prevention.

The list of significant challenges in aggression research remains daunting and the need to harness the full power of interdisciplinary approaches now appears more urgent than ever. Aside from our need to reconcile simple questions over terminology, a number of more serious impediments remain to be acknowledged. Concepts seem so intimately connected that we are tempted to view them as essentially overlapping, or to even use them synonymously (e.g., measures of an inherent tendency to fight, effectiveness in a contest, or the ability to socially dominate others). A common fallacy views these simply as separate perspectives onto the same, unitary phenomenon of aggression. For a synthesis to emerge we must accept aggression's multidimensional nature and recognize that the term "aggression" simply serves as an overarching label for an entangled complex of multiple, distinct components, causes, and functions. This volume comes at a critical juncture for defining a broader view of aggression and with it we hope to help define the structural elements that comprise the behavior in its full complexity.

The time is now right to bridge theoretical frameworks, combine experimental approaches, and relate significant findings across the many individual disciplines that are instrumental in the analysis of aggression. Center initiatives can serve as intellectual hubs for the comprehensive study of social conflict, violence, and related phenomena. Bringing together individual researchers from a broad range of disciplines to foster integrative and overarching themes, such centers provide a forum for a rich exchange of ideas, the development of human resources, a clearinghouse for notable discoveries, and to publicize their societal relevance through public outreach. The development of viewpoints spanning formerly separate disciplines, such as the one aimed for in this book, is cause for optimism that the future is not quite as far off as we had feared.

2

Evolutionary Aspects of Aggression: The Importance of Sexual Selection

Patrik Lindenfors*,† and Birgitta S. Tullberg*

*Department of Zoology, Stockholm University, Stockholm, Sweden
†Centre for the Study of Cultural Evolution, Stockholm University, Stockholm, Sweden

ABSTRACT

Aggressive behaviors in animals, for example, threat, attack, and defense, are commonly related to competition over resources, competition over mating opportunities, or fights for survival. In this chapter, we focus on aggressive competition over mating opportunities, since this competition explains much of the distribution of weaponry and large body size, but also because this type of competition sheds light on the sex skew in the use of violence in mammals, including humans. Darwin (1871) termed this type of natural selection, where differences in reproductive success are caused by competition over mates, sexual selection. Not all species have a pronounced competition over mates, however. Instead, this aspect of sociality is ultimately determined by ecological factors.

Advances in Genetics, Vol. 75
0065-2660/11 $35.00
DOI: 10.1016/B978-0-12-380858-5.00009-5

In species where competition over mates is rampant, this has evolutionary effects on weaponry and body size such that males commonly bear more vicious weapons and are larger than females. A review of sexual selection in mammals reveals how common aggressive competition over mating opportunities is in this group. Nearly half of all mammal species exhibit male-biased sexual size dimorphism, a pattern that is clearly linked to sexual selection. Sexual selection is also common in primates, where it has left clear historical imprints in body mass differences, in weaponry differences (canines), and also in brain structure differences. However, when comparing humans to our closest living primate relatives, it is clear that the degree of male sexual competition has decreased in the hominid lineage. Nevertheless, our species displays dimorphism, polygyny, and sex-specific use of violence typical of a sexually selected mammal. Understanding the biological background of aggressive behaviors is fundamental to understanding human aggression.   © 2011, Elsevier Inc.

I. INTRODUCTION

Why does aggression exist in nature? Darwin (1859, 1871) pointed out that the ultimate explanation for *any* trait has to be found in the effect that it has on survival and reproduction. From an evolutionary standpoint, individuals should thus mainly be expected to fight over resources, for survival and for mating opportunities, because these are what mainly affect how many genes that individual contributes to the gene pool of the coming generation. Another prediction is that the amount of aggression displayed in the encounters should increase with increasing value of the fought-over resource. Aggressive behaviors are associated with costs, and individuals are simply expected to take higher risks, that is, pay potentially higher costs, with increasing potential gains. In this chapter, we focus on aggression over mating opportunities—sexual selection—in mammals in general and in primates in particular. We focus on sexual selection because evidence suggests that it is the primary reason why animals fight with conspecifics and because it is the most likely explanation of some aspects of human aggression, such as why males tend to be more aggressive than females.

An important point about evolutionary explanations is the philosophical distinction between proximate and ultimate explanations. Take sex, for example. Do humans have sex because it feels good or in order to have children? Most sexual intercourse in current society probably has very little to do with actual procreation; on the contrary, there are many birth control methods available to make it possible to have sex without this resulting in a pregnancy. Despite the fact that protected sex happens "because" it feels good, the evolutionary explanation of sexual intercourse is "because" of procreation. This is where the crucial distinction between proximate and ultimate explanations comes into play. A proximate explanation is the explanation that is closest to the event that is to be explained. The higher, ultimate explanation is instead the deeper reason for why something happened.

In biology, the division in ultimate and proximate explanations has been extended to what is usually termed "Tinbergen's four questions" (Tinbergen, 1963); the four potential explanations of any behavior: (1) survival value or adaptive function, (2) phylogenetic history, (3) individual development, and (4) causal mechanisms such as hormonal mediation of behavior. The first two of Tinbergen's questions are ultimate whereas the latter two are proximate. To fully shed light on a biological phenomenon all four types of questions are needed and the answers complement each other. However, in this chapter, we focus entirely on ultimate, evolutionary answers to the question of why aggression exists and takes the form it does.

II. SEXUAL SELECTION

Natural selection is all about who gets to reproduce and who does not (Darwin, 1859). A central aspect of getting to reproduce is to survive until the opportunity to reproduce arises and to gain access to resources enabling you to do so, but another important aspect concerns direct competition in connection with the reproductive act itself. Darwin termed this second aspect "sexual selection": differences in reproductive success caused by competition over mates (Darwin, 1871).

Why did Darwin give a specific name to one part of natural selection: why not just stick to the umbrella term "natural selection"? Darwin had noted that there often seems to be a conflict of interest between traits that increase survival and traits that increase reproduction; many traits that give an advantage in reproduction have negative consequences for survival. A male peacock's large tail feathers are a prime example of such a trait. How can such a long colorful tail evolve when it makes the bearer simultaneously more visible and less adept at escaping predators? When thinking about this problem before having formulated the theory of sexual selection, Darwin wrote in a letter to his friend, the botanist Asa Gray, the famous line: "The sight of a feather in a peacock's tail, whenever I gaze at it, makes me sick!" (Darwin, 1860).

To clarify this second aspect of natural selection—selection that has to do with competition over mates—Darwin wrote a follow-up to "On the Origin of Species" (1859), "The Descent of Man and Selection in Relation to Sex" (1871). In this book, Darwin points out that there are two potential kinds of competition over mates, two forms of sexual selection. Either individuals of one sex (usually males) can fight with each other over mating opportunities (*intra*sexual selection) or, alternatively, individuals of one sex (usually females) can choose individuals of the other sex on the basis of some trait (*inter*sexual selection).

This second form of sexual selection is the explanation of the peacock's tail: peahens simply find it attractive and prefer to mate with the peacocks with the most elaborate tail. Further, the tail provides information on the genetic quality of the male—it is an honest signal (Petrie, 1994).

It is noteworthy that it was the idea of intersexual selection that caused the most furious debate in Darwin's time, foremost because it was judged utterly questionable that female aesthetical judgment could be the ultimate explanation for so many conspicuous characters in nature. However, partner choice is a more peaceful process than direct competition within a sex. Thus, because it induces so much aggression in nature, we focus in this chapter mainly on *intra*sexual selection; physical competition over mating opportunities. It should be pointed out that the two forms of sexual selection sometimes occur simultaneously, for instance, in lekking species where females choose as mating partners the winning males from physical competition (Andersson, 1994).

Sexual selection arises when one sex limits the reproductive success of the other. Most often it is females who are the limiting resource for the reproductive success of males due to a fundamental asymmetry between males and females in their defining characteristic, their gametes. Males are designated by their smaller, mobile gametes, called sperm cells. Females are designated by their larger, nutrition-carrying gametes, called eggs. Males can make more gametes than females, simply because sperm are energetically cheaper to make than eggs; thus, there is a fundamental reproductive difference between males and females. This initial asymmetry has consequences. Making sperm is cheap and easy, so this is not what limits the reproductive possibilities of males. Making eggs, on the other hand, is much costlier. Thus, sexual selection commonly—but not exclusively—affects males, because given an equal sex ratio, male reproductive success is limited by access to matings with females. Conversely, female reproductive success is limited by the number of eggs she can produce (Andersson, 1994). This sex specificity is so common that the reverse pattern, termed sex role reversal, is subject to intense interest from evolutionary biologists when it occurs (e.g., Ralls, 1976; Vincent *et al.*, 1992).

An important experimental verification of this theoretical insight was made by the geneticist Bateman (1948), who experimented on fruit flies. Bateman noted a pattern demonstrating that the number of offspring a male fruit fly can have is directly correlated with his number of matings. The same does not hold true for females, who have roughly the same number of offspring no matter how many times they mate (as long as it is at least once). This pattern is termed Bateman's principle. Later studies, however, have documented that a number of exceptions to Bateman's principle exist in nature (Birkhead, 2001). Individuals are not only sperm and eggs; there are a number of additional factors that need to be incorporated to understand what is going on in different species. In mammals, especially, one needs to incorporate two unique adaptations. While the energy investments in the mammal zygotes differ only marginally in relation to the body mass of most mammals, the cost to mammal females greatly exceeds that to males due to effects of pregnancy and lactation. This energy investment inequality has existed since the origin of the class Mammalia, 125 million years ago.

There are some exceptions to the general mammalian pattern, however. For instance, some mammal babies are so expensive to bring up to maturation that both sexes have to partake in the upbringing for it to be possible. In these species, where males and females work together to guard and rear the young, intrasexual competition occurs just prior to pair formation. In these species, the two sexes are usually morphologically alike. In other mammal species, however, competition between males over mating opportunities is fierce. In some species this affects the entire social life of the species, in that males physically exclude other males from the group. The result is a social system akin to a harem structure, with immature males roaming outside the social gathering or forming bachelor groups.

The importance of sexual selection in understanding aggression in mammals is most clearly illustrated by the presence and absence of weaponry. For example, male ungulates are commonly equipped with horns while females are not. Horns would be a good weapon to fend off predators, especially when you need to defend your young, or to fight off conspecific competitors. But most young are cared for by single mothers; the fathers—who have the weapons—are absent. Ungulate horns are commonly ready just in time for rutting season and are then shed (e.g., deer). Instead of predator defense, male ungulates mainly use their horns to fight each other (Caro et al., 2003; Stankowich and Caro, 2009). A similar case can be made for the large, sharp canines of primates (Thorén et al., 2006), and large body size in male mammals in general (Lindenfors et al., 2007a). Such sex-skewed distribution of size and weaponry, in combination with observations of fierce aggression, is what enables us to assert that most serious conflict and aggression in mammals is over mating opportunities.

Sexual selection acting primarily on one sex may have indirect but pronounced consequences for the relationship between the sexes. Thus, direct conflicts between males sometimes result in conflicts of interest between males and females. Early thoughts on this issue (Parker, 1979, Trivers, 1972; Williams, 1966) have received much empirical support (Arnqvist and Rowe, 2005), and it now almost seems the norm rather than an exception that there exists such a conflict and that this becomes more severe under strong intrasexual competition. This can lead to the interesting phenomena of sexually antagonistic coevolution where males and females become involved in an arms race, as traits in one sex entice the evolution of resistance in the other (Holland and Rice, 1998; Gavrilets et al., 2001; and others). On the other hand, intrasexual competition can lead to one sex dominating the other. With regard to aggression and physical prowess, the common situation in mammals, including primates (Hrdy, 1981), is that males are physically dominant over females.

III. MATING SYSTEMS

Not all animals have clear sexual differences (Fairbairn *et al.*, 2007). In birds, for example, many species of gulls and penguins are so alike that it is impossible to determine the sex except by close inspection of the genitals. At the other extreme are mallards, where the sexes are so different that Linnaeus classified them as two different species (Andersson, 1994). In mammals, we find the same variation even within a given mammalian order. Thus, within Pinnipedia we have, on the one hand, elephant seals where males may weigh up to five times as much as females and on the other hand, species such as Baikal seals where females are of similar weight as males (Lindenfors *et al.*, 2002). These differences in dimorphism are due to differences in the degree of sexual selection. But why are there differences in the degree of sexual selection between species to start with?

Fundamentally, this question is about factors affecting male and female social group size. These issues are commonly addressed by focusing on the ecological variables that determine the spatiotemporal distribution of females, based on the expectation that resources and predation account for variation in female reproductive success. By comparison, access to females is generally assumed to be the major factor influencing male reproductive success (Emlen and Oring, 1977; Trivers, 1972; Wilson, 1975). After risks and resources have determined the spatiotemporal distribution of females, the distribution of females is in turn expected to influence the degree of male intrasexual competition (Emlen and Oring, 1977). For instance, a group of concurrently fertile females opens the field for male competition and monopolization. The general framework is therefore that social evolution is driven by females.

The theoretical expectation that social evolution is ultimately driven by female distribution is empirically supported by comparative studies on primates, a group for which there is a significant correlation between evolution of male and female sociality (e.g., Altmann, 1990; Mitani *et al.*, 1996; Nunn, 1999). Further, a phylogenetic investigation has shown that the evolution of female group size precedes the evolution of male group size, that is, that evolutionary changes in male group size lag changes in female group size (Lindenfors *et al.*, 2004).

Ecological factors determine whether it is possible for a male to monopolize several mating opportunities. For example, elephant seal females give birth on beaches. With a limited number of suitable beaches available in the elephant seal range, females tend to crowd together when giving birth. Elephant seals mate soon after they have given birth, so at the time of mating females are gathered tightly on limited stretches of beach. Males can exclude other males from a stretch of beach and thereby secure matings with a large number of females. Successful males in this competition gain all matings, while the losers get none. Fighting among elephant seals over mating opportunities is thus a fierce and bloody affair, a scenario which has resulted in extreme size dimorphism. In other pinniped species, females give birth in isolated caves on the polar ice pack; thus, no opportunity exists to monopolize matings. Without evolutionary pressure for male fighting ability, the sexes are more equal in size (Lindenfors et al., 2002).

In conclusion, the ultimate cause for differences in mating systems can be traced back to ecological circumstances. The differences in mating systems in turn trigger differences in aggressive competition for mating opportunities which is what drives the evolution of sex differences in size and weaponry. These morphological sex differences are clear indicators of the severity of male–male aggression.

IV. WHEN TO FIGHT AND WHEN TO FLEE

Given that fighting is most often about mating opportunities, how are they predicted to pan out in terms of ferocity, number of behaviors involved, length of time, and so on? There are two things that determine the ferocity of fighting: the value of the object being fought over and the risks involved. The problem can be reduced to a cost-benefit analysis. A male can not give up at first instance to maximize his chances of survival, because that would result in total nonreproduction. Neither can he go "all-in" in just any aggressive encounter if there exists only a minute chance of success. Instead, males in competitive situations have to weigh the probabilities of success, injury, and survival against each other, while considering other factors such as energy expenditures and probabilities of success in future interactions with other competitors. It is important to note that animals make calculations and decisions about how to act, but such processes do not necessarily require the consciousness about the process usually ascribed to human decision-making. Rather, animals are believed to use cues with regard to the environment, as well as their own and the opponent's current status, and to use this information in an unconscious way when making decisions.

One consequence this accounting has had over evolutionary history is that competitive interactions often take the form of a "sequential assessment game". This prescribes that each competitor should attempt to assess his opponent's strength using as little energy as possible. Escalation should only be initiated by the competitor that feels he has the upper hand, or by either opponent if they cannot determine who is superior (Enquist and Leimar, 1983). Thus, a meeting between two deer males often starts out with a stage of roaring, which acts as a forcedly honest signal of body size. If this does not settle who is the larger/stronger, it is followed by "parallel walking," where each competitor tries to judge the size and strength of the other by walking back and forth in parallel. Only if it is still unclear who is the larger or stronger will the

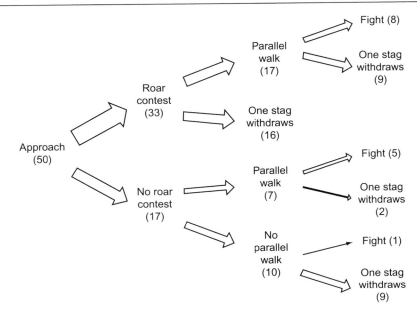

Figure 2.1. Sequential assessment in red deer (from Clutton-Brock and Albon, 1979).

competition escalate to actual fighting (Clutton-Brock and Albon, 1979; Fig. 2.1). Thanks to this "game," really fierce fights only happen between opponents of equal size—inferior competitors flee quickly to fight another day (and another opponent).

There is a common interest among fighters in trying to expend as little energy as possible while simultaneously minimizing the possibility of injury. For example, wolves and other canids ritualistically greet each other several times each day and display dominance or submission on a regular basis—they do not determine their relationship at every encounter. Some cichlid fish have a system akin to that of red deer, with different stages of escalation (Brick, 1999). Male lions fight savagely only if they stand a good chance of winning a pride of females. Research shows that they determine the quality of their rivals on cues from each other's manes (West and Packer, 2002). Lekking birds such as black grouse have distinctive courtship rituals where they make calls and visual displays, an odd mix of strength comparison and showing off, where females can pick winners according to some criterion, sometimes just by copying other females' choices (Andersson, 1994; Dugatkin and Godin, 1993; Wade and Pruett-Jones, 1990). Seldom do fights turn into vicious fighting, and when they do it is usually because either the contestants are judged by each other to be of equal strength, or because the benefit of winning—the value of the contested item—is much larger than the cost of losing. If the choice is reproduction or death, fights become deadly. This is why fights among elephant seals are so fierce and bloody. The chance to mate occurs only once per year and most males never even get close. For the successful males it is another story—in a study of Southern elephant seals, harem holders accounted for 89.6% of the recorded paternities (Fabiani *et al.*, 2004; Fig. 2.2).

The sequential assessment game is a variant of a game theoretical setup termed the "hawk-dove game" (see also Chapter 3). In this game, there are two possible strategies: always fight ("hawk") and always yield ("dove"), where it is assumed that the two competitors have equal fighting ability. An Evolutionarily

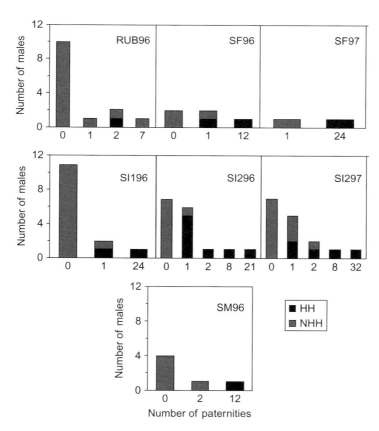

Figure 2.2. Number of paternities achieved by the harem holder (HH) and the other males (NHM) associated with each harem in seven different populations of Southern elephant seals (from Fabiani *et al.*, 2004).

Stable Strategy is a strategy which, if adopted by a population of players, cannot be invaded by any alternative strategy. It has been shown that the ESS is a mix of hawks and doves with proportions determined by the cost of fighting in relation to the benefit of winning (Maynard Smith, 1982). This game provides theoretical information that in a population of nonfighters it is profitable to be a fighter, and vice versa. If one extends the game to include a strategy called "assessor" that determines whether it will act as a "hawk" or a "dove" based on some criterion—for example, depending on priority at the resource—one can arrive at the sequential assessment game. The assessment strategy is also an ESS (Maynard Smith, 1982).

The prediction from these game theoretical models is that populations where individuals compete over resources or matings should consist of individuals utilizing different strategies depending on situation, where important factors are the current size and physical state of self and opponents and the value of resources (for instance, the number of females in the group being fought over). In this context, it should be noted that some animal populations have evolved alternatives to fighting strategies, usually known as sneaker strategies. Such males are usually much smaller than fighting males and can covertly sneak matings from females while the fighters are occupied with physical combat (Gross, 1996).

V. CASE STUDIES: SEXUAL DIMORPHISM

As mentioned above, animal groups differ in both the way and the degree to which they are exposed to sexual selection, and this will have great effects on the evolution of sex differences (Fairbairn et al., 2007). Although mammals as a group are characterized by a high degree of intrasexual selection (as compared with, for instance, birds, where intersexual selection seems to be more common), there is variation in the strength of sexual selection both within and among mammalian orders. An example of this variation is the pinnipeds (seals, sea lions, and walruses) where there exists a clear relationship between harem size and sexual size dimorphism (Lindenfors et al., 2002; Fig. 2.3).

In this section, we review some studies that have compared different mammalian groups with respect to the consequences of sexual selection on behavioral and morphological evolution. There are 4629 extant or recently extinct mammalian species, as listed by Wilson and Reeder (1993). In a survey of 1370 of these, Lindenfors et al. (2007a) showed that sexual selection is a prevalent selective force in mammals. With a cutoff point at a 10% size difference in either direction to "count" as sexual dimorphism, mammals were, on average, male-biased size dimorphic (average male/female mass ratio = 1.184; paired t-test $p \ll 0.001$; Table 2.1) with males being larger than females in 45% of

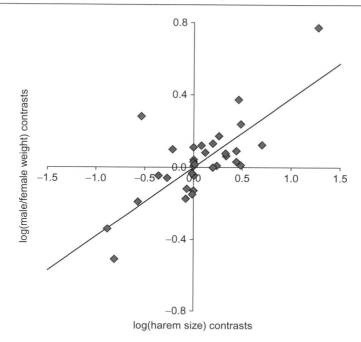

Figure 2.3. Regression line through the origin on harem size and body weight dimorphism. The data points are from a phylogenetic independent contrasts analysis. There is a significant relationship between harem size and sexual size dimorphism ($b=0.376$, $p=0.000$, $R^2=0.577$, $n=36$) (from Lindenfors et al., 2002).

extant species (Table 2.1). Systematists recognize 26 monophyletic mammalian orders (Wilson and Reeder (1993)). When investigating each order separately, the majority of orders also turned out to be significantly male-biased dimorphic (average male/female mass ratio >1.0 and $p<0.05$). Some orders exhibited no significant size dimorphism, and only one (Lagomorpha; hares, rabbits, and pikas) was significantly female-biased dimorphic on average (average male/female mass ratio <1.0, $p<0.05$; Table 2.1).

The mating system in a species describes the number of females a male can monopolize and thus affects the potential strength of sexual selection males are exposed to in that species. Using a phylogenetic tree for all mammals (Bininda-Emonds et al., 2007) together with information about mating system and sexual size dimorphism, Lindenfors et al. (2007a) tested whether sexual selection could explain the variation in sexual size dimorphism among mammals. Mating system was used as a three-state unordered categorical variable, an independent variable used to test for differences in dimorphism between

Table 2.1. Summary of the Patterns of Dimorphism Found in Mammals

Order	Number of recognized species	Number of species with body mass data	Average dimorphism	Sexual size dimorphism
	Mammalia			
All mammals	4629	1370	1.184	$p \ll 0.001$
	Subclass Prototheria			
Monotremata (Monotremes)	3	2	1.273	–
	Subclass Metatheria			
Didelphimorphia (American marsupials)	63	13	1.323	$p = 0.002$
Paucituberculata (Shrew oppossums)	5	2	1.840	–
Microbiotheria (Monito del monte)	1	1	1.044	–
Dasyuromorphia (Dasyuroids)	63	24	1.465	$p \ll 0.001$
Peramelemorphia (Bandicoots and bilbies)	21	9	1.496	$p = 0.015$
Notoryctemorphia (Marsupial moles)	2	0	–	–
Diprotodontia (Kangaroos, etc.)	117	63	1.306	$p \ll 0.001$
	Subclass Eutheria			
Insectivora (Insectivores)	428	59	1.048	$p = 0.081$
Macroscelidea (Elephant shrews)	15	5	0.964	$p = 0.142$
Scandentia (Tree shrews)	19	1	–	–
Dermoptera (Colugos)	2	0	–	–
Chiroptera (Bats)	925	354	0.999	$p = 0.091$
Primates (Primates)	233	198	1.247	$p \ll 0.001$
Xenarthra (Sloths, armadillos, and anteaters)	29	4	0.914	$p = 0.216$
Pholidota (Pangolins)	7	3	1.767	$p = 0.001$
Lagomorpha (Rabbits and pikas)	80	21	0.930	$p = 0.012$
Rodentia (Rodents)	2015	295	1.092	$p \ll 0.001$
Cetacea (Whales, dolphins, and porpoises)	78	10	1.414	$p = 0.082$
Carnivora (Carnivores)	271	180	1.476	$p \ll 0.001$
Tubulidentata (Aardwark)	1	0	–	–
Proboscidea (Elephants)	2	2	1.900	–
Hyracoidea (Hyraxes)	6	1	1.111	–
Sirenia (Dugongs and manatees)	5	0	–	–
Perissodactyla (Horses, rhinos, and tapirs)	18	8	1.164	$p = 0.156$
Artiodactyla (Antelopes, camels, pigs, etc.)	220	115	1.340	$p \ll 0.001$

Dimorphism is given as male mass/female mass. Mammals and the majority of mammalian orders are on average male-biased dimorphic (average dimorphism > 1.0 and $p < 0.05$), even if there exist a few orders with no significant dimorphism ($p > 0.05$) or female-biased dimorphism (Lagomorpha: average dimorphism < 1.0 and $p < 0.05$). p-Values represent the significance of paired t-tests where male body mass was paired with female body mass. Dashes indicate orders with too few data points for statistical analysis ($n < 3$ for tests of the presence of dimorphism, from Lindenfors et al., 2007a).

"more" and "less" sexually selected (polygynous) sister taxa. These tests revealed that a higher degree of sexual selection was associated with a higher degree of male-biased dimorphism. More polygynous taxa not only had larger males but also larger females than their less polygynous sister taxa. These results indicate that sexual selection is a significant explanatory factor of both sexual dimorphism as such, and of the general size increase in many mammalian lineages.

In primates, the mammal order humans belong to, the pattern is similar. Again using mating system as a three-state unordered categorical variable, testing for differences in dimorphism between "more" and "less" sexually selected sister taxa, a higher degree of sexual selection was associated with a higher degree of male-biased dimorphism. Again, more polygynous taxa also had larger males and females than their less polygynous sister taxa (Lindenfors, 2002; Lindenfors and Tullberg, 1998). Here, however, a novel method investigating temporal order of events revealed not only a correlation but also a causal link between sexual selection and sexual size dimorphism where changes in mating systems occurred before changes in the degree of sexual selection (Lindenfors and Tullberg, 1998). Using similar methods, sexual selection has also been shown to be an important determinant of sexual dimorphism in canine size in primates (Thorén et al., 2006), although primate canines are also of importance in predator defense (Harvey et al., 1978). Thus, both body size and canine size bear witness to an evolutionary history of male–male aggression in primates.

This selection history has its grounds in a sexual difference in behavior. While males compete more over matings than females, female reproduction is instead limited by resource allocation (Emlen and Oring, 1977). These differing demands should be expected to produce variation in the relative sizes of various brain structures, just as they are expected to produce differences in other morphological structures. However, data on brain structures in primates are not available for males and females separately. Instead, investigating *species* differences in brain structures and comparing them on basis of differences in the species-typical degree of sexual selection, research has shown that the degree of male intrasexual selection is positively correlated with several structures involved in autonomic functions and sensory-motor skills, and in pathways relating to aggression and aggression control (Lindenfors et al., 2007b).

The sizes of the mesencephalon, diencephalon (containing the hypothalamus), and amygdala, all involved in governing aggressive behaviors, are positively correlated with the degree of sexual selection, whereas the size of the septum, which has a role in facilitating aggression control, is negatively correlated with the degree of sexual selection. These correlations indicate that sexual selection affects physical combat skills. Moreover, male group size was positively correlated with the relative volume of the diencephalon and negatively correlated with relative septum size, further strengthening the conclusion that aggression is an evolutionarily important component of male–male interactions (Lindenfors *et al.*, 2007b). Thus, primate brain organization also reflects a history of male–male aggression.

VI. HUMANS AND THE MAMMALIAN PATTERN

So where do humans fit into this picture? Humans are one of the sexually size-dimorphic species in the primate order in Table 2.1. But compared to our closest relatives (chimpanzees, bonobos, gorillas, and orangutans) humans have the lowest degree of size dimorphism (Lindenfors, 2002), indicating a decreasing degree of sexual selection over human evolution. Nevertheless, in all measurements of length that have ever been carried out in human populations, males have been taller than females. The average dimorphism in humans from these surveys is 1.07 (Gustafsson and Lindenfors, 2004). Does this mean that we exhibit a tendency toward the polygyny that accompanies such size dimorphism?

According to the Ethnographic Atlas Codebook (Gray, 1999), a database of cultural characteristics for 1231 comparable cultures from around the world, polygyny is common in 48% of human societies. In another 37% polygyny is allowed, and in only 15% is monogamy the norm. Only four reported societies are considered polyandrous. From these data and the degree of human size dimorphism, one may draw the conclusion that humans are at least more polygynous than monogamous (see also Low, 2000), and also that Western cultures fall within the monogamous 15% of the world. Interestingly, intergroup differences in size dimorphism are not correlated with differences in the degree of polygyny (Gustafsson and Lindenfors, 2004), a clear indication that cultural evolution proceeds faster than biological evolution.

There are more indications that humans have an evolutionary history of sex differences as an explanatory factor in human aggression. For example, men commit most of the world's violent acts that are reported to the police. Men are consequently overrepresented in the world's prisons. Women typically make up only 10–15% of the prison population (Harrendorf *et al.*, 2010). Further, soldiering is most often an all-male vocation (personal observation). Humans are, however, the products of both biological and cultural inheritance (Boyd and Richerson, 2005). Here, we have presented only the biological side of the story, but we agree with Archer (2009) that sexual selection probably is the best explanation for the magnitude and nature of human sex differences in aggression. Humans fit into the mammalian scheme of things very well.

Acknowledgment

This research was supported through a generous research grant from the Swedish Research Council (P. L.).

References

Altmann, J. (1990). Primate males go where the females are. *Anim. Behav.* **39**, 193–195.

Andersson, M. (1994). Sexual Selection. Princeton University Press, New Jersey.

Archer, J. (2009). Does sexual selection explain human sex differences in aggression? *Behav. Brain Sci.* **32**, 249–311.

Arnqvist, G., and Rowe, L. (2005). Sexual Conflict. Princeton University Press, Princeton.

Bateman, A. J. (1948). Intra-sexual selection in Drosophila. *Heredity* **2**, 349–368.

Bininda-Emonds, O. R. P., *et al.* (2007). The delayed rise of present-day mammals. *Nature* **446**, 507–512.

Birkhead, T. (2001). Promiscuity: An Evolutionary History of Sperm Competition. Harvard University Press, Cambridge.

Boyd, R., and Richerson, P. J. (2005). The Origin and Evolution of Cultures. Oxford University Press, Oxford.

Brick, O. (1999). A test of the sequential assessment game: The effect of increased cost of sampling. *Behav. Ecol.* **10**, 726–732.

Caro, T. M., Graham, C. M., Stoner, C. J., and Flores, M. M. (2003). Correlates of horn and antler shape in bovids and cervids. *Behav. Ecol. Sociobiol.* **55**, 32–41.

Clutton-Brock, T. H., and Albon, S. D. (1979). The roaring of red deer and the evolution of honest advertisement. *Behaviour* **69**, 146–170.

Darwin, C. (1859). The Origin of Species by Means of Natural Selection. Penguin, London.

Darwin, C. (1860). Darwin Correspondence Project. Letter 2743—Darwin, C. R. to Gray, Asa, 3 Apr (1860).

Darwin, C. (1871). The Descent of Man and Selection in Relation to Sex. Murray, London.

Dugatkin, L. A., and Godin, J.-G. J. (1993). Female mate copying in the guppy (*Poecilia reticulata*): Age-dependent effects. *Behav. Ecol.* **4**, 289–292.

Emlen, S. T., and Oring, L. W. (1977). Ecology, sexual selection, and the evolution of mating systems. *Science* **197**, 215–223.

Enquist, M., and Leimar, O. (1983). Evolution of fighting behavior: Decision rules and assessment of relative strength. *J. Theor. Biol.* **102**, 387–410.

Fabiani, A., Galimberti, F., Sanvito, S., and Hoelzel, A. R. (2004). Extreme polygyny among southern elephant seals on Sea Lion Island, Falkland Islands. *Behav. Ecol.* **15**, 961–969.

Fairbairn, D. J., Blanckenhorn, W. U., and Székely, T. (eds.) (2007). *In* "Sex, Size and Gender Roles: Evolutionary Studies of Sexual Size Dimorphism". Oxford University Press, Oxford.

Gavrilets, S., Arnqvist, G., and Friberg, U. (2001). The evolution of female mate choice by sexual conflict. *Proc. R. Soc. Lond. B* **268**, 531–539.

Gray, J. P. (1999). A corrected ethnographic atlas. *World Cultures* **10**, 24–85.

Gross, M. R. (1996). Alternative reproductive strategies and tactics: Diversity within sexes. *Trends Ecol. Evol.* **11**, 92–97.

Gustafsson, A., and Lindenfors, P. (2004). Human size evolution: No allometric relationship between male and female stature. *J. Hum. Evol.* **47**, 253–266.

Harrendorf, S., Heiskanen, M., and Malby, S. (2010). International Statistics on Crime and Justice. United Nations Office on Drugs and Crime (UNODC), Helsinki, Finland.

Harvey, P. H., Kavanagh, M., and Clutton-Brock, T. H. (1978). Sexual dimorphism in primate teeth. *J. Zool.* **186**, 475–485.

Holland, B., and Rice, W. R. (1998). Chase-away sexual selection: Antagonistic seduction versus resistance. *Evolution* **52**, 1–7.

Hrdy, S. B. (1981). The Woman that Never Evolved. Harvard University Press, Cambridge.

Lindenfors, P. (2002). Sexually antagonistic selection on primate size. *J. Evol. Biol.* **15**, 595–607.

Lindenfors, P., and Tullberg, B. S. (1998). Phylogenetic analyses of primate size evolution: The consequences of sexual selection. *Biol. J. Linn. Soc.* **64,** 413–447.

Lindenfors, P., Tullberg, B., and Biuw, M. (2002). Phylogenetic analyses of sexual selection and sexual size dimorphism in pinnipeds. *Behav. Ecol. Sociobiol.* **52,** 188–193.

Lindenfors, P., Fröberg, L., and Nunn, C. L. (2004). Females drive primate social evolution. *Proc. R. Soc. Lond.* B **271,** S101–S103.

Lindenfors, P., Gittleman, J. L., and Jones, K. E. (2007a). Sexual size dimorphism in mammals. *In* "Sex, Size and Gender Roles: Evolutionary Studies of Sexual Size Dimorphism" (D. J. Fairbairn, W. U. Blanckenhorn, and T. Szekely, eds.), pp. 19–26. Oxford University Press, Oxford.

Lindenfors, P., Nunn, C. L., and Barton, R. A. (2007b). Primate brain architecture and selection in relation to sex. *BMC Biol.* **5,** 20.

Low, B. S. (2000). Why Sex Matters. Princeton University Press, Princeton.

Maynard Smith, J. (1982). Evolution and the Theory of Games. Cambridge University Press, Cambridge.

Mitani, J. C., Gros-Louis, J., and Manson, J. H. (1996). Number of males in primate groups: Comparative tests of competing hypotheses. *Am. J. Primatol.* **38,** 315–332.

Nunn, C. L. (1999). The number of males in primate social groups: A comparative test of the socioecological model. *Behav. Ecol. Sociobiol.* **46,** 1–13.

Parker, G. A. (1979). Sexual selection and sexual conflict. *In* "Sexual Selection and Reproductive Competition in Insects" (M. S. Blum and N. A. Blum, eds.), pp. 123–166. Academic Press, New York.

Petrie, M. (1994). Improved growth and survival of offspring of peacocks with more elaborate trains. *Nature* **371,** 598–599.

Ralls, K. (1976). Mammals in which females are larger than males. *Q. Rev. Biol.* **51,** 245–276.

Stankowich, T., and Caro, T. (2009). Evolution of weaponry in female bovids. *Proc. R. Soc. Lond.* B **276,** 4329–4334.

Thorén, S., Lindenfors, P., and Kappeler, P. M. (2006). Phylogenetic analyses of dimorphism in primates: Evidence for stronger selection on canine size than on body size. *Am. J. Phys. Anthropol.* **130,** 50–59.

Tinbergen, N. (1963). On aims and methods in ethology. *Z. Tierpsychol.* **20,** 410–433.

Trivers, R. L. (1972). Parental investment and sexual selection. *In* "Sexual Selection and the Descent of Man 1871–1971" (B. Campbell, ed.), pp. 136–179. Aldine, Chicago, IL.

Vincent, A., Ahnesjö, I., Berglund, A., and Rosenqvist, G. (1992). Pipefishes and seahorses: Are they all sex role reversed? *Trends Ecol. Evol.* **7,** 237–241.

Wade, M. J., and Pruett-Jones, S. G. (1990). Female copying increases the variance in male mating success. *Proc. Natl. Acad. Sci. USA* **87,** 5749–5753.

West, P. M., and Packer, C. (2002). Sexual selection, temperature, and the lion's mane. *Science* **297,** 1339–1343.

Williams, G. C. (1966). Adaptation and Natural Selection. Princeton University Press, Princeton.

Wilson, E. O. (1975). Sociobiology, the New Synthesis. Belknap, Cambridge, MA.

Wilson, D. E., and Reeder, D. M. (eds.) (1993). *In* "Mammal Species of the World: A Taxonomic and Geographic Reference". 2 nd edn. Smithsonian Institution Press, Washington and London.

3

Signaling Aggression

Moira J. van Staaden,* William A. Searcy,†
and Roger T. Hanlon‡

*Department of Biological Sciences and JP Scott Center for Neuroscience,
Mind & Behavior, Bowling Green State University, Bowling Green, Ohio,
USA
†Department of Biology, University of Miami, Coral Gables, Florida, USA
‡Marine Resources Center, Marine Biological Laboratory, Woods Hole,
Massachusetts, USA

0065-2660/11 $35.00
DOI: 10.1016/B978-0-12-380858-5.00008-3

ABSTRACT

From psychological and sociological standpoints, aggression is regarded as intentional behavior aimed at inflicting pain and manifested by hostility and attacking behaviors. In contrast, biologists define aggression as behavior associated with attack or escalation toward attack, omitting any stipulation about intentions and goals. Certain animal signals are strongly associated with escalation toward attack and have the same function as physical attack in intimidating opponents and winning contests, and ethologists therefore consider them an integral part of aggressive behavior. Aggressive signals have been molded by evolution to make them ever more effective in mediating interactions between the contestants. Early theoretical analyses of aggressive signaling suggested that signals could never be honest about fighting ability or aggressive intentions because weak individuals would exaggerate such signals whenever they were effective in influencing the behavior of opponents. More recent game theory models, however, demonstrate that given the right costs and constraints, aggressive signals are both reliable about strength and intentions and effective in influencing contest outcomes. Here, we review the role of signaling in lieu of physical violence, considering threat displays from an ethological perspective as an adaptive outcome of evolutionary selection pressures. Fighting prowess is conveyed by performance signals whose production is constrained by physical ability and thus limited to just some individuals, whereas aggressive intent is encoded in strategic signals that all signalers are able to produce. We illustrate recent advances in the study of aggressive signaling with case studies of charismatic taxa that employ a range of sensory modalities, viz. visual and chemical signaling in cephalopod behavior, and indicators of aggressive intent in the territorial calls of songbirds. © 2011, Elsevier Inc.

I. INTRODUCTION

Although physical fighting, including the killing of conspecifics, is widespread in nonhuman animals just as it is in humans, the majority of contests and disputes in nonhuman animals are settled without physical fighting. Rather than resorting to immediate physical combat, nonhuman animals often engage instead in extended bouts of signaling, making prominent display of their weapons (e.g., antlers, claws, and teeth), or running through a repertoire of highly stereotyped agonistic signals. With their high cognitive capacity, primates (humans included) are particularly good at reducing social tensions and resolving conflicts using agonistic signaling as opposed to sheer physical force (Cheney *et al.*, 1986).

Such aggressive signaling is found in virtually all of the multicellular taxa and can involve all communication modalities. Orthoptera (Alexander, 1961; Simmonds and Bailey, 1993) and many other insects (Clark and Moore, 1995; Jonsson *et al.*, 2011) use aggressive song to defend resources, and the use of territorial song in birds is well known (Searcy and Yasukawa, 1990; Stoddard *et al.*, 1988). Calls are employed to similar effect in the dramatic displays of large mammals or frog choruses (Bee *et al.*, 1999; Reby *et al.*, 2005; Wagner, 1992), and more subtly by other vertebrate taxa such as fish (Raffinger and Ladich, 2009). In these scenarios, signaling can be just as effective as physical attack in intimidating opponents and winning contested resources.

Chemical signals are widely used to signal resource defense and fighting ability, deposited either as scent marks in fixed locales by terrestrial species (Page and Jaeger, 2004) or contained in urine released during aggressive interactions in some aquatic organisms (Breithaupt and Eger, 2002). Visual signals are perhaps the most familiar and easily appreciated of aggressive displays, beginning with Darwin's (1871) graphic illustration of aggression and fear in the facial expression of the domestic dog. Visual signs of aggression include variable pigment patterns of many fish and cephalopods (DiMarco and Hanlon, 1997; Moretz and Morris, 2003), and the ritualized display of weapons (Huber and Kravitz, 1995; Lundrigan, 1996) or inedible objects as "props" (Murphy, 2008).

Phylogenetic comparative analyses demonstrate that many of these aggressive signals allowing opponents to resolve contests without physical harm evolved from nonsignaling behaviors through the process of ritualization (Scott *et al.*, 2010; Turner *et al.*, 2007). Whereas agonistic behavior runs the gamut from passivity, defense, and escape to full conflict, here, we reserve the terms aggressive/threatening behavior for that subset of agonistic behavior associated with the escalation toward physical fighting (Searcy and Beecher, 2009).

A. An ethological approach to aggression

The ethological approach to aggression derives historically from the traditional instincts and drives articulated by Lorenz (1978). Although the simple psychohydraulic model of motivation underlying this view proved inadequate in the long term, the idea that aggression is based on both internal state and external stimuli, and the proposed value of a comparative evolutionary approach, were both far-sighted and enduring. The classic *On Aggression* (Lorenz, 1963) which was written for a popular audience, highlighted aggression as a natural, evolved function, with a founding basis in other instincts, and a central role in animal communication. A more nuanced view is found in his work known as the Russian manuscript (Lorenz, 1995). In this, Lorenz discussed animals and humans separately, not because of any fundamental difference in their biology, but because he believed it necessary for the reader to have an adequate frame of reference.

Much current research on the biology of aggression focuses on identifying the physiological substrate to violence (i.e., on proximate cause and non-adaptive features). The ethological or sociobiological approach, in contrast, focuses attention on the ultimate causes and *adaptive* forms of aggressive behavior (e.g., Chen *et al*., 2002; Huber and Kravitz, 1995; Miczek, *et al*., 2007; Natarajan *et al*., 2009): how and why has evolution molded complex agonistic interactions built on reciprocal displays of threat or submission, affect or intent?

B. The classic game theory model

Evolutionary fitness is measured in terms of the number of offspring an individual produces over the course of its lifetime. In the evolutionary race to transmit their genes to the following generations at a higher frequency than that of their conspecifics, these individuals must compete for access to all the resources necessary to create and raise their progeny, including mates, dominance rights, and desirable territory. Winners in this intraspecific competition thus stand to gain both immediate personal advantages such as food, space, and safety, as well as long-term evolutionary fitness, that is, more offspring and therefore copies of their genes in subsequent generations. Simulation approaches from game theory have long provided a theoretical framework for analyzing and predicting the outcomes of competitive interactions. The classic "*Hawks*" and "*Doves*" game (Maynard Smith and Price, 1973) considers symmetrical contests between pairs of individuals who are equivalent in every respect (equal size, strength, fighting ability, etc.), differing only in behavioral/fighting strategy in intraspecific encounters. *Hawk* strategists are those who will always choose to fight when they encounter a conspecific at a contested resource. *Dove* strategists, in contrast, always retreat from an individual behaving as a *Hawk*, rather than engage them in combat. *Hawks* always best *Doves*, but they incur costs when they compete against other *Hawks*. The outcome between two *Dove* strategists is randomly determined. Each conflict consists of a series of agonistic moves (incorporating provocation, escalation, retaliation, etc.) with rewards or costs assigned to each contestant according to a particular payoff matrix (Table 3.1).

Populations are expected to converge on an evolutionarily stable strategy (ESS), strategies that once they are predominant cannot be invaded by any other strategy. The ESS depends critically on the ratio of what an individual stands to gain over what it stands to lose in a fight. Thus, in the common situation where the cost of injury exceeds the benefits of winning, populations are expected to adjust to balanced proportions of the two strategies with the majority of individuals behaving as *Doves*, while a smaller number of *Hawk* strategists persists. Only in extreme situations where the value of a resource greatly exceeds the cost of injury, will a *Hawk* strategy be superior and can become so widespread as to completely replace the *Dove* strategy. For instance,

Table 3.1. The Payoff Matrix for the *Hawk–Dove* Game Shows the Consequences that Result When a Player of a Given Strategy (Left Column) Encounters Another Player's Strategy

	Hawk	*Dove*
Hawk	Tie $[(V-C)/2]$	Win $[V]$
Dove	Lose $[0]$	Tie $[V/2]$

Choices are assumed to be rational where each individual would prefer to win, prefer to tie rather than lose, and prefer to lose over receiving injury. In this payoff matrix, V (value of the contested resource) and C (cost of an escalated fight) determines the net outcome when different strategies meet. In encounters between *Hawks*, the winner gains control over the value of the resource while the losing *Hawk* sustains an injury. In the common scenario, where the value of the resource is less than the cost of injury (i.e., $C > V$), average payoff in a *Hawk* meeting a *Hawk* is negative and less than that of a *Dove* meeting a *Hawk*. Only in rare situations, when the value of the resource exceeds the cost of injury, will *Hawk* be unequivocally the superior strategy.

intense fighting among male elephant seals results in the victorious male both monopolizing a section of the beach and gaining sole reproductive access to the harem of females which resides there. In the vast majority of cases, however, resources are rarely worth the risk of injury, and competing individuals would do best to resolve conflicts via ritualized displays.

C. Signaling games

The earliest game theoretical analyses of aggressive signaling were pessimistic about the evolutionary stability of such systems (Caryl, 1979; Maynard Smith, 1974, 1979). Their reasoning was that if we assume that signals can help in winning contests by conveying high levels of aggression or fighting ability, then it becomes advantageous for all individuals to give the highest levels of these signals. If all individuals signal maximally, then there is no information in the signal about either aggressive intentions or fighting ability. The first rigorous game theoretical model to demonstrate that reliable aggressive signaling could be evolutionarily stable was a mutual signaling game in which two interactants chose between two cost-free signals to create a stable global strategy (Enquist, 1985). This model demonstrated how threat displays reveal information about the strength or condition of the contestants via their choice of action in aggressive encounters. The players in this game each have a hidden state (strength or weakness) which determines their ability to win physical fights. An honest weak individual gives a signal conveying weakness, and abandons the contest if the other individual gives a signal conveying strength. A dishonest weak individual can successfully bluff other weak individuals by giving the signal of strength, but at a cost of sometimes being attacked by a better fighter if the opponent turns out to be strong. If the cost of being attacked by a stronger individual is high relative to the benefit of winning contests, then bluffing may not be advantageous, and honest signaling can be evolutionarily stable.

There followed a slew of variant *Hawk/Dove* models which attempted to accommodate the diversity of interactions between senders and receivers (e.g., Enquist and Leimar, 1983; Leimar and Enquist, 1984; Maynard Smith and Harper, 1988; Skyrms, 2009). These models of communication may be classified into five structures based on the relative timing of the (signal and/or response) choices made by the two players during the game (reviewed in Hurd and Enquist, 2005). Mutual signaling games, which most closely resemble agonistic interactions between animals, are increasingly being used as models (e.g., Kim, 1995; Számadó, 2000). In this structure, both players signal, and react to their opponent's signal, in biologically realistic ways. Genetic algorithms are also being used to examine non-ESS solutions to these games (Hamblin and Hurd, 2007). Alternative approaches employ simulation methods and neural networks (Noble, 2000; Wheeler and de Bourcier, 1995) to explore communication in animal contests.

D. Threat displays and why they are part of aggression

Aggression is costly to participants not only in terms of energy expenditure and the potential for injury but also because of opportunity costs. Time spent in physical conflict is time that is not available for other vital activities such as exploring, feeding, or mating. Thus, there are selective advantages to reducing aggression. Threat displays are a critical component of aggression because they modulate competitive social interactions among conspecifics. If signaling is effectively delivered by a sender and appropriately interpreted by the intended receiver it might be so subtle that the interaction is rendered virtually invisible to an outside observer. Alternatively, if sender and receiver perceive the competitive difference between them to be slight, the social interaction is prolonged, escalates in intensity, and may ultimately culminate in levels of overt conflict that result in physical damage or death of one or both interactants.

In such aggressive signaling contests, two kinds of information are important to receivers: information on the signaler's willingness to escalate (aggressiveness motivation) and on its fighting ability (resource-holding potential) (Searcy and Beecher, 2009). Classification schemes based on the type of interaction in which communication takes place and the nature of the signals used converge on the following signal categories (Hurd and Enquist, 2005; Maynard Smith and Harper, 2003; Vehrencamp, 2000).

Performance signals are signals constrained to a subset of signalers either by differences in the ability to perform them (Maynard Smith, 1982), or by possessing the information needed to produce them (Hurd and Enquist, 2005).

Performance displays ("index signals" of Maynard Smith and Harper, 2003) have excellent empirical support, as do models of their use (e.g., Enquist and Leimar, 1983; Leimar and Enquist, 1984). Examples include the lateral displays of many fish (Enquist and Jakobsson, 1986) or the pitch of calls in many frog and mammal species (e.g., Bee *et al.*, 1999; Reby *et al.*, 2005), both "unfakeable" signals as they are determined by the sender's size and fighting ability.

 Strategic signals are available to all signalers, and may be either *classic handicaps* or *conventional signals* (Hurd and Enquist, 2005). Classic handicaps have some inherent cost, independent of receiver response, and variation in the level of cost experienced by different individuals produces different optimum signaling levels (Grafen, 1990). Evidence for handicapped displays is theoretical (Zahavi, 1987) rather than empirical, though threat displays have been shown to advertise endurance in lizards (Brandt, 2002) and grasshoppers (Greenfield and Minckley, 1993). Conventional signals are arbitrary with respect to signal design and therefore dependent for meaning on an agreement between the signaler and receiver. Honesty of conventional signals in agonistic interactions is maintained by two forms of receiver-dependent stabilizing costs (Enquist, 1985; Guilford and Dawkins, 1995); receiver retaliation (Enquist, 1985) has empirical support (Molles and Vehrencamp, 2001) and vulnerability handicap (Zahavi, 1987) for which empirical support is contradictory (Laidre and Vehrencamp, 2008; Searcy *et al.*, 2006). Most threat displays appear to be conventional signaling systems. Examples include color patches and song-type sharing in birds (Molles and Vehrencamp, 2001; Vehrencamp, 2000). Aggressiveness motivation (or willingness to escalate) is most likely to be encoded this way (Hurd and Enquist, 2005).

E. Evolutionary issues

Empirical analysis of aggressive signaling is more complex than the classic ESS modeling approach would suggest. This is in large part attributable to the fact that evolution is not necessarily equilibrial (Houston and McNamara, 1999). An individual's success or failure in using signals depends upon how other individuals use and interpret those signals, that is, it is a trait under frequency-dependent selection (Maynard Smith, 1982). In addition to frequency dependence of the signal phenotype itself, selection pressures acting on signaler and receiver in a communicating dyad may be distinct if their genetic interests or risk profiles (Searcy and Nowicki, 2006) are not identical, or if signals have dual functions, affecting both aggression and mate choice (Wong and Candolin, 2005). Selection may also modify the responsiveness of other individuals to the signals (Arak and Enquist, 1995). Thus, like other significant evolutionary problems such as sexual selection and conflict, signaling strategies may lack stable equilibria and remain in constant evolutionary flux. Understanding the evolution of behavioral phenotypes under such nonequilibrial conditions requires dynamic approaches which have yet to be adequately deployed in the game-theoretical modeling of biological signaling (Hurd and Enquist, 2005).

F. The challenge of "incomplete honesty"

In animal contests, selection should favor displays providing reliable information about the fighting ability or aggressive intent of competitors. However, considerable theoretical work predicts that low levels of deception may occur within otherwise honest signaling systems (Adams and Mesterton-Gibbons, 1995; Számadó 2000). Strategic signals (i.e., ones of intent) are particularly prone to such corruption because they typically involve low production costs (Maynard Smith, 1974, 1979, 1982). Testing for such incomplete honesty is challenging because it is difficult to distinguish dishonest signals from natural variation in signal size (Moore *et al.*, 2009), and between a successful bluff and an honest signal, especially when signaled information is continuous rather than discrete. Hughes (2000) suggested that dishonesty could be detected by analysis of signal residuals, the residuals from a measure of the regression of signal structure on competitive ability. Whereas receivers take advantage of the strong relationship between signal and fighting ability, for example, signalers take advantage of the variation around this relationship. If individuals who exaggerate signals benefit from doing so, they should perform more repetitions of the signaling activity than those who do not exaggerate (Hughes, 2000). Empirical examples of incomplete honesty, though still comparatively rare, suggest this is not a fixed behavioral trait, and depends on context as well as signal residuals (Arnott and Elwood, 2010; Hughes, 2000; Lailvaux *et al.*, 2009).

G. Case studies in aggressive signaling

Using animal models and invasive techniques (e.g., drugs, hormones, brain lesions, and gene knockouts), we have made great strides in unraveling the mechanisms and internal states underlying aggression in controlled lab situations. This is true also with respect to aggressive signaling (see Chapter 5). Studies of nonmodel organisms are a necessary complement to this approach as these can provide the telling exceptions in field situations where more complex social/physical environments permit full expression of behaviors and analysis of adaptive function (see Logue *et al.*, 2010). Below, we present two case studies of taxa employing multimodal signaling systems to artfully modulate aggressive interactions in complex social systems.

II. BIRD SONG SIGNALS AGGRESSIVE INTENTIONS: SPEAK SOFTLY AND CARRY A BIG STICK

The use of song by songbirds provides an excellent illustration of how signals function in aggression in nonhuman animals. The songbirds (suborder Passeres) consist of over 4000 species of birds, which are distinguished in part by their intricate vocal musculature. This musculature functions most importantly in the production of the complex vocalizations from which the songbirds derive their name. Most species in the group are territorial and monogamous, and their songs are used in both territory defense and mate attraction (Catchpole and Slater, 2008; Searcy and Andersson, 1986). At least in temperate zone species, songs are given mainly by males and mainly during the breeding season. Some attributes of song and singing behavior have evolved to function in attracting females and persuading them to mate, but others have evolved to function in aggressive communication between males in the context of claiming and defending a territory.

Many of the signals employed by songbirds in aggressive communication can be illustrated using the signaling behavior of song sparrows (*Melospiza melodia*). Song sparrow songs (Fig. 3.1) are multiparted—that is, they contain multiple phrases differing in structure (Mulligan, 1963). Individual males sing several versions of the species' song, each consisting of a distinct and largely nonoverlapping set of phrases. These distinct versions are called *song types* (Fig. 3.1), and the collection of song types sung by one male is his *song repertoire*. Repertoire sizes vary geographically in song sparrows, with averages in the range of 8–12 song types per male (Peters *et al.*, 2000). Male song sparrows produce their repertoires with "eventual variety," meaning that they sing several to many repetitions of one song type before switching to another. The successive repetitions of a song type are themselves typically not identical, but instead show differences that are audible (Borror, 1965; Saunders, 1924) but of lower magnitude than differences between song types (Nowicki *et al.*, 1994). The minor variations of a song type are termed *song variants* (Fig. 3.1). Song sparrows respond to differences between song variants (Stoddard *et al.*, 1988) but less strongly than to differences between song types (Searcy *et al.*, 1995).

In some species of songbirds, different song types have different functions; for example, in wood warblers (Parulidae) some song types may be specialized for male–female communication and others for male–male signaling (Byers, 1996; Spector, 1992; Weary *et al.*, 1994; but see Beebee, 2004). In song sparrows, however, all song types are thought to be functionally equivalent, and in that sense "redundant." Even with redundant song types, however, certain signals can be produced with a repertoire of song types that are not possible with a single type. Some of these signals have been suggested to be aggressive.

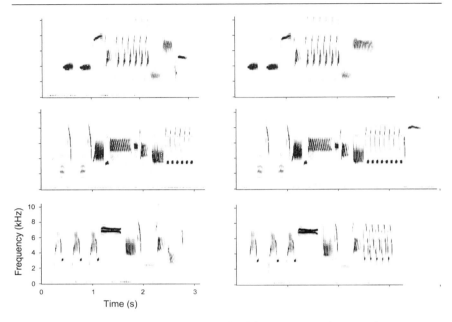

Figure 3.1. Spectrograms of two variants of each of three song types from a male song sparrow recorded in northwestern Pennsylvania. Each row shows two variants of one song type. Note that virtually every note differs between the different song types, whereas the two variants of any one song type tend to differ only in their endings.

Singing behaviors associated with aggressive contexts in song sparrows include:

1. Song-type switching. If a bird sings more than one song type, it can vary the frequency with which it switches between song types, and switching frequency becomes a possible signal. Song-type switching frequency has been suggested to be a conventional signal of aggression (Vehrencamp, 2000)—conventional in the sense that the meaning of the signal is arbitrary with respect to its form. In song sparrows, type-switching frequency increases in aggressive contexts, for example, during counter singing between territorial males or when an outside male intrudes on a territory (Kramer and Lemon, 1983; Kramer *et al.*, 1985; Searcy *et al.*, 2000). In other species, the opposite pattern holds—type-switching frequency decreases in aggressive contexts (Molles and Vehrencamp, 1999; Searcy and Yasukawa, 1990). The fact that either pattern can occur supports the arbitrariness of the signal (Vehrencamp, 2000).
2. Variant switching. In song sparrows, variant-switching frequency also increases in aggressive contexts, and the increase is if anything more consistent than the increase in type switching (Searcy *et al.*, 2000). Given

the evidence that male song sparrows attend to variant switching (Searcy et al., 1995; Stoddard et al., 1988), variant-switching frequency is another potential aggressive signal.

3. Song-type matching. Matching is a behavior in which one male replies to a rival with the same song type that the rival has just sung. Matching can occur by chance, but in song sparrows it has been shown that when wholly or partially shared songs are played to males on or near their territories, those males match the playback songs at levels significantly higher than chance (Anderson et al., 2005; Burt et al., 2002; Stoddard et al., 1992). Song sparrows match strangers more than neighbors (Stoddard et al., 1992), and are more aggressive in general toward strangers (Stoddard et al., 1990), providing further support for matching as an aggressive signal.

4. Song rate. The number of songs produced per unit time is a parameter that birds can vary even if they sing only a single song type. In some species of songbirds, territory owners consistently increase song rates in aggressive contexts (Vehrencamp, 2000). Song sparrows have shown this pattern in some experiments (Kramer et al., 1985) but not in others (Peters et al., 1980; Searcy et al., 2000).

5. Soft song. In her classic monograph on song sparrow behavior, Nice (1943) noted that during intense aggressive encounters, male song sparrows produce songs of especially low amplitude. In some other songbirds, such soft songs are produced during courtship as well as during aggression (Dabelsteen et al., 1998), but in song sparrows they apparently are given only in aggressive contexts. Anderson et al. (2008) found that the amplitude of soft songs was as much as 36 dB lower than the amplitude of the loudest normal or "broadcast" songs.

The five singing behaviors listed above are all associated with aggressive contexts in song sparrows, but signals used in aggressive contexts can convey submission or escape as well as attack, in which case they would be considered "agonistic" but not "aggressive." These alternative interpretations seem particularly likely a priori in the case of soft songs. To test whether a signal is aggressive rather than submissive, it is necessary to determine whether the signal predicts aggressive escalation (Searcy and Beecher, 2009). Aggressive escalation includes outright physical attack of course, but also includes other behaviors that lead up to attack, such as approach to a rival or giving signals that are higher in a hierarchy of aggressive signaling.

A test of the predictive power of singing behaviors was carried out for song sparrows by Searcy et al. (2006). In this study, a brief playback of song sparrow song was used to elicit aggressive signaling from a territory owner. After a 5-min period during which displays were recorded, a taxidermic mount of a song sparrow was revealed on the subject's territory, posed above the loudspeaker, in conjunction with another brief playback. The subject was then given a set period of time (14 min) to attack or not attack the mount. Of 95 males that were tested, 20 attacked and 75 did not. The display behavior of attackers and nonattackers was then compared, focusing on the five singing behaviors discussed above, plus *wing-waving*, a display in which a male fans one or both wings while remaining perched; this is the most prominent visual display given by song sparrows during aggressive contests. For the initial recording period, none of the display measures differed significantly between attackers and nonattackers, though the number of soft songs approached significance. A second analysis focused on the 1-min period directly before attack in the attacking subjects, using a matching time period in nonattackers as the control. Here, number of soft songs was significantly higher in attackers than nonattackers, whereas none of the other five measures differed (Fig. 3.2). In single-variable discriminant function analyses, the number of soft songs was the only display that discriminated between attackers and nonattackers; this display correctly predicted presence/absence of attack in 74% of the tested males. Soft song is thus a reliable signal of aggressive intentions in song sparrows.

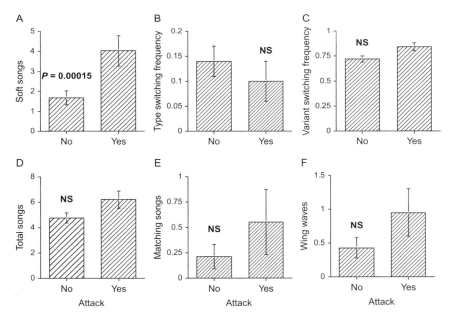

Figure 3.2. Display measures (mean ± s.e.) for the 1 min just prior to attack for male song sparrows that attacked compared to a matching 1-min period for nonattackers. The display measures are (A) number of soft songs, (B) type-switching frequency, (C) variant-switching frequency, (D) total songs, (E) number of matching songs, and (F) number of bouts of wing-waving. Only soft songs showed a significant difference between attackers and nonattackers. Redrawn from data in Searcy *et al.* (2006).

The use of soft, low-amplitude vocalizations as the most threatening of signals is somewhat counterintuitive, but this result has since been replicated in additional species. Ballentine *et al.* (2008) did a parallel study of aggressive signaling in swamp sparrows (*Melospiza georgiana*), a close relative of song sparrows, using methods similar to those of Searcy *et al.* (2006). Swamp sparrows have simpler songs than song sparrows, but again have repertoires of apparently redundant song types. In addition to songs, males give two types of calls in aggressive contexts, *buzzes* and *wheezes* (Ballentine *et al.*, 2008; Mowbray, 1997). In swamp sparrows as in song sparrows, wing-waving is the most prominent visual display given during aggressive encounters.

In 40 trials with swamp sparrows, 9 males attacked a taxidermic mount of a conspecific male and 31 did not. For the initial recording period, five of seven display measures did not differ between attackers and nonattackers; these were switching frequency, number of matching songs, number of broadcast songs, number of rasps, and number of wheezes. Two measures were significantly higher in attackers: number of soft songs and number of wing waves. In a forward, stepwise discriminant function analysis, soft songs entered first, followed by rasps, and these together correctly classified 83% of males as attackers or nonattackers. For the 1 min prior to attack, soft songs and wing waves were again the only two display measures that differed between attackers and nonattackers. For this time period, a discriminant function including soft songs and wing waves was the best predictor of attack, classifying 85% of males correctly.

Hof and Hazlett (2010) have recently performed a similar experiment with black-throated blue warblers (*Dendroica caerulescens*), which are also in the songbird suborder but in another family (Parulidae). In 54 trials with black-throated blue warblers, 19 males attacked the mount and 35 did not. Hof and Hazlett (2010) compared attackers and nonattackers for four display measures: type-switching frequency, total number of songs, number of soft songs, and number of ctuk calls. For both an initial recording period and the 1 min prior to attack, only the number of soft songs differed significantly between attackers and nonattackers, with attackers giving substantially more. In logistic regressions based on either time period, soft song was the only significant predictor of attack. In a logistic regression that incorporated displays for the entire trial, soft song correctly predicted attack behavior in a very impressive 93% of subjects.

In all three of the songbird species reviewed above, most of the displays given in aggressive contexts are not predictive of attack. One theory about such displays is that they were at one time predictors of attack, but that over evolutionary time their reliability was undermined by the spread of bluffing (Andersson, 1980). If an aggressive display is beneficial in intimidating opponents, such that the benefit of giving it is greater than any costs, then selection will favor its use in individuals that do not intend to attack as well as in those that do. Use of the display will then increase in frequency among individuals not intending attack, until at some point the signal ceases to be informative about attack likelihood. Another hypothesis is that these agonistic displays have evolved to convey messages other than imminent attack. Possible alternative messages include at one extreme retreat or submission, but another possibility is for a display to threaten a degree of aggressive escalation that falls short of attack. Song-type matching in song sparrows, for example, has been suggested to be part of a hierarchy of progressively more aggressive signals, which starts with singing a shared song, precedes to type matching, then to staying on the match, soft song, and finally attack (Beecher and Campbell, 2005; Searcy and Beecher, 2009). Because matching is low in this hierarchy of escalation, with several steps intervening between it and attack, matching would not be expected to be very informative about attack likelihood; nevertheless, it might still be predictive of the next level of escalation. Whether matching is predictive in this manner requires further testing.

Among the small number of songbird species that have been studied in this regard, soft song has emerged as an unusually reliable predictor of attack. Why a display whose distinguishing characteristic is low amplitude should be consistently favored for the highest level of aggressive signaling is not well understood. One hypothesis is that by using soft song during an encounter with an intruder, a territory owner lowers the chance of interference from other rival males by preventing them from eavesdropping on the interaction (McGregor and Dabelsteen, 1996), thereby concealing from them that an intrusion is taking place. In contradiction to this idea, Searcy and Nowicki (2006) found that, in song sparrows, more intrusions by third party males occurred during simulated interactions between an owner giving soft songs and an intruder giving loud songs than during interactions in which both owner and intruder gave loud songs. In other words, use of soft songs if anything increased interference by other rivals. A second hypothesis is that soft song is favored as an aggressive signal because its low amplitude makes its target unambiguous: only the male that is being confronted can discern the signal, so only he can be the target. Another way of stating this is that soft song is a performance signal subject to an informational constraint (Hurd and Enquist, 2005) that forces it to be honest at least with respect to the identity of its target.

If a display is a reliable signal of aggressive intentions, as is soft song, then theory predicts that it should be effective in changing the behavior of at least some opponents to the signaler's advantage (Enquist, 1985). In other words, a believable threat should intimidate some opponents, presumably the weaker ones, causing them to concede whatever resource is being contested. Effectiveness in this sense has not yet been demonstrated for soft song, in part because arranging tests of the effectiveness of displays in territorial defense is quite difficult (Searcy and Nowicki, 2000). Recent work with corn crakes (*Crex crex*), which are not songbirds and do not sing, shows that low amplitude calls predict attack, and suggests that these soft calls cause some receivers to retreat (Rek and Osiejuk, 2011). Effectiveness in intimidating opponents has been demonstrated in some other aggressive signaling systems (Dingle, 1969; Fugle *et al.*, 1984; Wagner, 1992).

III. VISUAL DISPLAYS SIGNAL AGGRESSIVE INTENT IN CEPHALOPODS: THE SWEET SMELL OF SUCCESS

Cephalopods—squid, octopus, and cuttlefish—are marine molluscs with large complex brains and highly diverse behavior (Hanlon and Messenger, 1996). They are highly visual animals, exemplified partly by their huge optic lobes that represent more than half of their central nervous system. These soft-bodied cephalopods are renowned for their rapid adaptive coloration: individuals of each species can instantly (<1 s) switch between any of 10–50 body patterns that are used for a wide range of communication and camouflage. The appearance of the animal can change so dramatically that they sometimes appear to be different species. This capability has been termed rapid adaptive polyphenism because the same genotype can produce multiple phenotypes.

Squids and cuttlefish have complex mating systems and their sexual selection mechanisms have been studied in some detail. During spawning, the operational sex ratio ranges from 2–4 M:F in some species to 4–11 M:F in others (e.g., Hall and Hanlon, 2002; Hanlon *et al.*, 1999, 2002; Jantzen and Havenhand, 2003). Thus, competition among males for mates is often intense and the visual signaling involved with male rivalry is diverse and dramatic in some cases. These agonistic visual displays are highly developed, and a few experimental studies have complemented field studies to determine the nature of aggression.

One of the most interesting aspects of agonistic behavior in cephalopods is its facultative nature. That is, small unpaired males seek extra-pair copulations using various "sneaking" tactics, but these are usually nonaggressive tactics that actively avoid confrontations with the paired males (for an unusual case involving sexual mimicry, see Hanlon *et al.*, 2005). However, if the large consort male leaves or is displaced (experimentally—in the field or lab), the small males immediately recognize the new behavioral context and become paired consorts to the female and will use agonistic displays to ward off other small males. This transition between sneaker/nonaggressive and consort/highly aggressive is quite remarkable for its speed and fluidity, and testifies to the cognitive abilities of these marine invertebrates. Many fishes and invertebrates have obligatory (i.e., genetic) sneaker morphs (Gross, 1996), but cephalopods accomplish this facultative switch with a large brain and extensive nervous system.

Early game theory models of agonistic behavior predicted that animals should not signal their probability of attack to their opponents. As Maynard Smith (1982) argued, if animals signaled their aggressive motivation during a fight, there would be strong selective pressure for animals to "bluff" and to signal the highest motivational state possible; such a system would likely be invaded by cheaters and become unreliable. However, some animals do signal intent (Hauser and Nelson, 1991), and below we provide an unusual example of this in cuttlefish.

As in birds, cephalopods signal aggressive intent but they do so with visual signals (chromatic skin patterns) as well as body postures (parallel positioning and arm postures). Two examples are given: one from cuttlefish (Order Sepioidea) and one from squid (Order Teuthoidea). In addition, a new finding is described in which a molecular trigger of aggression has been found in squid.

A. Cuttlefish agonistic bouts

In the Intense Zebra Display of the European cuttlefish, *Sepia officinalis*, the males turn on high-contrast stripes and dark eye ring and extend their large 4th arm toward the opponent (Fig. 3.3C). Such agonistic encounters between males can lead to aggressive grappling and biting. The experiments of Adamo and Hanlon (1996) showed that one visual component of the display—the *facial darkness*—was by far the most highly variable in expression, and was a good predictor of outcome in encounters in which one male withdrew. In non-escalated encounters, the male that ultimately withdrew always maintained a less dark face than its opponent (Fig. 3.3A). When the face of a displaying cuttlefish became lighter, the other male either remained in the Intense Zebra Display but did not approach closely or lightened the intensity of its own display within 15 s. When both males maintained a dark face, the agonistic encounters usually escalated to physical pushing, and sometimes to grappling and biting (Fig. 3.3B).

Why would males show an agonistic display to a rival male but simultaneously signal their intent not to be aggressive? Adamo and Hanlon (1996) pointed out that sexual recognition in cephalopods is poorly developed, and that the Intense Zebra Display (with 4th arm extended) identifies the signaler as a male. The authors suggest that male cuttlefish that are not prepared to attack an opponent still give the modified (i.e., light-faced) Intense Zebra Display to convey two messages: (1) that it is male, but (2) it is not prepared to escalate to aggressive physical contact. As the authors point out, when agonistic displays perform more than one function, signaling intent (i.e., signaling its likely subsequent behavior) can be an ESS. Unless the fight escalated to grappling and biting, there would be little cost to cheaters in this system since males that

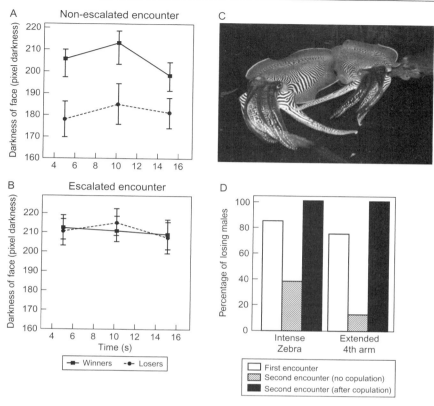

Figure 3.3. Cuttlefish signal intent to escalate a fight with a *dark face* component to their Intense Zebra Display. (A, B) Differences in facial darkness during a non-escalated versus escalated encounter. (C) Two males in Intense Zebra Display with different degrees of *facial darkness*. (D) When males that lost a fight copulated with a female, they became more aggressive in the successive fight. From Adamo and Hanlon (1996).

bluffed (i.e., gave a dark-faced Intense Zebra Display but had little fighting motivation and/or ability) could withdraw at the next stage of agonistic behavior with little penalty.

    In the same study, the authors allowed losing males to copulate with a female after a bout, and retested them with the male each had lost to. The former losers increased facial darkness dramatically in those encounters, showed a long-lasting Intense Zebra Display, and did not withdraw from an opponent (Fig. 3.3D), thus supporting the contention that facial darkness signals the animal's motivational state (i.e., tendency to attack).

B. Squid agonistic bouts

Male–male fights in *Loligo plei* are complex visual displays that include up to 21 behaviors. There is a hierarchy of agonistic signals that sometimes culminates in an aggressive physical lateral display and fin beating (Fig. 3.4A and B), which are then followed by *chase* or *flee*. DiMarco and Hanlon (1997) tested whether dominance was based upon the duration or frequency of these behaviors, but it was not. Instead, they found that certain visual features such as the *lateral flame markings* (Fig. 3.4B, top squid) could be expressed with high contrast and that this was a visual factor in escalation of the agonistic bout.

Two distinct tactics were exhibited by fighting males in this set of laboratory experiments: (1) long bouts with slow escalation from visual signaling to chasing and fleeing, or (2) short bouts with very rapid escalation from visual signaling to lateral displaying, aggressive physical fin beating, followed by chasing and fleeing (Fig. 3.4C). It is noteworthy that the second tactic occurred when a female was present (i.e., when a potential *resource value* was present). As shown in Fig. 3.4D, the presence of a female in various combinations had a dramatic effect on the nature and duration of the agonistic interactions. Longest bouts (mean 14 min) occurred when only two males were present. Bouts became progressively shorter when either two males and one female were assembled simultaneously, or two males were interacting and a female was then added (mean 9 and 3 min, respectively). But when a male and female were put in a tank and allowed to pair, and then a nonpaired male was added, tactic 2 was used and the highly aggressive interaction lasted only 30 s (a 28× difference over the simple two male scenario). As a control, when females were added to male/female pairs, there were no agonistic interactions (Fig. 3.4D).

In this squid species, the lateral display represents an escalation of aggression because it involves parallel posturing and the simultaneous expression of many high-contrast visual signals, which collectively give the impression of making the squid look larger (e.g., the mid-ventral ridge of the mantle protrudes vertically as in the dewlap extension of geckos). Fin beating is a physical, robust contest of pushing that can transmit information about strength and size of the competing individuals.

C. From molecules to aggression: Contact pheromone triggers strong aggression in squid

In the squid *Loligo pealei*, which conducts visual agonistic bouts similar to L. plei (above), it was found recently that females deposit a contact pheromone in the outer tunic of egg capsules that they lay on the sea floor. When males see the egg capsules (even in the absence of females), they are visually attracted to them and then physically contact the eggs, which leads to extremely aggressive fighting

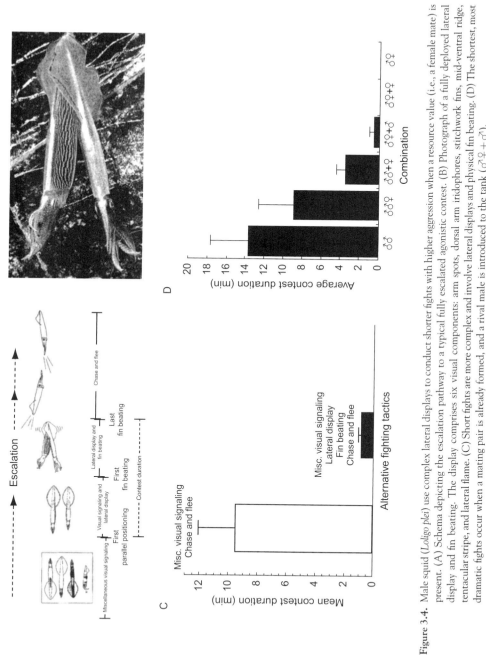

Figure 3.4. Male squid (*Loligo plei*) use complex lateral displays to conduct shorter fights with higher aggression when a resource value (i.e., a female mate) is present. (A) Schema depicting the escalation pathway to a typical fully escalated agonistic contest. (B) Photograph of a fully deployed lateral display and fin beating. The display comprises six visual components: arm spots, dorsal arm iridophores, stitchwork fins, mid-ventral ridge, tentacular stripe, and lateral flame. (C) Short fights are more complex and involve lateral displays and physical fin beating. (D) The shortest, most dramatic fights occur when a mating pair is already formed, and a rival male is introduced to the tank ($\male\female + \male$).

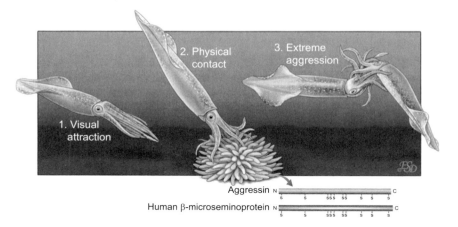

Figure 3.5. When male squids see egg capsules on the sea floor, then approach and touch them, this
leads to immediate and dramatic change from calm swimming to extreme fighting. The
contact pheromone ("aggressin" or *Loligo* β-MSP) is in the tunic of the egg capsules and
is similar in structure to that found in humans, mice, and other vertebrates. (For color
version of this figure, the reader is referred to the Web version of this chapter.)

within a minute or two (Fig. 3.5) (Cummins *et al.*, 2011). Thus, there is a two-
step sensory process: visual attraction to eggs followed by contact chemorecep-
tion that induces onset of aggression.

    In controlled experiments, the 10 kDa protein pheromone (termed
Loligo β-microseminoprotein, β-MSP) was isolated and coated onto a clear
glass flask containing egg capsules, and males that touched the glass (but not
the eggs) began to signal, fight, and bite each other violently within seconds.
Glass flasks without the pheromone coating failed to elicit those aggressive
behaviors. Thus, direct contact with the protein molecules immediately led to
the full cascade of complex aggressive fighting in the absence of females. Given
that aggression is often considered to be a result of multiple interactions of
physiology, hormones, sensory stimuli, etc., this finding reminds us that perhaps
in some cases there are straightforward pathways to aggression. In fact, the
proximate mechanisms that trigger or strengthen aggression are not well
known for many taxa (Wingfield *et al.*, 2005).

    There is a noteworthy vertebrate/mammalian connection to this
finding. As shown in Fig. 3.6, the β-MSPs are highly conserved throughout the
animal kingdom. The greatest known concentration of β-MSPs is in human and
rodent seminal fluid, yet regrettably the functions of β-MSPs are unknown in any
taxa except cephalopods, as explained above (Cummins *et al.*, 2011). As those
authors suggest, it would be worthwhile to look for an aggression function for

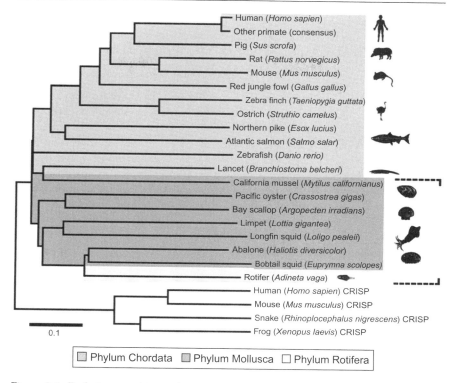

Figure 3.6. Evolutionary origins and conservation of β-microseminoproteins. The tree shows phylogenetic relationships among the protein sequences. The β-MSPs identified in the Cummins *et al.* (2011) study are within the dotted lines. (See Color Insert.)

β-MSPs in mammals and other vertebrates, given the molecular similarity and unique structure of these proteins, all of which seem to be most concentrated in exocrine glands in many taxa. Such findings remind us that multisensory cues are often involved in stimulating behaviors and that a good deal more research is needed before we understand subjects such as aggression.

D. Signaling aggression in humans

In humans, as in other species, signaler and receiver have both evolved to use variation in aggressive signal structure to their own advantage. In the case of human speech, fundamental vocal frequency is perceived to be associated with social cues for dominance and submissiveness (Bolinger, 1978; Huron *et al.*, 2009; Ohala, 1994), with vocal pitch height used to signal aggression (low pitch), or appeasement (high pitch). Moreover, a strong correlation with eyebrow position suggests an intermodal linkage between vocal and facial expressions (Huron *et al.*, 2009). Evidence implicates male dominance competition (Puts *et al.*, 2006), rather than intersexual selection (see Chapter 2), as the selective origin of this performance signal. Similarly, handgrip strength is correlated with level of aggression and appears to be an honest signal for quality in males (Gallup *et al.*, 2007). Mathematical models show, however, that the tradeoff of deceptive efficacy and dishonest signals of intent often favors signalers who produce imperfectly deceptive signals over perfectly honest or perfectly deceptive ones (Andrews, 2002). Competition among coalition groups (a characteristic shared with chimpanzees) initiated a social arms race, culminating in extraordinary human cognitive abilities (Flinn *et al.*, 2005), capable of parsing aggressive signals (Paul and Thelen, 1983), and competitive displays (Hawkes and Bird, 2002). This great capacity for signaling is outstripped only by the uniquely human ability to extend our phenotype with weaponry—with the unfortunate consequence that our potential to inflict damage frequently exceeds our ability to control aggression.

Rather than maximizing its absolute amount, natural selection enhances the overall *effectiveness* of aggression. In invertebrates, where individuals generally pursue a solitary existence, physical superiority primarily determines the eventual outcome of contests, and most fights are quickly resolved on the basis of prominent asymmetries in body or weapon size. In vertebrates, which must navigate the demands and opportunities of social living, aggressive success is largely contingent on the development of social competence. In this case, natural selection favors those with an ability to effectively anticipate their chances well in advance of a contest, and to signal strength while hiding any intentions to eventually withdraw. Generating and interpreting aggressive signals to form successful alliances and to inherit status from high-ranking kin, is thus key to winning both short-term contests and long-term evolutionary success.

Acknowledgments

Production of this chapter was partially supported by funding from NSF grant DUE-0757001 (to M. v. S.). W. A. S. thanks his collaborators on the sparrow signaling research including Steve Nowicki, Rindy Anderson, Barb Ballentine, Mike Beecher, and Susan Peters. R. T. H. is grateful for partial funding from NSF grant IBN-0415519 and many wonderful colleagues who participated in these experiments and field observations.

References

Adamo, S. A., and Hanlon, R. T. (1996). Do cuttlefish (Cephalopoda) signal their intentions to conspecifics during agonistic encounters? *Anim. Behav.* **52**, 73–81.
Adams, E. S., and Mesterton-Gibbons, M. (1995). The cost of threat displays and the stability of deceptive communication. *J. Theor. Biol.* **175**, 405–421.

Alexander, R. D. (1961). Aggressiveness, territoriality, and sexual behavior in field crickets (Orthoptera: Gryllidae). *Behaviour* **17**, 130–223.

Anderson, R. C., Searcy, W. A., and Nowicki, S. (2005). Partial song matching in an eastern population of song sparrows, *Melospiza melodia*. *Anim. Behav.* **69**, 189–196.

Anderson, R. C., Searcy, W. A., Peters, S., and Nowicki, S. (2008). Soft song in song sparrows: Acoustic structure and implications for signal function. *Ethology* **114**, 662–676.

Andersson, M. (1980). Why are there so many threat displays? *J. Theor. Biol.* **86**, 773–781.

Andrews, P. W. (2002). The influence of postreliance detection on the deceptive efficacy of dishonest signals of intent: Understanding facial clues to deceit as the outcome of signaling tradeoffs. *Evol. Hum. Behav.* **23**, 103–121.

Arak, A., and Enquist, M. (1995). Conflict, receiver bias and the evolution of signal form. *Phil. Trans. R. Soc. Lond. B* **349**, 337–344.

Arnott, G., and Elwood, R. W. (2010). Signal residuals and hermit crab displays: Flaunt it if you have it! *Anim. Behav.* **79**, 137–143.

Ballentine, B., Searcy, W. A., and Nowicki, S. (2008). Reliable aggressive signalling in swamp sparrows. *Anim. Behav.* **75**, 693–703.

Bee, M. A., Perrill, S. A., and Owen, P. C. (1999). Size assessment in simulated territorial encounters between male green frogs (*Rana clamitans*). *Behav. Ecol. Sociobiol.* **45**, 177–184.

Beebee, M. D. (2004). The function of multiple singing modes: Experimental tests in yellow warblers, *Dendroica petechia*. *Anim. Behav.* **67**, 1089–1097.

Beecher, M. D., and Campbell, S. E. (2005). The role of unshared songs in singing interactions between neighbouring song sparrows. *Anim. Behav.* **70**, 1297–1304.

Bolinger, D. L. (1978). Intonation across languages. *In* "Universals of Human Language", Vol. 2: Phonology (J. H. Greenberg, C. A. Ferguson, and E. A. Moravcsik, eds.), pp. 471–524. Stanford University Press, Palo Alto, CA.

Borror, D. J. (1965). Song variation in Maine song sparrows. *Wilson Bull.* **77**, 5–37.

Brandt, Y. (2002). Lizard threat display handicaps endurance. *Proc. R. Soc. Lond. B* **270**, 1061–1068.

Breithaupt, T., and Eger, P. (2002). Urine makes the difference: Chemical communication in fighting crayfish made visible. *J. Exp. Biol.* **205**, 1221–1231.

Burt, J. M., Bard, S. C., Campbell, S. E., and Beecher, M. D. (2002). Alternative forms of song matching in song sparrows. *Anim. Behav.* **63**, 1143–1151.

Byers, B. E. (1996). Messages encoded in the songs of chestnut-sided warblers. *Anim. Behav.* **52**, 691–705.

Caryl, P. G. (1979). Communication by agonistic displays: What can games theory contribute to ethology? *Behaviour* **68**, 136–169.

Catchpole, C. K., and Slater, P. J. B. (2008). Bird Song: Biological Themes and Variations. Cambridge University Press, Cambridge.

Chen, S., Lee, A. Y., Bowens, N. M., Huber, R., and Kravitz, E. A. (2002). Fighting fruit flies: A model system for the study of aggression. *Proc. Natl. Acad. Sci. USA* **99**, 5664–5668.

Cheney, D., Seyfarth, R., and Smuts, B. (1986). Social relationships and social cognition in nonhuman primates. *Science* **234**, 1361–1366.

Clark, D. C., and Moore, A. J. (1995). Genetic aspects of communication during male-male competition in the Madagascar hissing cockroach: Honest signalling of size. *Heredity* **75**, 198–205.

Cummins, S. F., Boal, J. G., Buresch, K. C., Kuanpradit, C., Sobhon, P., Holm, J. B., Degnan, B. M., Nagle, G. T., and Hanlon, R. T. (2011). Extreme aggression in male squid induced by a β-MSP-like pheromone. *Curr. Biol.* **21**, 322–327.

Dabelsteen, T., McGregor, P. K., Lampe, H. M., Langmore, N. E., and Holland, J. (1998). Quiet song in song birds: An overlooked phenomenon. *Bioacoustics* **9**, 89–105.

Darwin, C. (1871). The Descent of Man and Selection in Relation to Sex. Murray, London.

DiMarco, F. P., and Hanlon, R. T. (1997). Agonistic behavior in the squid *Loligo plei* (Loliginidae, Teuthoidea): Fighting tactics and the effects of size and resource value. *Ethology* **103,** 89–108.

Dingle, H. (1969). A statistical and information analysis of aggressive communication in the mantis shrimp *Gonodactylus bredini* Manning. *Anim. Behav.* **17,** 561–575.

Enquist, M. (1985). Communication during aggressive interactions with particular reference to variation in choice of behaviour. *Anim. Behav.* **33,** 1152–1161.

Enquist, M., and Jakobsson, S. (1986). Decision making and assessment in the fighting behaviour of *Nannacara anomala* (Cichlidae, Pisces). *Ethology* **72,** 143–153.

Enquist, M., and Leimar, O. (1983). Evolution of fighting behaviour: Decision rules and assessment of relative strength. *J. Theor. Biol.* **102,** 387–410.

Flinn, M. V., Geary, D. C., and Ward, C. V. (2005). Ecological dominance, social competition, and coalitionary arms races: Why humans evolved extraordinary intelligence. *Evol. Hum. Behav.* **26,** 10–46.

Fugle, G. N., Rothstein, S. I., Osenberg, C. W., and McGinley, M. A. (1984). Signals of status in wintering white-crowned sparrows, *Zonotrichia leucophrys gambelii. Anim. Behav.* **32,** 86–93.

Gallup, A. G., White, D. D., and Gallup, G. G. (2007). Handgrip strength predicts sexual behavior, body morphology, and aggression in male college students. *Evol. Hum. Behav.* **28,** 423–429.

Grafen, A. (1990). Biological signals as handicaps. *J. Theor. Biol.* **144,** 517–546.

Greenfield, M. D., and Minckley, R. L. (1993). Acoustic dueling in tarbush grasshoppers: Settlement of territorial contests via alternation of reliable signals. *Ethology* **95,** 309–326.

Gross, M. R. (1996). Alternative reproductive strategies and tactics: Diversity within sexes. *Trends Ecol. Evol.* **11,** 92–98.

Guilford, T., and Dawkins, M. S. (1995). What are conventional signals? *Anim. Behav.* **49,** 1689–1695.

Hall, K. C., and Hanlon, R. T. (2002). Principal features of the mating system of a large spawning aggregation of the giant Australian cuttlefish *Sepia apama* (Mollusca: Cephalopoda). *Mar. Biol.* **140,** 533–545.

Hamblin, S., and Hurd, P. L. (2007). Genetic algorithms and non-ESS solutions to game theory models. *Anim. Behav.* **74,** 1005–1018.

Hanlon, R. T., and Messenger, J. B. (1996). Cephalopod Behaviour. Cambridge University Press, Cambridge.

Hanlon, R. T., Maxwell, M. R., Shashar, N., Loew, E. R., and Boyle, K. L. (1999). An ethogram of body patterning behavior in the biomedically and commercially valuable squid *Loligo pealei* off Cape Cod, Massachusetts. *Biol. Bull.* **197,** 49–62.

Hanlon, R. T., Smale, M. J., and Sauer, W. H. H. (2002). The mating system of the squid *Loligo vulgaris reynaudii* (Cephalopoda, Mollusca) off South Africa: Fighting, guarding, sneaking, mating and egg laying behavior. *Bull. Mar. Sci.* **71,** 331–345.

Hanlon, R. T., Naud, M. J., Shaw, P. W., and Havenhand, J. N. (2005). Behavioural ecology: Transient sexual mimicry leads to fertilization. *Nature* **430,** 212.

Hauser, M. D., and Nelson, D. A. (1991). 'Intentional' signalling in animal communication. *Trends Ecol. Evol.* **6,** 186–189.

Hawkes, K., and Bird, R. B. (2002). Showing off, handicap signaling, and the evolution of men's work. *Evol. Anthropol.* **11,** 58–67.

Hof, D., and Hazlett, N. (2010). Low-amplitude song predicts attack in a North American wood warbler. *Anim. Behav.* **80,** 821–828.

Houston, A. I., and McNamara, J. M. (1999). Models of Adaptive Behaviour, an Approach Based on State. Cambridge University Press, Cambridge.

Huber, R., and Kravitz, E. A. (1995). A quantitative study of agonistic behavior and dominance in juvenile American lobsters (*Homarus americanus*). *Brain Behav. Evol.* **46,** 72–83.

Hughes, M. (2000). Deception with honest signals: Signal function for signalers and receivers. *Behav. Ecol.* **6**, 614–623.

Hurd, P. L., and Enquist, M. (2005). A strategic taxonomy of biological communication. *Anim. Behav.* **70**, 1155–1170.

Huron, D., Dahl, S., and Johnson, R. (2009). Facial expression and vocal pitch height: Evidence of an intermodal association. *EMR.* **4**, 93–100.

Jantzen, T. M., and Havenhand, J. N. (2003). Reproductive behavior in the squid *Sepioteuthis australis* from South Australia: Interactions on the spawning grounds. *Biol. Bull.* **204**, 305–317.

Jonsson, T., Kravitz, E. A., and Heinrich, R. (2011). Sound production during agonistic behavior of male Drosophila melanogaster. *Fly (Austin)* **5**, 29–38.

Kim, Y.-G. (1995). Status signalling games in animal contests. *J. Theor. Biol.* **176**, 221–231.

Kramer, H. G., and Lemon, R. E. (1983). Dynamics of territorial singing between neighboring song sparrows (*Melospiza melodia*). *Behaviour* **85**, 198–223.

Kramer, H. G., Lemon, R. E., and Morris, M. J. (1985). Song switching and agonistic stimulation in the song sparrow (*Melospiza melodia*): Five tests. *Anim. Behav.* **33**, 135–149.

Laidre, M. E., and Vehrencamp, S. L. (2008). Is bird song a reliable signal of aggressive intent? *Behav. Ecol. Sociobiol.* **62**, 1207–1211.

Lailvaux, S. P., Reaney, L. T., and Backwell, P. R. Y. (2009). Dishonesty signalling of fighting ability and multiple performance traits in the fiddler crab *Uca mjoebergi*. *Funct. Ecol.* **23**, 359–366.

Leimar, O., and Enquist, M. (1984). Effects of asymmetries in owner-intruder conflicts. *J. Theor. Biol.* **111**, 475–491.

Logue, D. M., Abiola, I., Raines, D., Bailey, N., Zuk, M., and Cade, W. H. (2010). Does signaling mitigate the cost of agonistic interactions? A test in a cricket that has lost its song. *Proc. R. Soc. Lond. B* **277**, 2571–2575.

Lorenz, K. (1963). Das sogenannte Böse. Methuen Publishing, London.

Lorenz, K. (1978). Vergleichende Verhaltensforschung Grundlagen der Ethologie. Springer-Verlag, Wien.

Lorenz, K. (1995). The Natural Science of the Human Species: An Introduction to Comparative Behavioral Research—The Russian Manuscript (1944–1948). MIT Press, Cambridge, MA.

Lundrigan, B. (1996). Morphology of horns and fighting behavior in the family Bovidae. *J. Mammal.* **77**, 462–475.

Maynard Smith, J. (1974). The theory of games and the evolution of animal conflicts. *J. Theor. Biol.* **47**, 209–221.

Maynard Smith, J. (1979). Game theory and the evolution of behaviour. *Proc. R. Soc. Lond. B* **205**, 475–488.

Maynard Smith, J. (1982). Do animals convey information about their intentions? *J. Theor. Biol.* **97**, 1–5.

Maynard Smith, J., and Harper, D. (1988). The evolution of aggression: Can selection generate variability? *Phil. Trans. R. Soc. Lond. B* **319**, 557–570.

Maynard Smith, J., and Harper, D. (2003). Animal Signals. Oxford University Press, Oxford.

Maynard Smith, J., and Price, G. R. (1973). The logic of animal conflict. *Nature* **246**, 15–18.

McGregor, P. K., and Dabelsteen, T. (1996). Communication networks. In "Ecology and Evolution of Acoustic Communication in Birds" (D. E. Kroodsma and E. H. Miller, eds.), pp. 409–425. Cornell University Press, Ithaca, NY.

Miczek, K. A., De Almeida, R. M. M., Kravitz, E. A., Rissman, E. F., De Boer, S. F., and Raine, A. (2007). Neurobiology of escalated aggression and violence. *J. Neurosci.* **27**, 11803–11806.

Molles, L. E., and Vehrencamp, S. L. (1999). Repertoire size, repertoire overlap, and singing modes in the Banded Wren (*Thryothorus pleurostictus*). *Auk* **116**, 667–689.

Molles, L. E., and Vehrencamp, S. L. (2001). Songbird cheaters pay a retaliation cost: Evidence for auditory conventional signals. *Proc. R. Soc. Lond.* B **268**, 2013–2019.

Moore, J. C., Obbard, D. J., Reuter, C., West, S. A., and Cook, J. M. (2009). Male morphology and dishonest signalling in a fig wasp. *Anim. Behav.* **78**, 147–153.

Moretz, J. A., and Morris, M. R. (2003). Evolutionarily labile responses to a signal of aggressive intent. *Proc. R. Soc. Lond.* B **270**, 2271–2277.

Mowbray, T. B. (1997). Swamp sparrow (*Melospiza georgiana*). *In* "The Birds of North America" (A. Poole and F. Gill, eds.), No. 509. Academy of Natural Sciences, Philadelphia.

Mulligan, J. A. (1963). A Description of Song Sparrow Song Based on Instrumental Analysis Proceedings of the XIII International. Ornithological Congress. pp. 272–284.

Murphy, T. G. (2008). Display of an inedible prop as a signal of aggression? Adaptive significance of leaf-display by the turquoise-browed motmot, *Eumomota superciliosa*. *Ethology* **114**, 16–21.

Natarajan, D., de Vries, H., Saaltink, D. J., de Boer, S. F., and Koolhaas, J. M. (2009). Delineation of violence from functional aggression in mice: An ethological approach. *Behav. Genet.* **39**, 73–90.

Nice, M. M. (1943). Studies in the life history of the song sparrow. II. The behavior of the song sparrow and other passerines. *Trans. Linn. Soc. N.Y.* **6**, 1–328.

Noble, J. (2000). Talk is cheap: Evolved strategies for communication and action in asymmetrical animal contests. *In* "SAB00" (J.-A. Meyer, A. Berthoz, D. Floreano, H. Roitblat, and S. Wilson, eds.), pp. 481–490. MIT Press, Massachusetts.

Nowicki, S., Podos, J., and Valdes, F. (1994). Temporal patterning of within-song type and between-song type variation in song repertoires. *Behav. Ecol. Sociobiol.* **34**, 329–335.

Ohala, J. (1994). The frequency code underlies the sound-symbolic use of voice pitch. *In* "Sound Symbolism" (L. Hinton, J. Nichols, and J. Ohala, eds.), pp. 325–347. Cambridge University Press, Cambridge.

Page, R. B., and Jaeger, R. G. (2004). Multimodal signals, imperfect information, and identification of sex in red-backed salamanders (*Plethodon cinereus*). *Behav. Ecol. Sociobiol.* **56**, 132–139.

Paul, S. C., and Thelen, M. H. (1983). The use of strategies and messages to alter aggressive interactions. *Aggress. Behav.* **9**, 183–193.

Peters, S., Searcy, W. A., Beecher, M. D., and Nowicki, S. (2000). Geographic variation in the organization of song sparrow repertoires. *Auk* **117**, 936–942.

Puts, D. A., Gaulin, S. J. C., and Verdolini, K. (2006). Dominance and the evolution of sexual dimorphism in human voice pitch. *Evol. Hum. Behav.* **27**, 283–296.

Raffinger, E., and Ladich, F. (2009). Acoustic threat displays and agonistic behaviour in the red-finned loach *Yasuhikotakia modesta*. *J. Ethol.* **27**, 239–247.

Reby, D., McComb, K., Cargnelutti, B., Darwin, C., Fitch, W. T., and Clutton-Brock, T. (2005). Red deer stags use formants as assessment cues during intrasexual agonistic interactions. *Proc. R. Soc. Lond.* B **272**, 941–947.

Rek, P., and Osiejuk, T. S. (2011). Nonpasserine bird produces soft calls and pays retaliation cost. *Behav. Ecol.* **22**, 657–662.

Saunders, A. A. (1924). Recognizing individual birds by song. *Auk* **41**, 242–259.

Scott, J. L., et al. (2010). The evolutionary origins of ritualized acoustic signals in caterpillars. *Nat. Commun.* **1**, 4. doi: 10.1038/ncomms1002.

Searcy, W. A., and Andersson, M. (1986). Sexual selection and the evolution of song. *Annu. Rev. Ecol. Syst.* **17**, 507–533.

Searcy, W. A., and Beecher, M. D. (2009). Song as an aggressive signal in songbirds. *Anim. Behav.* **78**, 1281–1292.

Searcy, W. A., and Nowicki, S. (2000). Male-male competition and female choice in the evolution of vocal signaling. *In* "Animal Signals: Signalling and Signal Design in Animal Communication" (Y. Espmark, T. Amundsen, and G. Rosenqvist, eds.), pp. 301–315. Tapir Academic Press, Trondheim.

Searcy, W. A., and Nowicki, S. (2006). Signal interception and the use of soft song in aggressive interactions. *Ethology* **112**, 865–872.

Searcy, W. A., and Yasukawa, K. (1990). Use of the song repertoire in intersexual and intrasexual contexts by male red-winged blackbirds. *Behav. Ecol. Sociobiol.* **27**, 123–128.

Searcy, W. A., Podos, J., Peters, S., and Nowicki, S. (1995). Discrimination of song types and variants in song sparrows. *Anim. Behav.* **49**, 1219–1226.

Searcy, W. A., Nowicki, S., and Hogan, C. (2000). Song type variants and aggressive context. *Behav. Ecol. Sociobiol.* **48**, 358–363.

Searcy, W. A., Anderson, R. C., and Nowicki, S. (2006). Bird song as a signal of aggressive intent. *Behav. Ecol. Sociobiol.* **60**, 234–241.

Simmonds, L. W., and Bailey, W. J. (1993). Agonistic communication between males of a zaprochi-line katydid (Orthoptera: Tettigoniidae). *Behav. Ecol.* **4**, 364–368.

Skyrms, B. (2009). Evolution of signalling systems with multiple senders and receivers. *Phil. Trans. R. Soc. B* **364**, 771–779.

Spector, D. A. (1992). Wood-warbler song systems: A review of paruline song systems. *Curr. Ornithol.* **9**, 199–239.

Stoddard, P. K., Beecher, M. D., and Willis, M. S. (1988). Response of territorial male song sparrows to song types and variations. *Behav. Ecol. Sociobiol.* **22**, 125–130.

Stoddard, P. K., Beecher, M. D., Horning, C. L., and Willis, M. S. (1990). Strong neighbor-stranger discrimination in song sparrows. *Condor* **92**, 1051–1056.

Stoddard, P. K., Beecher, M. D., Campbell, S. E., and Horning, C. L. (1992). Song-type matching in the song sparrow. *Can. J. Zool.* **70**, 1440–1444.

Számadó, S. (2000). Cheating as a mixed strategy in a simple model of aggressive communication. *Anim. Behav.* **59**, 221–230.

Turner, C. R., Derylo, M., de Santana, C. D., Alves-Gomes, J. A., and Smith, G. T. (2007). Phylogenetic comparative analysis of electric communication signals in ghost knifefishes (Gymnotiformes: Apteronotidae). *J. Exp. Biol.* **210**, 4104–4122.

Vehrencamp, S. L. (2000). Handicap, index, and conventional signal elements of bird song. *In* "Animal Signals: Signalling and Signal Design in Animal Communication" (Y. Espmark, T. Amundsen, and G. Rosenqvist, eds.), pp. 277–300. Tapir Academic Press, Trondheim.

Wagner, W. E. (1992). Deceptive or honest signalling of fighting ability? A test of alternative hypotheses for the function of changes in call dominant frequency by male cricket frogs. *Anim. Behav.* **44**, 449–462.

Weary, D. M., Lemon, R. E., and Perrault, S. (1994). Different responses to different song types in American redstarts. *Auk* **111**, 730–734.

Wheeler, M., and De Bourcier, P. (1995). How not to murder your neighbor: Using synthetic behavioral ecology to study aggressive signaling. *Adapt. Behav.* **3**, 273–309.

Wingfield, J. C., Moore, I. T., Goymann, W., Wacker, D. W., and Sperry, T. (2005). Contexts and ethology of vertebrate aggression: Implications for the evolution of hormone-behavior interactions. *In* "Biology of Aggression" (R. J. Nelson, ed.), pp. 179–211. Oxford University Press, Oxford.

Wong, B. B. M., and Candolin, U. (2005). How is female mate choice affected by male competition? *Biol. Rev.* **80**, 559–571.

Zahavi, A. (1987). The theory of signal selection and some of its implications. *In* "International Symposium of Biological Evolution" (V. P. Delfino, ed.), pp. 305–327. Adriatica Editrice, Bari.

4

Self-Structuring Properties of Dominance Hierarchies: A New Perspective

Ivan D. Chase* and Kristine Seitz†
*Department of Sociology, Stony Brook University, Stony Brook,
New York, USA
†Department of Biology, Stony Brook University, Stony Brook,
New York, USA

Advances in Genetics, Vol. 75
0065-2660/11 $35.00
DOI: 10.1016/B978-0-12-380858-5.00001-0

ABSTRACT

Using aggressive behavior, animals of many species establish dominance hierarchies in both nature and the laboratory. Rank in these hierarchies influences many aspects of animals' lives including their health, physiology, weight gain, genetic expression, and ability to reproduce and raise viable offspring. In this chapter, we define dominance relationships and dominance hierarchies, discuss several model species used in dominance studies, and consider factors that predict the outcomes of dominance encounters in dyads and small groups of animals. Researchers have shown that individual differences in attributes, as well as in states (recent behavioral experiences), influence the outcomes of dominance encounters in dyads. Attributes include physical, physiological, and genetic characteristics while states include recent experiences such as winning or losing earlier contests. However, surprisingly, we marshal experimental and theoretical evidence to demonstrate that these differences have significantly less or no ability to predict the outcomes of dominance encounters for animals in groups as small as three or four individuals. Given these results, we pose an alternative research question: How do animals of so many species form hierarchies with characteristic linear structures despite the relatively low predictability based upon individual differences? In answer to this question, we review the evidence for an alternative approach suggesting that dominance hierarchies are self-structuring. That is, we suggest that linear forms of organization in hierarchies emerge from several kinds of behavioral processes, or sequences of interaction, that are common across many different species of animals from ants to chickens and fish and even some primates. This new approach inspires a variety of further questions for research.   © 2011, Elsevier Inc.

I. INTRODUCTION

Both humans and animals use aggression in many contexts as discussed in this volume. In this chapter, we talk about aggression as it is used in the social context of establishing dominance relationships and dominance hierarchies. A broad range of species—from insects to humans—form these types of relationships and hierarchies, and where they occur, hierarchy rank has wide-ranging and serious consequences for individuals (Addison and Simmel, 1970; Barkan *et al.*, 1986; Goessmann *et al.*, 2000; Hausfater *et al.*, 1982; Heinze, 1990; Nelissen, 1985; Post, 1992; Savin-Williams, 1980; Vannini and Sardini, 1971; Wilson, 1975). These consequences include variation in physiology, stress, health, growth rate, access to sexual partners, ability to raise viable offspring, and even in the thickness of nerves leading from the hypothalamus to the pituitary gland (Clutton-Brock *et al.*, 1984; Ellis, 1995; Francis *et al.*, 1993; Holekamp and Smale, 1993; Post, 1992; Sapolsky and Share, 1994).

Our discussion centers on dominance in less complex animals such as chickens, fish, and nonprimate mammals. We begin with behavioral definitions of dominance relationships and hierarchies. We then move on to the factors that predict dominance in pairs of animals including differences in both traits and behavioral states. Traits include genetic, physical, physiological, and "personality" attributes; behavioral states include the influences of winning and losing contests, as well as observing the contests of other individuals. Next, we review the research that demonstrates, very surprisingly, that factors affecting dyadic encounters are poor predictors of outcome in dominance encounters within groups of any size, even as small as three or four individuals. Consequently, researchers should not expect that findings from dominance in pairs of animals will easily generalize to small groups of animals.

Finally, we suggest that rather than trying to predict the ranks of individuals in groups, a more appropriate research question is to ask how these hierarchies often come to have a linear structure (defined below) across many species. We discuss recent experimental findings that conclude that these linear hierarchies are self-structuring or self-organizing. Self-structuring hierarchies are common across a range of species. They emerge from the repeated use of several characteristic, small-scale behavioral processes, or sequences of behavior. We describe how these processes generate linear hierarchies and discuss some unresolved experimental questions suggested by these behavioral processes.

II. DEFINITIONS

A. Dominance relationships

While researchers have proposed a variety of definitions, in this chapter, we will be using a strictly behavioral measure of dominance relationships. More specifically, we will define a dominance relationship as characterized by an asymmetry of aggressive behaviors by one animal toward another. Typically, both noncontact and contact behaviors are involved. Common noncontact type behaviors are chasing, displacement, or the threat of aggressive contact. Threat behaviors vary by species and include gill flaring in fish; certain vocalizations and gestures in primates; and short, rapid movements toward the threatened individual in a variety of species. Aggressive contact behaviors also vary by species and include pecking in the case of chickens and other birds; biting in fish, rodents, and primates; and grasping with claws in crayfish, lobsters, hermit crabs, and other crustaceans.

There is no established standard that defines the number of acts in a row that are recorded to determine dominance in different species. However, the general rule is to use a sufficient number such that, when it is reached, a stable relationship has been revealed with very little likelihood of the animal declared subordinate beginning to attack the one declared dominant. For example, in determining the presence of dominance relationships, Chase *et al.* (2002) declared that one cichlid fish was dominant over another if it bit or chased the other fish six times in a row without the other fish initiating an aggressive action in return.

Other methods for determining dominance, such as recording which of two animals obtains a desired piece of food, do not necessarily reflect the kind of relatively stable social relationships that are indicated by asymmetries in aggressive behavior in pairs of animals. For example, when two monkeys are competing for a peanut, the winner may be an otherwise subordinate animal that is quicker and ready to withstand the chasing and harassing it will receive from the normally dominant animal with which it lives in order to secure the peanut. Researchers also sometimes determine dominance by observing which individual delivers the majority, rather than an uninterrupted sequence, of aggressive acts over a period of observation. A measure of dominance such as this may be appropriate in some species, such as pigeons and young children, who do not always form completely asymmetric relationships. Such measures, however, can be misleading if used when animals meet for the first time. When animals meet initially, there is often a trading back and forth of aggressive actions before one individual begins to deliver all the actions and clearly becomes dominant over the other. Figure 4.1 is a music notation graph showing the interactions of two

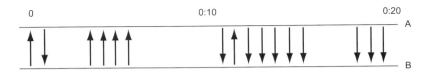

Figure 4.1. Graphic display of the record of interaction between two hens using music notation. Horizontal lines represent individuals, ordered by eventual dominance rank. Each aggressive act between individuals is indicated by a vertical arrow from the line representing the initiator to the line representing the receiver. The time in minutes since the assembly of the pair appears above the graph. Letters at the ends of the horizontal lines identify the hens. See Chase (2006) for more information about the uses of music notation in visualizing interactions in groups of animals and humans.

chickens setting up a relationship (Chase 2006). If only the first 10 min of interactions are considered, then chicken B is dominant over A since B delivers 83% of the aggressive interactions and chicken A only 17%. However, as time goes on, the back and forth actions stop, and chicken A is soon delivering all aggressive acts. If a researcher had only considered which individual initiated the majority of acts during the initial phases of the interaction, an incorrect indication of dominance would have resulted.

B. Dominance hierarchies

A dominance hierarchy is the overall collection, or network, of dominance relationships among the pairs of individuals in a group. In many small groups of animals and human children of around eight or ten members or less, dominance hierarchies often take a classical *linear* form (Addison and Simmel, 1970; Barkan *et al.*, 1986; Goessmann *et al.*, 2000; Hausfater *et al.*, 1982; Heinze, 1990; Nelissen, 1985; Post, 1992; Savin-Williams, 1980; Vannini and Sardini, 1971; Wilson, 1975). In a linear hierarchy, there is one individual who dominates all the other group members, a second who dominates all but the top individual, and so on, down to the last individual who dominates no one. In larger groups, there is often the skeleton of a linear structure, but even with extensive observations, researchers do not see interactions between some pairs, especially those that seem distant in rank. In hierarchies that are not linear, there are inconsistencies in rank showing intransitive relationships (A dominates B, B dominates C, but C dominates A). The hierarchies in some animals, especially those with more complex social organization such as dolphins, chimpanzees, hyenas, baboons, and macaques, are often too complex to be simply classified as linear (Holekamp and Smale, 1993; Kummer, 1984; Möller *et al.*, 2001, 2006; Surbeck *et al.*, 2011; Widdig *et al.*, 2001, among many others). Our discussion here will concentrate on those animals forming more linear hierarchies.

III. ANIMAL MODELS

Animal behaviorists have shown that a huge range of animals establish dominance relationships and dominance hierarchies in the wild and in the laboratory. These include insects such as fruit flies and some ants, wasps and cockroaches; crustaceans such as hermit crabs, crayfish, and lobsters; reptiles such as anoles; many species of fish, birds, and mammals; and even human preschoolers and adolescents (Addison and Simmel, 1970; Barkan *et al.*, 1986; Clark, 1998; Goessmann *et al.*, 2000; Hausfater *et al.*, 1982; Heinze, 1990; Nelissen, 1985; Post, 1992; Queller *et al.*, 2000; Savin-Williams, 1980; Vannini and Sardini, 1971; Wilson 1975).

In biomedical research, an animal model is usually an animal species used for research on a human disease or other condition. Dominance researchers do not often use specific species in the strict sense of models for human conditions, but instead as models for dominance in animals more broadly. The partial exception to this is that some researchers have studied dominance in primates to discover information about how dominance processes may work in human groups (see below). Below is a brief description of animals or animal groups used frequently in dominance research and a representative, but by no means comprehensive, sampling of work done with these animals.

A. Chickens

Chickens were among the first animals to be studied for their dominance relationships. Schjelderup-Ebbe (1922) introduced dominance hierarchies into the modern study of animal behavior and coined the term "peck order" (or *Hackordnung* in the original German in which he wrote). Schjelderup-Ebbe (1922) was among the first researchers to observe the highly linear structure of pecking orders. He noted that a number of factors influenced rank in the flock, including stress, prior experience, overall health, mating condition, and age. He further concluded that dominance is based, not just on the size and strength of the combatants but also on the perception of fellow flock members (Schjelderup-Ebbe, 1922).

Other early researchers used chickens to explore the relationship between stress and dominance. Sactuary (1932), for example, showed that hens that mysteriously molted out of season and went out of laying condition had been relegated to the lower ranks of the flock. Thus Sactuary (1932) linked rank to both fitness (ability to produce offspring) and stress levels.

Some more recent investigations in chickens have focused on the relationship between dominance, aggression, and selective breeding. These lines of inquiry began with the rise of the factory farms, in which aggression leading to deaths and lowered egg production is of great concern (Craig and Muir, 1996; Craig *et al.*, 1965, 1969, 1975, among others).

B. Fish

Fish have recently become one of the most popular vertebrate models for dominance research. The fish model system is comparable to chickens, in that fish possess easily observed dominance behaviors (chases, bites, and threat behaviors), recognize members of their group as individuals, and can maintain their dominance hierarchies for extended periods of time. Fish, however, are easier to maintain in the laboratory than chickens. They have been used for a variety of studies relating to dominance which we can only briefly cover here.

One common type of study of fish explores how differences in behavioral states and individual attributes affect the outcome of dominance contests. For example, researchers have used fish to investigate the so-called *winner, loser,* and *bystander* effects (defined below) (Chase *et al.*, 1994, 2002; Hsu and Wolf, 1999; Hsu *et al.*, 2006; Oliveira *et al.*, 2009, among others); *prior residency*, a kind of home field advantage assumed to confer benefits in social contexts (Beaugrand and Beaugrand, 1991); and the effects of differences in size and prior social experience (Beaugrand and Cotnoir, 1996).

Besides studying factors affecting the outcome of contests, researchers have utilized fish in selection experiments studying the heritability of *dominance-related* aggression (Bakker, 1985, 1986; Francis, 1984, 1987) and in investigations into physiological and genetic components of dominance (see Sloman and Armstrong 2002 for a review of physiological aspects).

C. Crustaceans

Crustaceans are a unique model system that allows researchers to study chemical signaling behaviors, and the anatomy of their nervous systems enables researchers to study the neural underpinnings of dominance relationship formation (Moore and Bergman, 2005). One of the most commonly studied chemical signals in this group of species is the release of urine during agonistic behaviors. In lobsters, this signal has been implicated in the maintenance of dominance hierarchies (Breithaupt and Atema, 2000; Karavanich and Atema, 1998), and in crayfish, it has been shown to reduce aggression in opponents during dominance bouts (Breithaupt and Eger, 2002). Yeh *et al.* (1997) showed that changes in dominance status altered levels of serotonin in the crayfish, *Procambarus clarkii*. These changes caused modifications in the command neuron involved in escape behaviors in this species and were found to be reversible and linked to changes in the population of serotonin receptors.

D. Primates

The dominance systems of primates are often more complex than the other model systems just discussed. As a result, their dominance behavior can more easily be generalized to humans. One of the most important lines of research in primates has shown how stress affects the hormonal responses of animals of different ranks. Sapolsky (1982) studied wild olive baboons *(Papio anubis)* and found that high-ranking males showed a low initial level of cortisol, but in response to stress, they had faster and larger spikes of cortisol than their less successful counterparts. Sapolsky (1982) suggested that the high ranking member's cortisol responses might be more adaptive to their social environments, because their usual, lower cortisol levels conferred immunological and other health benefits. A general review of the influence of hierarchies on primate health can be found in Sapolsky (2005).

IV. FACTORS AFFECTING DOMINANCE RELATIONSHIPS IN PAIRS OF ANIMALS

In attempting to predict the outcomes of dominance contests in pairs of animals, researchers have used two broad classes of variables: differences in attributes or traits and differences in behavioral conditions or states. Attributes are relatively long-lasting characteristics, while states are shorter-term conditions often influenced by behavioral events.

A. Physical differences

Differences in physical attributes often have a considerable impact on the outcome of dominance encounters. One common characteristic used is differences in the sizes of the organisms, which can be broken down into two categories: differences in weights and differences in lengths or heights. Researchers have found that larger, heavier animals usually dominate animals that are smaller and lighter (Frey and Miller, 1972; Houpt et al., 1978; Nakano and Furukawa-Tanaka, 1994; Knights, 1987; Lott and Galland, 1987). However, when size differences are smaller, other factors can influence the outcomes of contests. For example, in male green swordtail fish, a difference of 20–30% in the size of the lateral surface area of fish meeting in dyadic dominance contests generally resulted in the larger fish becoming dominant over the smaller (Beaugrand et al., 1996). Contests between fish with size differences of 10–20%, however, showed that other factors such as prior social experiences (winning or losing) and prior residency influence the outcome of contests. Size differences below 10% do not influence dominance contests at all. Instead, the social experience (discussed below) of the fish is usually the deciding factor in dominance contests (Beaugrand et al., 1996). Similar to standard length, the effects of weight on dominance success can be ameliorated by other factors such as having won or lost a prior contest (Beacham, 1988; Schulte-Hostedde and Millar, 2002).

Although size and weight are, perhaps, the physical attributes most widely studied for their effect on dominance contests, other physical features and conditions have also been implicated in dominance success. One example that has been studied extensively is the dominance badge, an area of color on the body of an animal that acts to indicate dominance to conspecifics (see Senar, 1999 for a review). Other examples include the state of molt and the size of combs in chickens (Collias, 1943), the size of genital papilla in fish (Schwanck, 1980), and the bill size in birds (Shaw, 1986).

1. Behavioral profile or personality

Repeatable behavioral type or *personality* can be defined as suites of behaviors that differ among individuals but are consistently repeatable in multiple contexts over time (Bell *et al.*, 2009; Boon *et al.*, 2007; Groothius and Carere, 2005; Martin and Réale, 2008; Sih *et al.*, 2004; Sinn and Moltschaniwskyj, 2005; Svartberg *et al.*, 2005). Researchers have demonstrated that behavioral profiles can predict the outcome of dyadic dominance encounters in a variety of species with high accuracy. For example, in fish, brown trout that scored higher in boldness were more likely to dominate those that scored lower (Sundström *et al.*, 2004), and rainbow trout that had shorter or more proactive responses to stress were more likely to dominate those with longer or reactive responses (Øverli *et al.*, 2004; Schjolden *et al.*, 2005). In birds, mountain chickadees classified as high-exploring individuals (those that visited more sites within a strange area) dominated low-exploring individuals, and in great tits, fast explorers dominated slow explorers (Fox *et al.*, 2009; Verbeek *et al.*, 1996). Interestingly, Verbeek *et al.* (1999) found that the same behavioral profiles in great tits that predicted dominance in dyads gave opposite results in groups of five to eight great tits. Here, the slow explorers had higher average dominance scores.

B. Physiology

Whether or not physiological differences can predict the outcome of dyadic dominance encounters is an extremely vexed question. In the mid-twentieth century, researchers thought that differences in testosterone, among other hormones, were reliable determinants of dominance (For recent work see Huber *et al.*, 1997). However, subsequent research demonstrated that the causal direction is often reversed—in many species, the ranks of individuals in hierarchies has a strong influence on the levels of their hormones and other physiologically active chemicals rather than vice versa (see, e.g., Eaton and Resko, 1974; Sapolsky, 1982; Trainor and Hofmann, 2007). Further, Sloman and Armstrong (2002), in a general review, suggest that at least for fish, the physiological effects of dominance encounters in simple laboratory settings may be stronger than those observed in more complex laboratory or natural habitats.

These caveats notwithstanding, there is a considerable recent literature on physiological predictors of dominance in the dyadic encounters of fish, chiefly in trout and salmon. For example, Metcalfe *et al.* (1995), Cutts *et al.* (1999), and McCarthy (2001) find that fish with higher relative metabolic rates dominate those with lower relative rates. In tests of responses to stress, Øverli *et al.* (2004) and Schjolden *et al.* (2005) report that trout with lower levels of cortisol defeat those with higher levels. In experiments in which Arctic charr are dosed with

L-dopa, the immediate precursor of the neurotransmitter dopamine, and trout are dosed with growth hormone, treated fish dominated control fish at significant rates (Johnsson and Björnsson, 1994; Winberg and Nilsson, 1992).

We obviously need further research to untangle the complicated chains of cause and effect among various physiological variables and outcomes in dominance relationships.

C. Genetics

The inheritance of dominance and aggressiveness has been a topic of interest since the field's inception. In artificial selection experiments, Craig et al. (1965) were able to produce hens with diverging dominance abilities. In dyadic contests, hens of high dominance ability usually defeated those of low dominance ability. Even based on these early findings, however, Craig et al. (1965) concluded that variations caused by interactions between genes (nonadditive genetic variation) and environmental factors were likely to be important in the inheritance of dominance. Similar studies of paradise fish (Francis, 1984, 1987), cockroaches (Moore, 1990), and deer mice (Dewsbury, 1990) confirmed that dominance could be artificially selected in a variety of species. Artificial selection, however, can only imply a genetic basis for dominance and cannot identify which genes are responsible for dominance or subordination. An alternative explanation for at least some of these results is that the social environment of mothers (including their levels of aggression and dominance ranks) can expose prenatal young to androgens that can influence their offsprings' behavior. Such maternal influences operate independently of genotype and have been implicated in the inheritance of rank-related behavior. For example, the level of female aggression affects the amount of maternally derived testosterone in tree swallow eggs, which, in turn, influences the growth and dominance of the hatchlings (Whittingham and Schwabl, 2002). In mammals, higher ranking female hyenas (*Crocuta crocuta*) have higher levels of *in utero* androgens causing their cubs to more aggressive than those of lower ranking females (Dloniak et al., 2006).

To begin to tease apart these and other influences on rank order, the newest technological advances in molecular biology are being employed to investigate which genes influence social behavior. *Sociogenomics*, the study of social systems at a molecular level, can offer us insights and information never before available to behavioral scientists. A variety of unique insights have arisen from this new way of studying social dominance and are revealing two major themes: one theme is that genes involved in nonsocial behaviors are often also implicated in social behaviors; the second is that genes are highly sensitive to social influences, and regulation of gene expression by social factors heavily influences behavior (Robinson et al., 2005).

An outstanding example of the interplay between genetic expression and social factors occurs in the cichlid fish *Astatolapia burtoni*. In this fish, dominant males are brightly colored and actively defend territories for mating. Subordinate males are nonreproductive, move about in schools and mimic females' cryptic coloration. Subordinate males, however, grow faster than dominant males, giving subordinates the opportunity to depose dominant males from their territories. These phenotypes are plastic and males may switch back and forth between phenotypes several times in a life span, depending on the availability of suitable territories to defend (Burmeister *et al.*, 2005; Renn *et al.*, 2008).

Burmeister *et al.* (2005) investigated the neural mechanisms linking social environment to physiological changes associated with dominance. They found that when a subordinate male perceives an opportunity to move to a territory and become dominant, he begins to produce dominant coloration and some initial behavioral changes in as little as a day (Burmeister *et al.*, 2005). It takes about 7 days, however, for males ascending to dominance to produce the same amount of gonadotropin-releasing hormone (GnRH1) as a dominant male, during which time the size of testes and GnRH1 neurons increase to sizes comparable to dominants (Burmeister *et al.*, 2005). In *A. burtoni*, GnRH1 is produced by neurons in the anterior parvocellular preoptic nucleus (aPPn), the most anterior part of the preoptic area in teleosts. To study the behavioral and *the* genomic response to social opportunity, researchers chose to focus on the gene *egr-1*, which codes for a transcription factor involved in neuronal plasticity and links membrane depolarization to late-response target genes. Expression of this gene was compared in socially ascending males and dominant and subordinate males in stable hierarchies (Burmeister *et al.*, 2005). Their results show that socially ascending males had a twofold induction of *egr-1* in the aPPn, compared to the both dominant and subordinate males in stable positions. Expression levels in other parts of the brain did not differ with social status or opportunity (Burmeister *et al.*, 2005). Stable dominant males do not show this spike in *egr-1*, suggesting that this change is a response to social opportunity rather than a response to dominance itself. Although socially ascending males also show a difference in physical activity (e.g., more threat displays), it is not clear whether there is a simple relationship between *egr-1* expression and increased motor activity. Instead, Burmeister *et al.* (2005) conclude that the relationship of social context, expression of *egr-1*, and activity differences have a complex relationship that cannot be adequately explained by the simple functional motor and sensory aspects of the experience (Burmeister *et al.*, 2005).

Renn *et al.* (2008) studied how social context affects physiology, but in this case, a microarray was used to investigate coregulated gene sets that might differentiate dominant males, subordinate males, and brooding females. The results show that there are, indeed, gene sets that are common to each of these phenotypes, with males (both dominant and subordinate) and females having the largest (16%) difference in gene expression (Renn *et al.*, 2008). Twenty-one genes were found to be upregulated in the subordinate male phenotype, and it was hypothesized that downregulation of these genes would lead to the dominant phenotype. Additionally, subordinate males and brooding females were found to have a similar expression pattern for 16 genes, possibly suggesting a type of subordination module (Renn *et al.*, 2008). Interestingly, although gene sets for phenotypic traits were found, results also showed that there was as much variation in gene expression between individuals of the same phenotype as there was between the phenotypes themselves. This suggests that widely variant gene expression in individuals can still yield reliable, easily identifiable phenotypes (Renn *et al.*, 2008).

Trainor and Hofmann (2007) investigated the neuropeptide hormone somatostatin, and its receptors, for possible involvement in social behavior. This hormone and its numerous receptor subtypes have been shown to play a role in the inhibition of growth hormone secretion, among other more diverse effects. The relationship between somatostatin gene expression and body size differed between dominant and subordinate individuals. In dominant males, the gene expression of one subgroup of receptors in the hypothalamus was negatively associated with body size. In subordinate fish, however, gene expression was positively correlated with body size. This suggests that growth in this animal may be socially mediated at the genetic level (Trainor and Hofmann, 2007).

D. Behavioral states: Winner, loser and bystander effects

In addition to differences in attributes or traits, considerable research also demonstrates that differences in states can influence the outcomes of dominance encounters in pairs of animals. Most of this research has examined what are known as winner, loser, and bystander effects. In a winner effect, an animal that has won an earlier contest with one individual has an increased probability of winning a second contest with another individual. In a loser effect, an animal that has lost a dominance encounter with one individual has an increased probability of losing a subsequent contest with another individual. In a bystander effect, an animal that has observed a dominance contest between two others alters its behavior, compared to a nonobserver, when it meets either of the animals that it observed interacting.

Researchers have discovered loser effects in a broad range of species, and there is some evidence that these effects may last for several days (see, e.g., Chase *et al.*, 1994; Hsu *et al.*, 2006). Winner effects seem to be less common across species and not as strong as loser effects. Further, some species seem not to have them at all (Chase *et al.*, 1994; Rutte *et al.*, 2006). In particular, Fuxjagera and Marlera (2010) show that winner effects can be documented in some species but can be nonexistent in other, closely related species. Where winner effects occur, there is some evidence that they are of much shorter duration than loser effects, lasting perhaps less than an hour or so after an individual's initial winning experience (Bergman *et al.*, 2003; Chase *et al.*, 1994). Oliveira *et al.* (2009) have shown that winner effects can be ameliorated with antiandrogen drugs, but loser effects are unchanged. Clearly, additional work is needed in this area to elucidate the role these effects have on the formation of dominance relationships in pairs of animals.

Research indicates that animals in many species are attentive observers of other individuals and that they use the information gained in their observations in shaping their future behavior with those observed. For example, a bystander fish may be less aggressive when it meets another fish it has observed winning a contest. Bystander effects have been observed in a broad range of species (see, e.g., Oliveira *et al.*, 1998; Oliveira *et al.*, 2001; Danchin *et al.*, 2004; Peake and McGregor, 2004).

V. FORMATION OF DOMINANCE RELATIONSHIPS AND DOMINANCE HIERARCHIES IN GROUPS

Given the research that we have just reviewed, it would seem natural to assume that the same factors that strongly predict the outcomes of dominance encounters in pairs of animals by themselves should also work for dominance encounters between pairs of animals in groups. That is, the factors that predict dominance in isolated pairs should also predict dominance for socially embedded pairs. Predicting the outcome of dominance encounters for all the embedded pairs in a group would allow us to rank individuals within the dominance hierarchy and reveal the hierarchical structure. Surprisingly, while individual differences in attributes and states do have some influence on rank in hierarchies, that influence is significantly less in groups than it is in dyadic pairs. Consequently, other factors must be at work in determining individual ranks within hierarchies and in generating hierarchical structure. Unraveling that paradox—how individuals can be clearly differentiated by rank, but with that differentiation not strongly based upon individual differences—is a great challenge in the study of dominance hierarchies.

In the next section, we present evidence demonstrating that individual differences can neither adequately predict the places of individuals within hierarchies nor explain their overall linear structure. Following that, we describe a new approach that we believe can account for the common formation of linear hierarchies across a variety of species.

A. Differences in individual attributes and hierarchy formation

The prior attributes hypothesis proposes that differences in the characteristics that animals possess before forming a hierarchy predetermine their resulting hierarchy ranks. Figure 4.2 illustrates this hypothesis in graphical form. In the hypothesis, the individual ranking highest on attributes takes the top position when the hierarchy forms; the individual ranking second-highest takes the *next-to-the-top* position; and so on. Rank based on prior attributes could be determined by any set of characteristics: physical ones such as weight, personality ones such as aggressiveness or boldness, genetic ones such as overall genotype or specific genetic markers, social ones such as the conditions of rearing or family background, physiological ones such as various hormone levels, and so on.

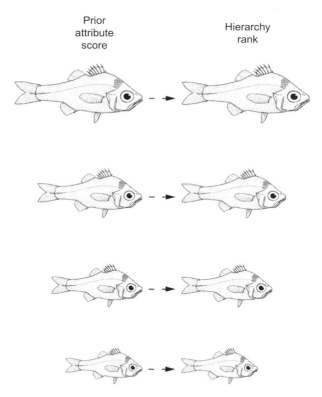

Figure 4.2. Graphical illustration of the prior attributes hypothesis. Size indicates relative prior attribute value; larger size indicates higher rank on prior attributes. Figure adapted from Figure 24.1, p. 570 in "Dominance hierarchies" by Ivan D. Chase and W. Brent Lindquist from *Oxford Handbook of Analytical Sociology* ed. by Hedström, P. and Bearman, P. (2009), by permission of Oxford University Press.

The problem in testing the prior attributes hypothesis is that although an experimenter might know some of the traits that influence dominance, he or she might not know all those involved or the size of the contribution of a specific trait to dominance outcomes. In order to get around this problem, Chase *et al.* (2002) designed an experiment to test the prior attributes hypothesis without knowing which attributes or the sizes of their contributions that might be involved in hierarchy formation.

In their experiment, they brought together groups of four cichlid fish and let them form hierarchies, separated the fish for two weeks, which was sufficient time for them to forget one another as individuals, and then reassembled them to form second hierarchies. The plan was to let the fish form a hierarchy and then, to the extent possible, "rewind their tape," removing all memory of recent social experience before letting them form a second hierarchy. If prior attributes, whatever they might have been, determined the ranks of the fish in the first hierarchy, the attributes, provided they were reasonably stable, should also have determined the ranks of the fish in the second hierarchies. Consequently, by the prior attributes hypothesis, the positions of the fish in the first and second hierarchies should have been the same for all, or at least most, of the groups.

The results of this experiment are shown in Fig. 4.3 and Table 4.1. Instead of a high proportion of groups having identical first and second hierarchies, the experimenters found that only about a quarter (26%) of the groups did so. In nearly three-quarters of the groups, two, three, or even all four fish had different ranks in the two hierarchies. Prior attributes did, however, have a moderate influence on the ranks of the fish within the hierarchy—more groups formed identical first and second hierarchies than expected by chance alone, and there was, on average, moderate rank order correlations between the ranks of individuals in the two hierarchies. However, the lack of a high proportion of groups with identical first and second hierarchies indicated that some other factor played a substantial part in the formation of linear hierarchies and the ranks of individuals within them.

One question that can be raised about the interpretation of these results is, what if the fish changed after the first hierarchy so that their attribute ranks were different before they formed their second hierarchy? For example, for simplicity consider just one attribute called dominance ability. What if the rank on dominance ability before the first hierarchy had been A, B, C, D, but before they formed the second hierarchy their order had changed to give a new ranking D, C, A, B? The difference in dominance ability could still have accounted for the linear hierarchies but the fish would have different ranks in the second hierarchy than the first. This question is discussed in some detail in Chase *et al.*, (2002), which argues against this counter-explanation. In addition to that discussion, some more recent experimental work also suggests a lack of

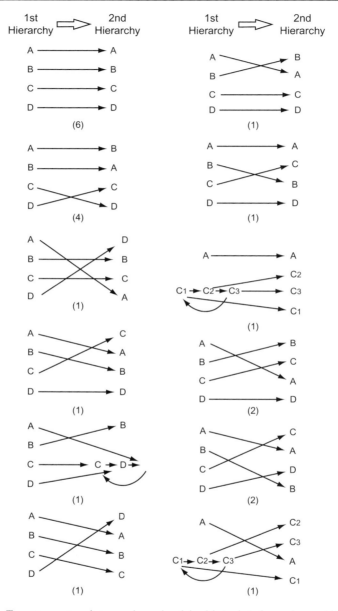

Figure 4.3. Transition pattern between the ranks of the fish in their first and second hierarchies. The total number of groups in the experiment was 22; the number of groups showing a particular transition pattern is indicated in parentheses below each pattern. Fish that have an intransitive dominance relationship (A dominates B, B dominates C, and C dominates A) share the same rank. Intransitive relationships are discussed below.

Table 4.1. Percentage of Groups with Different Numbers of Fish Changing Ranks Between their First and Second Hierarchies ($n = 22$)

No. of fish changing ranks	Percentage of groups
0	27.3
2	36.4
3	18.2
4	18.2

support for this counter-explanation. When isolated pairs of fish are tested under the same experimental conditions as the groups forming and reforming hierarchies, they have an extremely high rate (94%) of forming the same dominance relationship each time. If A dominated B when they met the first time, A virtually always dominated B the second time they met. However, at 76%, the replication rate for socially embedded pairs in the hierarchy groups was significantly less than that of isolated pairs. If the rank of the fish on attributes changed between meetings so that they did not always dominate the same fish when their hierarchies formed the second time, then likewise there should have been a similarly low rate of replication in relationships when the isolated pairs met for the second time. But this did not occur, so the changes in the relationships and ranks of the fish must be accounted for by other factors rather than changes in ranks on attributes.

B. Influence of social factors on linear hierarchy formation

In order to discover what factors were necessary for the formation of linear hierarchies, Chase *et al.* (2002) carried out a second experiment with cichlid fish. In this experiment, they set out to test that social processes, behavioral processes that could only take place in a group context, were crucial for linear hierarchy formation and that they contributed more to linearity than prior attributes. In this experiment, they allowed groups of four and five fish to form hierarchies by two means: round-robin competition and group assembly. In round-robin competition, the fish in a group met only as isolated pairs, out of sight of the other members of their group. The sequence of meeting was as follows: first, A and B met, then C and D, A and C, D and B, and so on. In this way, differences in individual attributes could determine which one of a pair would dominate. If, say, A was superior in dominance attributes to B, A could dominate when they met as a pair. However, social processes such as C getting information by observing the outcome of a contest between A and B and then using that information in interacting with either A or B was not possible, since each pair formed a relationship in isolation from other group members. In group assembly, all the members of a group met in an aquarium at the same time. This allowed fish to use whatever social process they were capable of in forming their hierarchies.

Table 4.2 shows the results of this experiment. When groups of fish established their hierarchies using round-robin competition, only about half of them formed linear hierarchies (but see McGhee and Travis 2010 for contrasting results). When they established their hierarchies using group assembly, nearly all of the hierarchies were linear. Behavior that only occurred in a group context ensured the development of linear hierarchies, while differences in individual attributes did not. However, differences in individual attributes still had some influence on the production of linear hierarchies: the proportion of linear hierarchies with round-robin competition was higher in groups of five fish than would be expected by chance alone.

Although the experiments just described demonstrated that differences in the attributes of individuals were of some importance in generating linear hierarchy structures, social processes of some sort were necessary to ensure the formation of the structures. Another way to look at these findings is that, given the attributes of individuals, there was still considerable randomness in their positions in the hierarchies. It was far from total randomness, but the amount of chance in dominance rank was still substantial. In spite of this degree of randomness, the hierarchy structures themselves were almost always linear. What social processes could ensure the common formation of these linear forms of social organization in spite of the lack of predictability concerning the individuals in the structure?

Chase (1982) proposed that winner, loser, and bystander effects together might be the social processes largely accounting for the formation of linear hierarchies across a variety of species. The basic idea was that even if you started with a group of animals of equal prior attributes, they could eventually be differentiated in terms of dominance through feedback during the course of their interactions. For example, assume that there are both winner and loser effects for some species and that A and B have the first interaction when a group is assembled. If A defeats B, A

Table 4.2. Percentage of Linear Structures Expected in Random Hierarchies and Observed in Round-Robin Competition and Group Assembly in Groups of Four and Five Fish

	Method of forming hierarchy		
Size of group	Random (%)	Round robin (%)	Group assembly (%)
4	37.5	56.2 ($n=16$)	92.0 ($n=25$)
5	11.7	50.0 ($n=12$)	90.9 ($n=11$)

increases her probability of winning her next encounter and B decreases hers. A next meets C, defeats her, and again increases her probability of winning, while B encounters D, loses, and decreases her probability further.

A number of researchers have developed mathematical models and computer simulations to show that feedback from winner and loser effects, either working by themselves or together, can, at least in theory, produce highly linear hierarchies (e.g., Bonabeau et al., 1999; Dugatkin, 1997; Hemelrijk, 1999; Hock and Huber, 2006; 2007; 2009; Skvoretz and Fararo, 1996; Skvoretz et al., 1996). However, these models were not tested against actual interaction records of real animals forming dominance hierarchies. When Lindquist and Chase (2009) did evaluate three (Bonabeau et al., 1999; Dugatkin, 1997; Hemelrijk, 1999) of the most prominent of these models and simulations using detailed data records for hens establishing hierarchies (Chase 1982), they found little support for the idea that winner and loser effects were responsible for the formation of linear dominance structures. In addition, when Lindquist and Chase (2009) examined the background assumptions on which these models and simulations were based, they found little support for these assumptions in the experimental literature. Assumptions in the models and simulations include animals not remembering one another as individuals, outcomes of earlier dominance contests during hierarchy formation not influencing the outcomes of later contests, and most important, winner and loser effects actually occurring in groups forming hierarchies. In fact, the literature indicated that the actual experimental findings were in virtually all cases directly opposite to the assumptions of the models and simulations. For example, animals setting up hierarchies do remember one another as individuals—even in the case of fruit flies (Yurkovic et al., 2006). The outcome of earlier contests do influence the later ones, and perhaps most striking of all, winner and loser effects do not seem to occur in groups forming hierarchies (Brown and Colgan, 1986; Chase et al., 2003; Cheney and Seyfarth, 1990; D'Eath and Keeling, 2003; D'Ettore and Heinze, 2005; Gherardi and Atema, 2005; Karavanich and Atema, 1998; Lai et al., 2005; McLeman et al., 2005; Tibbetts, 2002; Todd et al., 1967; Yurkovic et al., 2006).

In particular, Chase et al. (2003) investigated several basic aspects of dominance relationships in isolated versus socially embedded pairs in groups of three and four fish. These aspects of relationships included winner and loser effects, stability of relationships over time, and the ability of pairs to replicate a relationship after a separation of two weeks (as described above). Specifically, while there was a strong loser effect in isolated pairs of fish, this effect was not above chance for socially embedded pairs. There was no winner effect in either isolated or socially embedded pairs. In addition, dominance relationships were much less stable over time for pairs within groups (a significant proportion of these relationships reversed over 24 h) compared to no relationships reversing in isolated pairs, and a significantly smaller proportion of pairs within groups did not form the same relationships when they met a second time after a separation of two weeks as compared to isolated pairs. In summary, all the aspects of relationships the researchers tested either disappeared or were significantly reduced for socially embedded pairs as compared to pairs by themselves. This experiment provides a strong warning of the danger of simply assuming that experimental results for isolated pairs can be automatically generalized to animals that are part of groups—even those as small as three or four individuals.

Given the lack of fit between these three prominent winner/loser models and actual data on the formation of hierarchies in real animals, and the almost total absence of experimental support for the basic assumptions of the models, it seems reasonable to suggest that winner and loser effects cannot account for the common occurrence of linear hierarchies in animal groups. But could some other models, based upon states, still satisfactorily explain linear structures? For example, what about bystander states? These states have also been used in models, both for animals and humans, as a way to account for linear hierarchies (e.g., see Dugatkin and Earley, 2003; Skvoretz and Fararo, 1996; Skvoretz et al., 1996). Although Lindquist and Chase (2009) did not look at bystander effects per se, they did point out that a bystander effect can often be decomposed into winner and loser effects; for example, a bystander increases its probability of winning over an individual that it has observed losing a contest and decreases its probability of defeating an individual that it has observed winning a contest. In cases of this sort, Lindquist and Chase's (2009) results also indicate the difficulty of bystander effects in explaining the common presence of linear structures.

While it is impossible to prove categorically that no differences in states for individuals could ever account for the forms of hierarchy structures, winner, loser, and bystander effects are the best candidates that have been proposed so far. Given the lack of support for them, it appears doubtful, at least to us, that differences in the states of animals can be adequate explanations for the social organization of hierarchies.

VI. A NEW APPROACH TO EXPLAINING THE FORMATION OF LINEAR HIERARCHIES: BEHAVIORAL PROCESSES

Given the apparent absence of support for differences in individual attributes and states as satisfactory explanations for linear hierarchies, we now review an alternative view: that the social organization of hierarchies can be explained by characteristic behavioral processes that commonly occur across many species during hierarchy formation.

Chase's (1982) "jigsaw puzzle" model presented the original version of this idea. The jigsaw puzzle model suggested that like the picture in a real jigsaw puzzle, a linear hierarchy forms when the "right" small pieces, in this case of social interaction, are put together in the correct manner. In this way, the model sees the dominance hierarchy as self-organizing or self-structuring. More specifically, the model indicates four possible sequences for the formation of the first two dominance relationships in subgroups of three animals making up a larger group. These four sequences, shown in Fig. 4.4, have different implications for the formation of linear hierarchies. The two patterns on the left, Double Dominance and Double Subordinance, guarantee transitive dominance relationships, regardless of the direction that the missing third relationship in those sequences takes when it fills in later. In a transitive relationship, individual X dominates individual Y, individual Y dominates individual Z, and individual X also dominates individual Z. For example, in Double Dominance, if B later comes to dominate C, the transitive relationship is A dominates B, B dominates C, and A dominates C. If C later comes to dominate B, the transitive relationship is A dominates C, C dominates B, and A dominates B. The fact that Double Dominance and Double Subordinance guarantee transitive relationships is very important because transitive relationships are the building blocks of linear hierarchies. By mathematical definition, if all the subgroups of three individuals in a larger group have transitive relationships, the hierarchy is necessarily linear. Thus the presence of only Double Dominance and Double Subordinance sequences in the subgroups of three animals (component triads) making up a larger group guarantees that a hierarchy will be linear, even before the missing third relationships in the component triads have formed.

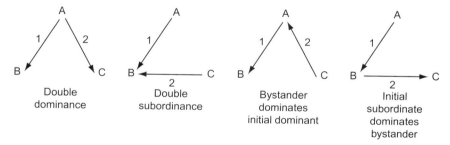

Figure 4.4. The four possible sequences for the first two dominance relationships in a component triad. Arrows show the direction of dominance relationships between the members of a triad. The number 1 indicates the first relationship to form in a triad and 2 indicates the second relationship. In all triads, A is the initial dominant, B is the initial subordinate, and C is the bystander. Figure adapted from Figure 24.2, p. 574, in "Dominance hierarchies" by Ivan D. Chase and W. Brent Lindquist from *Oxford Handbook of Analytical Sociology* ed. by Hedström, P. and Bearman, P. (2009), by permission of Oxford University Press.

On the other hand, if a hierarchy is nonlinear, it contains at least one component triad with an intransitive dominance relationship; the more intransitive triads, the further the hierarchy is from linearity. In an intransitive relationship, X dominates Y, Y dominates Z, but Z dominates X. The two sequences on the right of the figure, Bystander Dominates Initial Dominant and Initial Subordinate Dominates Bystander, can lead to either transitive or intransitive relationships depending upon how the third dominance relationship eventually fills in. For example, in Bystander Dominates Initial Dominant, the relationship is transitive if C later dominates B (C dominates A, A dominates B, and C dominates B), but intransitive if B later dominates C (A dominates B, B dominates C, C dominates A). If a group had one or more Bystander Dominates Initial Dominant and Initial Subordinate Dominates Bystander sequences, the chance for a nonlinear hierarchy would be increased when the third relationships in the triads eventually formed.

In a study of groups of three hens forming hierarchies, Chase (1982) found that almost all relationships were established using Double Dominance and Double Subordinance sequences. In 23 groups of three hens, 91% of the groups used Double Dominance and Double Subordinance together, while only 8% used the sequences not ensuring transitivity. In a second study of 14 groups of four hens (Chase, 1982), approximately 87% of the component triads in the groups (each group of four had four subgroups of three for 56 triads altogether) used Double Subordinance and Double Dominance, while approximately 13% had sequences not guaranteeing transitivity. These results are in contrast to those expected by chance, in which each sequence would have occurred about 25% of the time (or 50% combined for the two ensuring transitivity and 50% for the two not doing so).

Thus, the jigsaw puzzle model indicated that the hens established dominance relationships largely through behavioral sequences that guaranteed transitivity in their triads, and transitivity in their triads in turn guaranteed overall linear hierarchy structures. After Chase's (1982) application of the model to hens, other researchers used the original model and some modifications of it to examine dominance interactions in a broad range of species including rhesus monkeys, Japanese macaques, Harris sparrows, crayfish, and ants (Barchas and Mendoza, 1984; Chase and Rohwer, 1987; Eaton, 1984; Goessmann *et al.*, 2000; Heinze, personal communication; Mendoza and Barchas, 1983). In spite of the great differences among these species in phylogenetics, intelligence, and ways of making a living, all showed highly significant use of sequences ensuring transitivity, although this was somewhat lower in crayfish and large groups of Harris sparrows (Chase and Rohwer, 1987; Goessmann *et al.*, 2000). These results suggest that behavioral sequences ensuring transitive dominance relationships may be common for many species that form dominance hierarchies. Further research to confirm this possibility would be helpful.

A. Modifications of the jigsaw puzzle model

In recent work, Lindquist and Chase (2009) reanalyzed Chase's (1982) original data for groups of four hens forming dominance hierarchies and discovered two additional behavioral processes promoting the efficient formation of linear hierarchies in addition to those described above under the jigsaw puzzle model. The first additional process is the relative lack of back and forth fighting in pairs of animals within a group establishing a dominance hierarchy. Lindquist and Chase (2009) referred to back and forth fighting as *pair flips*—first A attacks B, then B attacks A, and so forth. Consider two groups forming a dominance hierarchy: in one group, there are many pair flips before they eventually form a stable linear hierarchy. In the second group, there are no counterattacks, and the group forms a stable linear hierarchy without pair flips. The formation of dominance relationships and their linear hierarchy is much more "efficient" in the second group.

In their analysis, Lindquist and Chase (2009) found that the hens formed their relationships with a high level of efficiency—one approaching that of the hypothetical second group just mentioned. Of the 7257 aggressive acts recorded over the 12 h of observation for each of the 14 groups of four hens (168 h of observation, total), only 138 interactions (1.9%) involved pair flips.

The second additional behavioral process was the rapid conversion of intransitive dominance relationships to transitive ones. As indicated above, the original application of the jigsaw puzzle model to the hen data showed that the great majority (87%) of the behavioral sequences in the component triads for the groups of four hens were those ensuring transitivity. However, a more detailed reanalysis of the data indicated that a few triads did initially form intransitive relationships and that several others that initially had transitive relationships later developed intransitive ones. For example, if the triad ABC initially had a transitive relationship $A > B$, $B > C$, and $A > C$, it would become intransitive if C reversed its relationship with A to give $A > B$, $B > C$, and $C > A$.

In their analysis, Chase and Linquist, (2009) found that, in virtually all cases, intransitive relationships were unstable and quickly converted to transitive ones, returning hierarchies to linearity after brief episodes of nonlinearity. The average number of dominance interactions (pecks, feather pulls, etc.) that occurred among group members before an intransitive triad was converted back to a transitive one was approximately 5.2. In contrast, the average number of interactions occurring before a transitive triad was converted into an intransitive one or into a different transitive one was approximately 54.3 interactions or over 10 times as many. Research by Chase and Rohwer (1987) on Harris sparrows supports these findings: they also found a strong tendency for intransitive dominance relationships to be unstable and to convert to transitive ones. Further experimental work is needed to determine whether the instability of intransitive relationships and their reformation as transitive ones are found in other species.

B. Experimental evidence concerning animal cognitive abilities and processes of interaction

In order for animals to carry out the kinds of behavioral processes that we have just discussed, they must possess an array of quite complex cognitive abilities. These include the abilities to remember one another as individuals, to make inferences about their own interactions and the interactions of others, and to use those inferences in adjusting their future dominance behavior. We have already discussed the extensive experimental evidence concerning the ability of animals to identify and remember one another as individuals, and to make inferences about certain kinds of interactions. In particular, we know that a broad range of species can infer transitivity (Bond *et al.*, 2003; Davis, 1992; Gillian, 1981; Grosenick *et al.*, 2007; Lazareva, 2004; Paz-y-Mino *et al.*, 2004; Roberts and Phelps, 1994; Steirn *et al.*, 1995; von Fersen *et al.*, 1991). For example, if B has dominated C, and C observes A dominating B, C will act less aggressively toward A when they meet than C will toward an animal that it has simply seen dominate another individual.

As far as we are aware, however, there have been no experimental studies to show that animals can infer or act upon intransitivity. Such studies could confirm the findings of Lindquist and Chase (2009) and Chase and Rohwer (1987) mentioned above and extend our knowledge of the behavioral processes leading to the formation of linear hierarchies.

VII. CONCLUSION

We have pointed out that there are two types of approaches that researchers can use to explain the formation of dominance relationships and dominance hierarchies in small groups of animals. The first approach uses individuals as the unit of analysis—either concentrating on differences in their traits before hierarchy formation or differences in the states that they develop during group formation. The theoretical and experimental evidence that we have reviewed indicates that explanations based upon differences in the attributes and states of individuals often work quite well to predict the outcome of dominance encounters in isolated pairs of animals. But that evidence also demonstrates, surprisingly, that these same differences in individuals are less able to predict the outcomes of relationships in socially embedded pairs or the overall ranks of individuals in their groups.

In order to resolve this problem, we have suggested that we need to ask a new research question—not what determines the ranks of individuals in hierarchies, but how linear hierarchies themselves develop. As the beginning of an answer to this question, we have discussed the support for a new approach that uses behavioral processes to account for hierarchy structures. For some, such an approach may seem to be a kind of "cheating," an avoidance of discovering things about individuals that really do explain their "successes" and "failures" in winning dominance contests within hierarchies. However, because the greater complexities of groups introduce an unavoidable chance element into the predictions about individuals within hierarchies, then perforce we need an approach that does not depend upon those individuals as units of analysis. A rough analogy is the way we look at the organization of tosses of a coin. Because of randomness, we do not try to predict the outcomes of individual coin tosses. Instead we move to a higher level of the phenomenon: we say something about the organization, or form of the distribution, of a great many coin flips. The behavioral process explanation of hierarchy structure is our attempt to get around the chance elements in what individuals do in hierarchies, and to say something about the organization of hierarchies in spite of that randomness.

Recognizing when individual-based and process-based approaches are best applied in studies of dominance has fundamental importance in choosing the kinds of data we collect, how we analyze those data, the explanations that we develop for hierarchies, and for the cognitive capacities, for both humans and animals, that we envision as underlying dominance behavior.

Acknowledgments

We thank Hronn Axelsdottir and Paul St. Denis for their help with the graphics in this chapter and Robert Huber for inviting us to participate in this volume.

References

Addison, W. E., and Simmel, E. C. (1970). The relationship between dominance and leadership in a flock of ewes. *Bull. Psychon. Soc.* **15**, 303–305.

Bakker, T. C. M. (1985). Two-way selection for aggression in juvenile, female and male sticklebacks (*Gasterosteus aculeatus*), with some notes on hormonal factors. *Behaviour* **93**, 69–81.

Bakker, T. C. M. (1986). Aggression in sticklebacks (*Gasterosteus aculeatus* L.): A behavior genetic study. *Behaviour* **98**, 1–144.

Barchas, P. R., and Mendoza, S. P. (1984). Emergent hierarchical relationships in rhesus macaques: An application of Chase's model. *In* "Social Hierarchies: Essays Toward a Sociophysiological Perspective" (P. R. Barchas, ed.), pp. 23–44. Greenwood, Westport, Conn.

Barkan, C. P. L., Craig, J. L., Strahl, S. D., Stewart, A. M., and Brown, J. L. (1986). Social dominance in communal Mexican Jays *Aphelocoma ultramarine*. *Anim. Behav.* **34**, 175–187.

Beacham, J. L. (1988). The relative importance of body size and aggressiveness as determinants of dominance in pumpkinseed sunfish, *Lepomis gibbosus*. *Anim. Behav.* **36**, 621–623.

Beaugrand, J., and Beaugrand, M. (1991). Prior residency and the stability of dominance relationships in pairs of green swordtail fish *Xiphophorus helleri* (Pisces, Poeciliidae). *Behav. Process.* **24**, 169–175.

Beaugrand, J. P., and Cotnoir, P. (1996). The role of individual differences in the formation of triadic dominance orders of male green swordtail fish (*Xiphophorus helleri*). *Behav. Process.* **38**, 287–296.

Beaugrand, J. P., Payette, D., and Goulet, C. (1996). Conflict outcome in male green swordtail fish dyads (*Xiphophorus helleri*): Interaction of body size, prior dominance/subordination experience, and prior residency. *Behaviour* **133**, 303–319.

Bell, A. M., Hankison, S. J., and Laskowski, K. L. (2009). The repeatability of behavior: A meta-analysis. *Anim. Behav.* **77**, 771–783.

Bergman, D. A., Kozlowski, C., McIntyre, J. C., Huber, R., Daws, A. G., and Moore, P. A. (2003). Temporal dynamics and communication of winner-effects in the crayfish, *Orconectes rusticus*. *Behaviour* **140**, 805–825.

Bonabeau, E., Theraulaz, G., and Deneubourg, J. L. (1999). Dominance orders in animal societies: The self-organization hypothesis revisited. *Bull. Math. Biol.* **61**, 727–757.

Bond, A. B., Kamil, A. C., and Balda, R. P. (2003). Social complexity and transitive inference in corvids. *Anim. Behav.* **65**, 479–487.

Boon, A. K., Réale, D., and Boutin, S. (2007). The interaction between personality, offspring fitness and food abundance in North American red squirrels. *Ecol. Lett.* **10**, 1094–1104.

Breithaupt, T., and Atema, J. (2000). The timing of chemical signaling with urine in dominance fights of male lobsters (*Homarus americanus*). *Behav. Ecol. Sociobiol.* **49**, 67–78.

Breithaupt, T., and Eger, P. (2002). Urine makes the difference: Chemical communication in fighting crayfish made visible. *J. Exp. Biol.* **205**, 1221–1231.

Brown, J. A., and Colgan, P. W. (1986). Individual and species recognition in centrachid fishes: Evidence and hypotheses. *Behav. Ecol. Sociobiol.* **19**, 373–379.

Burmeister, S. S., Jarvis, E. D., and Fernald, R. D. (2005). Rapid behavioral and genomic responses to social opportunity. *PLoS Biol.* **3**, e363.

Chase, I. D. (1982). Dynamics of hierarchy formation: Tthe sequential development of dominance relationships. *Behaviour* **80**, 218–240.

Chase, I. D. (2006). Music notation: A new method for visualizing social interaction in animals and humans. *Front. Zool.* **3**, 18.

Chase, I. D., and Rohwer, S. (1987). Two methods for quantifying the development of dominance hierarchies in large groups with applications to Harris' sparrows. *Anim. Behav.* **35**, 1113–1128.

Chase, I. D., Bartolomeo, C., and Dugatkin, L. A. (1994). Aggressive interactions and inter-contest interval: How long do winners keep winning? *Anim. Behav.* **48**, 393–400.

Chase, I. D., Tovey, C., Spangler-Martin, D., and Manfredonia, M. (2002). Individual differences versus social dynamics in the formation of animal dominance hierarchies. *Proc. Natl. Acad. Sci. USA* **99**, 5744–5749.

Chase, I. D., Tovey, C., and Murch, P. (2003). Two's company, three's a crowd: Differences in dominance relationships in isolated versus socially embedded pairs of fish. *Behaviour* **140**, 1193–1217.

Chase, I. D., and Lindquist, W. B. (2009). Dominance hierarchies. *In* "Oxford Handbook of Analytical Sociology" (P. Hedström and P. Bearman, eds.), pp. 566–591. Oxford University Press, New York.

Cheney, D. L., and Seyfarth, R. M. (1990). How Monkeys See the World: Inside the Mind of Another Species. University of Chicago Press, Chicago.

Clark, D. C. (1998). Male mating success in the presence of a conspecific opponent in a Madagascar hissing cockroach, *Gromphadorhina portentosa*. *Ethology* **104**, 877–888.

Clutton-Brock, T. H., Albon, S. D., and Guinness, F. E. (1984). Maternal dominance, breeding success and birth sex-ratios in red deer. *Nature* **308**, 358–360.

Collias, N. E. (1943). Statistical analysis of factors which make for success in initial encounters between hens. *Am. Nat.* **77**, 519–538.

Craig, J. V., and Muir, W. M. (1996). Group selection for adaptation to multiple-hen cages: Beak related mortality, feathering, and body weight responses. *Poult. Sci.* **75**, 294–302.

Craig, J. V., Ortman, L. L., and Guhl, A. M. (1965). Genetic selection for social dominance ability in chickens. *Anim. Behav.* **13**, 114–131.

Craig, J. V., Biswas, D. K., and Guhl, A. M. (1969). Agonistic behavior influenced by strangeness, crowding and heredity in female domestic fowl (*Gallus gallus*). *Anim. Behav.* **17**, 498–506.

Craig, J. V., Jan, M., Polley, C. R., Bhagwat, A. L., and Dayton, A. D. (1975). Changes in relative aggressiveness and social dominance associated with selection for early egg production in chickens. *Poult. Sci.* **54**, 1647–1658.

Cutts, C. J., Metcalf, N. B., and Taylor, A. C. (1999). Competitive asymmetries in territorial juvenile Atlantic salmon, *Salmo salar*. *Oikos* **86**, 479–486.

D'Eath, R. B., and Keeling, L. J. (2003). Social discrimination and aggression by laying hens in large groups: From peck orders to social tolerance. *Appl. Anim. Behav Sci.* **84**, 197–212.

D'Ettore, P., and Heinze, J. (2005). Individual recognition in ant queens. *Curr. Biol.* **15**, 2170–2174.

Danchin, E., Giraldeau, L. A., and Valone, T. J. (2004). Public information: From noisy neighbors to cultural evolution. *Science* **305**, 487–491.

Davis, H. (1992). Transitive Inference in Rats (*Rattus norvegicus*). *J. Comp. Psychol.* **106, 342**–349.

Dloniak, S. M., French, J. A., and Holekamp, K. E. (2006). Rank-related maternal effects of androgens on behavior in wild spotted hyaenas. *Nature* **440**, 1190–1193.

Dugatkin, L. A. (1997). Winner and loser effects and the structure of dominance hierarchies. *Behav. Ecol.* **8**(6), 583–587.

Dugatkin, L. A., and Earley, R. L. (2003). Group fusion: The impact of winner, loser, and by-stander effects on hierarchy formation in large groups. *Behav. Ecol.* **14**, 367–373.

Eaton, G. G. (1984). Aggression in adult male primates: A comparison of confined Japanese macaques and free-ranging olive baboons. *Int. J. Primatol.* **5**, 145–160.

Eaton, G. G., and Resko, J. A. (1974). Plasma testosterone and male dominance in a Japanese macaque (*Macaca fuscata*) troop compared with repeated measures of testosterone in laboratory males. *Horm. Behav.* **5**, 251–259.

Ellis, L. (1995). Dominance and reproductive success among nonhuman animals: A cross-species comparison. *Ethol. Sociobiol.* **16**, 257–333.

Fox, R. A., Ladage, L. D., Roth, T. C. I. I., and Pravosudov, V. V. (2009). Behavioural profile predicts dominance status in mountain chickadees. Poecile gambeli. *Anim. Behav.* **77**, 1441–1448.

Francis, R. C. (1984). The effects of bidirectional selection for social dominance on agonistic behavior and sex ratios in paradise fish (*Macropodus opercularis*). *Behavior* **90**, 25–44.

Francis, R. C. (1987). The interaction of genotype and experience in the dominance success of paradise fish (*Macropodus opercularis*). *Biol. Behav.* **12**, 1–11.

Francis, R. C., Soma, K., and Fernald, R. D. (1993). Social regulation of the brain—Pituitary gonadal axis. *Proc. Natl. Acad. Sci. USA* **90**, 7794–7798.

Frey, D. F., and Miller, R. J. (1972). The establishment of dominance relationships in blue gourami, *Trichogaster trichopterus* (Pallas). *Behaviour* **42**, 8–62.

Fuxjagera, M. J., and Marlera, C. (2010). How and why the winner effect forms: Influences of contest environment and species differences. *Behav. Ecol.* **21**, 37–45.

Gherardi, F., and Atema, J. (2005). Memory of social partners in hermit crab dominance. *Ethology* **111**(3), 271–285.

Gillian, D. J. (1981). Reasoning in the chimpanzee. II. Transitive inference. *J. Exp. Psychol. Anim. Behav. Process.* **7**, 87–108.

Goessmann, C., Hemelrijk, C., and Huber, R. (2000). The formation and maintenance of crayfish hierarchies: Behavioral and self-structuring properties. *Behav. Ecol. Sociobiol.* **48**, 418–428.

Groothius, T. G. G., and Carere, C. (2005). Avian personalities: Characterization and epigenesis. *Neurosci. Biobehav. Rev.* **29**, 137–150.

Grosenick, L., Clement, T. S., and Fernald, R. D. (2007). Fish can infer social rank by observation alone. *Nature* **445**, 429–432.

Hausfater, G., Altmann, J., and Altmann, S. (1982). Long-term consistency of dominance relations among female baboons (*Papio-cynocephalus*). *Science* **217**, 752–755.

Heinze, J. (1990). Dominance behavior among ant females. *Naturwissenschaften* **77**, 41–43.

Hemelrijk, C. K. (1999). An individual-oriented model of the emergence of despotic and egalitarian societies. *Proc.R. Soc. Lond. B* **266**, 361–369.

Hock, K., and Huber, R. (2006). Modeling the acquisition of social rank in crayfish: Winner and loser effects and self-structuring properties. *Behaviour* **143**, 325–346.

Hock, K., and Huber, R. (2007). Effects of fighting decisions on formation and structure of dominance hierarchies. *Mar. Freshw. Behav. Phy.* **40**, 153–169.

Hock, K., and Huber, R. (2009). Models of winner and loser effects: A cost-benefit analysis. *Behaviour* **146**, 69–87.

Holekamp, K. E., and Smale, L. (1993). Ontogeny of dominance in free-living spotted hyenas: Juvenile rank relations with other immature individuals. *Anim. Behav.* **46**, 451–466.

Houpt, K. A., Law, K., and Martinisi, V. (1978). Dominance hierarchies in domestic horses. *Appl. Anim. Ethol.* **4**, 273–283.

Hsu, Y., and Wolf, L. L. (1999). The winner and loser effect: Integrating multiple experiences. *Anim. Behav.* **57**, 903–910.

Hsu, Y. Y., Earley, R. L., and Wolf, L. L. (2006). Modulation of aggressive behavior by fighting experience: Mechanisms and contest outcomes. *Biol. Rev.* **81**, 33–74.

Huber, R., Smith, K., Delago, A., Isaksson, K., and Kravitz, E. A. (1997). Serotonin and aggressive motivation in crustaceans: Altering the decision to retreat. *Proc. Natl. Acad. Sci. USA* **94**, 5939–5942.

Johnsson, J. I., and Björnsson, B. T. (1994). Growth hormone increases growth rate, appetite and dominance in juvenile rainbow trout, *Oncorhynchus mykiss*. *Anim. Behav.* **48**, 177–186.

Karavanich, C., and Atema, J. (1998). Olfactory recognition of urine signals in dominance fights between male lobster, *Homarus americanus*. *Behaviour* **135**(6), 719–730.

Knights, B. (1987). Agonistic behaviour and growth in the European eel *Anguilla anguilla* L. in relation to warm-water aquaculture. *J. Fish Biol.* **31**, 265–276.

Kummer, H. (1984). From laboratory to desert and back: A social system of hamadryas baboons. *Anim. Behav.* **32**(4), 965–971.

Lai, W.-S., Ramiro, L.-L. R., Yu, H. A., and Johnston, R. E. (2005). Recognition of familiar individuals in golden hamsters: A new method and functional neuroanatomy. *J. Neurosci.* **25**, 11239–11247.

Lazareva, O. F. (2004). Transitive responding in hooded crow requires linearly ordered stimuli. *J. Exp. Anal. Behav.* **82**, 1–19.

Lindquist, W. B., and Chase, I. (2009). Data-based analysis of winner-loser models of hierarchy formation in animals. *Bull. Math. Biol.* **71**, 556–584.

Lott, D. F., and Galland, J. C. (1987). Body mass as a factor influencing dominance status in American bison cows. *J. Mammal.* **63**, 683–685.

Martin, J. G. A., and Réale, D. (2008). Temperament, risk assessment and habituation to novelty in eastern chipmunks, *Tamias striatus*. *Anim. Behav.* **75**, 309–318.

McCarthy, I. D. (2001). Competitive ability is related to metabolic asymmetry in juvenile rainbow trout. *J. Fish Biol.* **59**, 1002–1014.

McGhee, K. E., and Travis, J. (2010). Repeatable behavioural type and stable dominance rank in the bluefin killifish. *Anim. Behav.* **79**, 497–507.

McLeman, M. A., Mendl, M., Jones, R. B., White, R., and Wathes, C. M. (2005). Discrimination of Conspecifics by Juvenile Domestic Pigs, Sus scrofa. *Anim. Behav.* **70**, 451–461.

Mendoza, S. P., and Barchas, P. R. (1983). Behavioral processes leading to linear status hierarchies following group formation in rhesus monkeys. *J. Hum. Evol.* **12**, 185–192.

Metcalfe, N. B., Taylor, A. C., and Thorpe, J. E. (1995). Metabolic rate, social status and life history strategies in Atlantic salmon. *Anim. Behav.* **49**, 431–436.

Möller, L. M., Beheregaray, L. B., Allen, S. J., and Harcourt, R. G. (2006). Association patterns and kinship in female Indo-Pacific bottlenose dolphins (*Tursiops aduncus*) of southeastern Australia. *Behav. Ecol. Sociobiol.* **61**, 109–117.

Möller, L. M., Beheregaray, L. B., Hart, R. G., and Krützen, M. (2001). Alliance membership and kinship in wild male bottlenose dolphins (*Tursiops aduncus*) of southeastern Australia. *Proc. R. Soc. Lond. B.* **268**(1479), 1941–1947.

Moore, A. J. (1990). The inheritance of social dominance, mating behavior and attractiveness to mates in male *Nauphoeta cinera*. *Anim. Behav.* **42**, 497–498.

Moore, P. A., and Bergman, D. A. (2005). The smell of success and failure: The role of intrinsic and extrinsic chemical signals on the social behavior of crayfish. *Integr. Comp. Biol.* **45**, 650–657.

Nakano, S., and Furukawa-Tanaka, T. (1994). Intra- and interspecific dominance hierarchies and variation in foraging tactics of two species of stream-dwelling chars. *Ecol. Res.* **9**, 9–20.

Nelissen, M. H. J. (1985). Structure of the dominance hierarchy and dominance determining group factors in *Melanochromic auratus* (Pisces, Cichlidae). *Behaviour* **94**, 85–107.

Oliveira, R. F., McGregor, P. K., and Latruffe, C. (1998). Know thine enemy: Fighting fish gather information from observing conspecific interactions. *Proc. Biol. Sci.* **265**, 1045–1049.

Oliveira, R. F., Lopes, M., Carneiro, L. A., and Canário, A. V. M. (2001). Watching fights raises fish hormone levels. *Nature* **409**, 475.

Oliveira, R. F., Silva, A., and Canário, V. M. (2009). Why do winners keep winning? Androgen mediation of winner but not loser effects in cichlid fish. *Proc. R. Soc. B* **276**(1665), 2249–2256.

Øverli, Ø., Korzan, W. J., Höglund, E., Winberg, S., Bollig, H., Watt, M., Forster, G. L., Barton, Bruce A., Øverli, E., Renner, K. J., *et al.* (2004). Stress coping style predicts aggression and social dominance in rainbow trout. *Horm. Behav.* **45**, 235–241.

Paz-y-Mino, G., *et al.* (2004). Pinyon jays use transitive inference to predict social dominance. *Nature* **430**, 778–781.

Peake, T. M., and McGregor, P. K. (2004). Information and aggression in fishes. *Learn. Behav.* **32**, 114–121.

Post, W. (1992). Dominance and mating success in male boat-tailed grackles. *Anim. Behav.* **44**, 917–929.

Queller, D., Zacchi, F., Cervo, R., Tupillazi, S., Henshaw, M. T., Santorelli, L. A., and Strassman, J. E. (2000). Unrelated helpers in a social insect. *Nature* **405**, 784–787.

Renn, S. C. P., Aubin-Horth, N., and Hofmann, H. A. (2008). Fish and chips: Functional genomics of social plasticity in an African cichlid fish. *J. Exp. Biol.* **211**, 3041–3056.

Roberts, W. A., and Phelps, M. T. (1994). Transitive inference in rats: A test of the spatial coding hypothesis. *Psychol. Sci.* **5**, 368–374.

Robinson, G. E., Grozinger, C. M., and Whitfield, C. W. (2005). Sociogenomics: Social life in molecular terms. *Nature Rev.* **6**, 257–270.

Rutte, C., Taborsky, M., and brinkhof, M. W. G. (2006). What sets the odds of winning and losing? *Trends Ecol. Evol.* **21**, 16–21.

Sactuary, W. C. (1932). A study in avian behavior to determine the nature and persistency of the order of dominance in the domestic fowl and to relate these to certain physiological reactions. Masters Thesis at the Massachusetts State College at Amherst (unpublished).

Sapolsky, R. M. (1982). The endocrine stress-response and social status in wild the baboon. *Horm. Behav.* **16**(3), 279–292.

Sapolsky, R. (2005). The influence of social hierarchy on primate health. *Science* **308**, 648–652.

Sapolsky, R. M., and Share, L. J. (1994). Rank-related differences in cardiovascular function among wild baboons: Role of sensitivity to glucocorticoids. *Am. J. Primatol.* **32**, 261–275.

Savin-Williams, R. C. (1980). Dominance hierarchies in groups of middle to late adolescent males. *J. Youth Adolesc.* **9**, 75–85.

Schjelderup-Ebbe, T. (1922). Contributions to the social psychology of the domestic chicken. Social Hierarchy and Dominance, Benchmark Papers in Animal Behavior, vol. 3. pp. 7–94. Hutchinson and Ross, Inc, Dowden, Stroudsburg, PA.

Schjolden, J., Stoskhus, A., and Winberg, S. (2005). Does individual variation in stress responses and agonistic behavior reflect divergent stress coping strategies in juvenile rainbow trout? *Physiol. Biochem. Zool.* **78**, 715–723.

Schulte-Hostedde, A. I., and Millar, J. S. (2002). 'Little Chipmunk' Syndrome? Male Body Size and Dominance in Captive Yellow-Pine Chipmunks (*Tamias amoenus*). *Ethology* **108**, 127–137.

Schwanck, E. (1980). The effect of size and hormonal state on the establishment of dominance in young males of *Tilapia mariae* (Pisces: Cichlidae). *Behav. Processes* **5**, 45–53.

Senar, J. C. (1999). Plumage coloration as a signal of social status. *Proc. Int. Ornithol. Congr* **22**, 1669–1686.

Shaw, P. (1986). The relationship between dominance behaviour, bill size and age group in Greater Sheathbills, Chionis alba. *Ibis* **128**, 48–56.

Sih, A., Bell, A. M., Johnson, J. C., and Ziemba, R. M. (2004). Behavioral syndromes: An integrative overview. *Q. Rev. Biol.* **79**, 241–277.

Sinn, D. L., and Moltschaniwskyj, N. A. (2005). Personality traits in dumpling squid (*Euprymna tasmanica*): Context-specific traits and their correlation with biological characteristics. *J. Comp. Pyschol.* **119**, 99–110.

Skvoretz, J., and Fararo, T. J. (1996). Status and participation in task droups: A dynamic network model. *Am. J. Sociol.* **101**(5), 1366–1414.

Skvoretz, J., Faust, K., and Fararo, T. J. (1996). Social structure, networks, and e-state structuralism models. *J. Math. Sociol.* **21**, 57–76.

Sloman, K. A., and Armstrong, J. D. (2002). Physiological effects of dominance hierarchies: Laboratory artefacts or natural phenomena? *J. Fish Biol.* **61**, 1–23.

Steirn, J. N., Weaver, J. E., and Zentall, T. R. (1995). Transitive inference in pigeons: Simplified procedures and a test of value transfer theory. *Anim. Learn. Behav.* **23**, 76–82.

Sundström, L. F., Petersson, E., Höjesjö, J., Johnsson, J. I., and Jörvi, T. (2004). Hatchery selection promotes boldness in newly hatched brown trout (*Salmo trutta*): Implications for dominance. *Behav. Ecol.* **15**, 192–198.

Surbeck, M., Mundry, R., and Hohmann, G. (2011). Mothers matter! Maternal support, dominance status and mating success in male bonobos (*Pan paniscus*). *Proc. R. Soc. Lond. B* **278**(1705), 590–598.

Svartberg, K., Tapper, I., Temrin, H., Radesäter, T., and Thorman, S. (2005). Consistency of personality traits in dogs. *Anim. Behav.* **69**, 283–291.

Tibbetts, E. A. (2002). Visual signals of individual identity in the wasp, *Polistes fuscatus*. *Proc. R. Soc. Lond. B* **269**, 1423–1428.

Todd, J. H., Atema, J., and Bardach, J. E. (1967). Chemical communication in social behavior of a fish, the yellow bullhead (*Ictalurus natalis*). *Science* **158**, 672–673.

Trainor, B. C., and Hofmann, H. A. (2007). Somatostatin and somatostatin receptor gene expression in dominant and subordinate males of an African cichlid fish. *Behav. Brain Res.* **179**, 314–320.

Vannini, M., and Sardini, A. (1971). Aggressivity and dominance in river crab *Potomon fluviantile* (Herbst). *Monitore Zool. Ital.* **5**, 173–213.

Verbeek, M. E. M., Boon, Anne, and Drent, P. J. (1996). Exploration, aggressive behaviour and dominance in pair-wise confrontations of juvenile male great tits. *Behaviour* **133**, 945–963.

Verbeek, M. E. M., De Goede, P., Drent, P. J., and Wiepkema, P. R. (1999). Individual behavioural characteristics and dominance in aviary groups of great tits. *Behaviour* **136**, 23–48.

von Fersen, L., Wynne, C. D. L., and Delius, J. D. (1991). Transitive inference formation in pigeons. *J. Exp. Psychol. Anim. Behav. Process.* **17,** 334–341.

Whittingham, L. A., and Schwabl, H. (2002). Maternal testosterone in tree swallow eggs varies with female aggression. *Anim. Behav.* **63,** 63–67.

Widdig, A., Nürnberg, P., Krawczak, M., Streich, W. J., and Bercovitch, F. B. (2001). Paternal relatedness and age proximity regulate social relationships among adult female rhesus macaques. *Proc. Natl. Acad. Sci. USA* **98**(24), 13769–13773.

Wilson, E. O. (1975). Sociobiology. Harvard University Press, Cambridge.

Winberg, S., and Nilsson, G. E. (1992). Induction of social dominance by L-dopa treatment in Arctic charr. *Neuroreport* **3,** 243–246.

Yeh, S., Musolf, B. E., and Edwards, D. H. (1997). Neuronal adaptations to changes in the social dominance status of crayfish. *J. Neurosci.* **17**(2), 697–708.

Yurkovic, A., Wang, O., Basu, A. C., and Kravitz, E. A. (2006). Learning and memory associated with aggression in *Drosophila melanogaster. Proc. Natl. Acad. Sci. USA* **103,** 17519–127524.

5

Neurogenomic Mechanisms of Aggression in Songbirds

Donna L. Maney* and James L. Goodson[†]
*Department of Psychology, Emory University, Atlanta, Georgia, USA
[†]Department of Biology, Indiana University, Bloomington, Indiana, USA

Advances in Genetics, Vol. 75
0065-2660/11 $35.00
DOI: 10.1016/B978-0-12-380858-5.00002-2

ABSTRACT

Our understanding of the biological basis of aggression in all vertebrates, including humans, has been built largely upon discoveries first made in birds. A voluminous literature now indicates that hormonal mechanisms are shared between humans and a number of avian species. Research on genetics mechanisms in birds has lagged behind the more typical laboratory species because the necessary tools have been lacking until recently. Over the past 30 years, three major technical advances have propelled forward our understanding of the hormonal, neural, and genetic bases of aggression in birds: (1) the development of assays to measure plasma levels of hormones in free-living individuals, or "field endocrinology"; (2) the immunohistochemical labeling of immediate early gene products to map neural responses to social stimuli; and (3) the sequencing of the zebra finch genome, which makes available a tremendous set of genomic tools for studying gene sequences, expression, and chromosomal structure in species for which we already have large datasets on aggressive behavior. This combination of hormonal, neuroendocrine, and genetic tools has established songbirds as powerful models for understanding the neural basis and evolution of aggression in vertebrates. In this chapter, we discuss the contributions of field endocrinology toward a theoretical framework linking aggression with sex steroids, explore evidence that the neural substrates of aggression are conserved across vertebrate species, and describe a promising new songbird model for studying the molecular genetic mechanisms underlying aggression.   © 2011, Elsevier Inc.

I. AGGRESSION IN CONTEXT

Biomedical studies of aggression and its genetic basis are most often focused on pathology, yet aggressive behaviors and related agonistic displays are essential, adaptive components of social behavior that enable animals to secure and defend food, mates, and territories. For many species, aggression is also required to protect offspring from would-be predators. Thus, given that effective aggression is often essential for gene propagation, we can expect that it will be under strong selection to meet species-typical and population-specific demands. Further, for any given species, aggression will be adaptive in some contexts but not others, and it may therefore be the case that the neural and neurogenomic mechanisms of aggression vary in relation to the functional goals of the behavior. Numerous findings support this view, including evidence that parental aggression and male–male aggression are regulated by different suites of neuroendocrine mechanisms in rodents (Gammie, 2005; Trainor et al., 2008; Veenema et al., 2007) and that neuropeptides differentially influence territorial aggression and aggressive competition for mates in songbirds (Goodson and Kabelik, 2009). Indeed, the idea that aggression is differentially regulated across distinct functional contexts, and distinct motivational states, has been around for more than 40 years (Moyer, 1968). This functional perspective suggests that ethological approaches will be particularly useful for identifying integrated suites of neurogenomic mechanisms that regulate aggression in any given context and will provide a powerful framework for distinguishing pathology from normal, adaptive variation (Blanchard and Blanchard, 2003, 2005).

In this review, we focus on aggression in the context of competition for resources, for example, defending a breeding territory or a position in a dominance hierarchy within a social group. This type of aggression, particularly in a reproductive context, is part of a suite of related behaviors that characterize a "life history strategy" maximizing short-term gains as opposed to longer term investments (Maynard-Smith, 1977; Trivers, 1972). Short-term relationships with mates, high aggression among same-sex individuals, and low parental care typify this strategy. At the other end of this continuum is a strategy characterized by commitment to one mate, avoidance of injury, and a high level of parental care. The two strategies are difficult to employ simultaneously, resulting in a trade-off between time spent on territorial aggression versus parenting. This trade-off has become a classic principle in ethology and is universal among vertebrates, including humans (Trivers, 1972).

Disruptive selection that drives the sequestration of territorial and parental behavior into alternative strategies is most likely to act on genes with widespread effects—particularly those with multiple functions. Genes encoding the action or regulation of hormones are obvious examples of such genes (Finch and Rose, 1995; Hau, 2007; Ketterson and Nolan, 1992; McGlothlin and Ketterson, 2007; Miles et al., 2007; Moore, 1991; Nijhout, 2003; Rhen and Crews, 2002; Sinervo and Svensson, 2002). A growing literature suggests that trade-offs between parenting and territorial aggression are associated with gonadal steroids; in many species of fish, birds, rodents, and primates, including humans, high levels of circulating androgens are associated with increased intrasexual competition manifested as aggression or mating effort, whereas low levels are associated with increased parenting effort (e.g., Ketterson and Nolan, 1994; McGlothlin et al., 2007). In humans, paternal care and fatherhood have been repeatedly shown to correspond to low levels of testosterone (T) (Fleming et al., 2002; Gray, 2003; Gray et al., 2002; Storey et al., 2000; Wynne-Edwards, 2001), whereas high T levels are associated with male–male aggression and competition (Bernhardt et al., 1998; Book et al., 2001; Booth et al., 1989). These opposing strategies can be generalized as a tendency to prioritize shorter term goals (mating) versus longer term goals (parental investment); at the former end of this continuum in humans, associations have been reported between T and antisocial activities such as alcoholism, drug use, reckless driving, failure to plan ahead, risk-taking, and assaults (Aromaki et al., 1999; Dabbs and Morris, 1990; Udry, 1990).

Strategies to balance effort toward short-term versus long-term goals may therefore involve a limited number of hormones and genes. In this review, we attempt to bring together behavior, reproductive endocrinology, and genetics by focusing on species in which all three have been characterized in some detail.

Since the scientific study of behavior began, birds have been the most commonly studied animals in relation to territoriality, dominance, and agonistic communication. Their popularity primarily reflects their unique accessibility—including location and use of vocal and visual communication channels—and the fact that territorial birds are readily captured using mist nets and playback of song. For the biomedical researcher attempting to model social behavior in humans, birds may seem to represent a rather distant taxonomic group. But in fact, birds have provided the test bed for some of the most influential theories in the history of aggression research, and it is no exaggeration to say that our understanding of the hormonal mechanisms of aggression in all vertebrates, including humans, has been built in large part upon discoveries that were first made in birds (Archer, 2006; Goodson et al., 2005a,b,c; Konishi et al., 1989). For example, pioneering studies in birds established the theoretical framework currently used by researchers to understand how hormones mediate a trade-off between aggression and parenting in mammals (Wingfield et al., 1990). This theoretical framework, which has been called the "challenge hypothesis," is based on the idea that the role of gonadal steroids in aggression is modulated by social context. It predicts that when males are challenged in a reproductive context, T levels rise to facilitate territorial aggression and suppress parental behavior. Since it was first proposed by Wingfield et al. (1990), the challenge hypothesis has found support in a wide variety of nonavian vertebrate taxa including fish, reptiles, and primates, including humans (reviewed by Archer, 2006). The parallels between songbirds (particularly New World sparrows) and humans with regard to the social modulation of gonadal steroids and their effects on aggressive and parental behavior are voluminous and are summarized in Table 5.1. The underlying neuroendocrine mechanisms are nearly identical in birds and humans and are based on the function of the hypothalamo-pituitary-gonadal (HPG) axis, which is universally recognized as being highly conserved across all vertebrates (reviewed by Adkins-Regan, 2005).

Despite the contributions of avian research to our understanding of human behavior, genomic resources in birds have lagged well behind those in mammals—although this situation is now rapidly changing. In the sections that follow, we first explore the neural and endocrine literature on songbird aggression, and then describe a relatively new research program that is focused on the neurogenomics of territorial aggression in white-throated sparrows (*Zonotrichia albicollis*), a species that exhibits morphological and behavioral polymorphisms associated with a chromosomal inversion (Thomas et al., 2008; Thorneycroft, 1975). Importantly, the morphs differ not only in their territorial aggression, but also in parental

Table 5.1. Evidence of Shared Mechanisms of Competitive Aggression in Birds and Humans

Prediction	Evidence in New World sparrows	Evidence in humans
Males respond to competition with increased plasma T	Wingfield (1985), Wingfield and Hahn (1994), Wingfield and Wada (1989), Wingfield et al. (1990)	Meta-analysis of 23 studies in Archer (2006)
The plasma T response to challenge increases aggression	Archawaranon et al. (1991), Wingfield (1984b, 1994b)	Meta-analysis of 11 studies in Archer (2006)
Plasma T levels are lower among paternal males	Wingfield (1984a), Wingfield and Farner (1978), Wingfield and Goldsmith (1990), Wingfield et al. (1990)	Berg and Wynne-Edwards (2001), Fleming et al. (2002), Gray et al. (2002), Storey et al. (2000)
Aggressive dominance is correlated with plasma T levels	Archawaranon and Wiley (1988), Schlinger (1987), Wiley et al. (1993)	Meta-analysis of 13 studies in Archer (2006)
Plasma T is associated with alternative life history strategies regarding territoriality versus parenting	Hau (2007), Ketterson and Nolan (1992), McGlothlin et al. (2007), Schoech et al. (1998), Spinney et al. (2006), Wingfield (1984a,b,c)	Dabbs and Morris (1990), Dabbs et al. (1997), Daitzman and Zuckerman (1980), Gray et al. (2002), Julian and McHenry (1989)

The endocrine underpinnings of competitive aggression are broadly similar in New World sparrows (here limited to the *Zonotrichia*, *Melospiza*, and *Junco* genera) and humans (based primarily on Archer, 2006). Only a fraction of the relevant literature is cited here.

behavior, and thus this species offers an extraordinary opportunity to examine neurogenomic mechanisms that integrate aggression with other aspects of social phenotype and context-specific behavior.

II. HORMONAL MECHANISMS OF AGGRESSION

A. Territoriality in the breeding season

There are about 10,000 species of birds, almost half of which are songbirds. Territoriality runs the gamut, with members of some species nesting colonially or cooperatively, others defending territories of several hectares. Perhaps the best-studied territorial species are the seasonally breeding New World sparrows, which include song sparrows (*Melospiza melodia*), field sparrows (*Spizella pusilla*), white-crowned sparrows (*Zonotrichia leucophrys*), and dark-eyed juncos (*Junco hyemalis*; see Arcese et al., 2002; Chilton et al., 1995; Carey et al., 2008; Nolan et al., 2002 for reviews). In migratory populations, the males arrive at the breeding grounds a week or so before the females and stake out territories containing food sources and nest sites. The females then arrive, basing their mate choices on the quality of the males as well as their territories. It is therefore important, in fact critical, for males to establish high-quality territories early in the breeding season. Once a male has attracted a mate, she will help defend the territory.

The most ubiquitous and frequent behavior used for territory defense by songbirds is, not surprisingly, song. Each species' song is distinct, and within a species there is enough variation that individuals can recognize each other's songs (Krebs, 1971). Some species sing different types of song in different contexts; for example, the "complex song" of the field sparrow is considered more aggressive than the "simple song" (Carey *et al.*, 2008; Nelson and Croner, 1991), and the chestnut-sided warbler (*Dendroica pensylvanica*) sings a different song to an intruder than he does to a potential mate (Kroodsma *et al.*, 1989; Lein, 1978). Although most of the singing is done by males, females of some species do sing during agonistic encounters (e.g., Baptista *et al.*, 1993; Falls and Kopachena, 2010). Males typically choose a centrally located perch and sing loudly at regular intervals, making their presence known to would-be mates and intruders. In a now-classic study, Krebs (1976) showed that experimental removal of territorial male great tits (*Parus major*) resulted in rapid takeover of the vacated territories by other males; however, broadcasting a former resident's song from a loudspeaker in his territory significantly delayed that takeover (see also Falls, 1988). Although most song is typically sung from a prominent perch in the center, it is also commonly used near territory boundaries, particularly directed at neighbors, as the territory is established. Males learn to recognize their neighbors' songs and will tolerate them at a distance; however, hearing an unfamiliar song will generally trigger an investigation and aggressive response (Falls, 1969; Goldman, 1973, Krebs, 1971; Kroodsma, 1976).

In addition to song, territorial sparrows are likely to exhibit a number of other displays during territory establishment and maintenance. Birds of both sexes may puff out their feathers, in particular raising those on the head to form a crest, or quiver their wings while pointing a closed bill at the intruder (Elekonich, 2000; Nice, 1943). They may peck furiously at nearby objects. If the intruder is unfazed, the resident then resorts to more direct physical threats, flying directly over the intruder, chasing him, and eventually attacking him. Opponents may fly at each other with feet stretched forward and may even fall to the ground as they engage and struggle. Physical fights are rare, however, and generally limited to the early breeding season before territory boundaries are firmly established.

Territorial responses can be studied in the field by observing naturally occurring behavior or by staging a "simulated territorial intrusion" (STI). In this procedure, experimenters place a decoy "intruder," often accompanied by song played through a loudspeaker, onto a resident's territory; the resident's behavioral response is then quantified. Taxidermic or painted models may be used as decoys, or a live, unfamiliar male in cage may be presented. The most robust responses are obtained by presenting both decoy and recorded song so that the resident receives both visual and auditory cues (Wingfield and Wada, 1989). The behavioral data that are collected typically include latency to respond, songs, flights directed at the decoy threat displays (e.g., wing quivers), distance from the decoy at the closest approach, and the time spent within a certain distance, for example 5 m, of the decoy (Wingfield, 1984b, 1985; Wingfield and Hahn, 1994).

B. Hormones and territoriality

In the 1970s and 1980s, John Wingfield and Donald Farner revolutionized the study of behavior in songbirds by developing methods to measure gonadal hormones in small plasma samples collected from free-living individuals (Wingfield and Farner, 1976). The ensuing research in "field endocrinology" (Wingfield et al., 1990; Walker et al., 2005) elucidated patterns of gonadal hormone secretion over the reproductive cycle in a wide variety of avian species. In general, the stages of breeding associated with high levels of aggression coincide with high plasma levels of T. In song sparrows, for example T peaks during territory establishment when agonistic encounters are most frequent (reviewed by Wingfield et al., 2001), rises again during egg-laying, and slowly declines until the incubation phase when territory disputes are rare (Fig. 5.1A). In some multiple-brooded species, for example house sparrows (*Passer domesticus*), competition for nest holes appears to drive an increase in plasma T during each egg-laying period (Hegner and Wingfield, 1986; Fig. 5.1B).

The temporal correlation between high plasma T and territorial behavior suggests that the two are related, and experimental manipulation of either T or the competitive environment shows that the relationship is bidirectional. Male song sparrows implanted subcutaneously with T-filled silastic capsules during territory establishment showed a more aggressive response to STI than males given empty capsules, and won territories that were twice the size (Wingfield, 1984b,c). Perhaps more interesting, however, was the effect on these males' neighbors. The residents occupying territories adjacent to the treated males also showed increases in plasma T, suggesting that having to defend their territories against their more aggressive, T-treated neighbors stimulated HPG activity. In a separate study, male song sparrows were removed from their territories, allowing new residents to take over. In this case, both the new residents and their neighbors experienced high T levels compared with unmanipulated controls (Wingfield et al., 1987). Together, these studies show not only that T increases aggression, but also that engaging in agonistic encounters increases plasma T. The HPG response is rapid; plasma T rises significantly

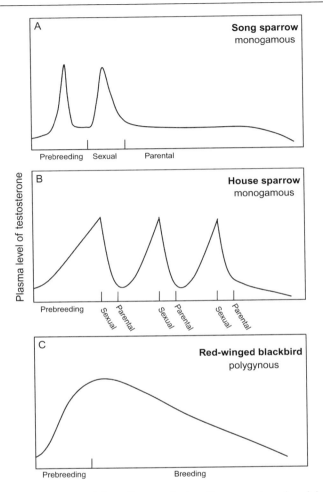

Figure 5.1. Plasma testosterone (T) profiles over the breeding season in males of three passerine species. (A) In song sparrows, T peaks during territory establishment (prebreeding) and again during laying of the first clutch when females are receptive (sexual), and then falls during incubation and feeding (parental). (B) In house sparrows, T peaks during periods of intense competition for nest sites, prior to each of multiple broods. (C) In red-winged blackbirds (*Agelaius phoeniceus*), males provide little parental care and spend more time on territorial defense; T remains relatively high until the end of the breeding season. Redrawn from data in Wingfield (1984a) and Hegner and Wingfield (1986).

within 10 min of STI onset in male song sparrows (Wingfield *et al.*, 1987) and can remain high for several days. This prolonged hormonal response probably heightens vigilance in anticipation of a sustained challenge (Wingfield, 1994a).

C. Aggression outside the breeding season

In seasonally breeding birds, the gonads regress and become quiescent during the nonbreeding season. A testis that measures more than 10 mm across during breeding may shrink to less than 1 mm in the fall. Ovaries likewise regress such that follicles are barely visible. As a result, plasma levels of gonadal steroids fall to low or even nondetectable levels (Wingfield and Farner, 1993). The gonads remain in this state until the long days of spring stimulate photoreceptors in the mediobasal hypothalamus, triggering HPG hormone secretion and gonadal recrudescence (reviewed by Dawson *et al.*, 2001).

For many species, the fall in plasma gonadal steroid levels at the end of the breeding season coincides with abandonment of breeding territories and the onset of flocking behavior. For others, quiescence of the HPG axis does not seem to affect territorial behavior at all. Both of these scenarios are considered below.

1. Aggression in flocks

In species that do not defend territories year-round, the conclusion of territoriality each year may give rise to flocking behavior. In flocks, birds maximize food-finding while minimizing predation risk (Hamilton, 1971). Competition for food, roosting sites, and other resources within flocks creates many opportunities for agonistic interactions, and some of the same behaviors used to defend territories, for example, song and threat displays, are also seen in this context. Other common aggressive displays in flocks include displacements, wherein one individual approaches another and causes it to move away, and hold-offs, wherein an individual refuses to be displaced. Each of these behaviors is easily observed in free-living groups, for example, in the vicinity of a popular food source (e.g., Ficken *et al.*, 1978; Harrington, 1973; Rohwer and Rohwer, 1978), or in captivity (e.g., Schlinger, 1987; Watt, *et al.*, 1984).

In some species, winter flocks adopt a highly organized social structure. Many of us are familiar, for example, with the dominance hierarchies, or "pecking order," established by chickens (*Gallus gallus domesticus*; Allee, 1936, 1942). Similar social arrangements have been observed in wild and captive groups of sparrows, for example, dark-eyed juncos (*Junco hyemalis*; Sabine, 1959), white-throated sparrows (Schneider, 1984; Watt *et al.*, 1984), Harris sparrows (*Zonotrichia querula*; Rohwer, 1975) and to a lesser extent, song sparrows (Nice, 1943). Within a group, each pair of individuals has a stable relationship such that one is dominant to the other. The subordinate will allow the dominant to displace it, deferring access to resources, and the dominant is unlikely to tolerate the close proximity of the subordinate. When all members of the group are considered together, there is in most cases a linear order of dominance; in other words, there is an alpha bird that dominates all others, a beta that dominates all but the alpha, and so on down to a bird that is subordinate to all. Exceptions, such as triangular relationships, are fairly common. Reversals, in which the hierarchy is challenged and altered, do occur, but overall the arrangement is fixed and stable (Sabine, 1959). In abiding by this social structure, the members of the group avoid the potential injuries and high energy expenditure that would result from constant aggressive encounters; when the hierarchy is stable, actual fighting is extremely rare. Only newcomers are subject to aggressive behavior, which subsides as they are assimilated into the group and assume a fixed rank (Tompkins, 1933).

Whereas physical contact and escalated fights are not normally required for the maintenance of a stable hierarchy, its initial establishment is associated with frequent aggressive interactions. When unfamiliar birds are forming a group, or when newcomers arrive, rank is settled by the outcome of agonistic encounters. Age and sex can predict rank in some cases. For example in white-crowned sparrows and related species, males tend to dominate females and older birds dominate younger ones such that the hierarchy within a group follows the general rule: adult males > adult females > immature males > immature females (Keys and Rothstein, 1991). In some species, such as house sparrows and Harris sparrows, plumage characteristics can predict rank; dominant individuals have dark "bibs" that appear to serve as status signals (Møller, 1987; Rohwer, 1975). Manipulation of bib coloration does not, however, alter status; rather, it results in heightened aggression toward the altered individuals, whose new plumage color-ation is inconsistent with their behavior (Rohwer and Rohwer, 1978).

The relationships between dominance rank and HPG function have been explored in many species of birds. This body of work, which spans almost 75 years, collectively shows that dominance rank is predicted by plasma T levels only in groups that are newly forming. In other words, when unfamiliar birds come together to form a social group, their T levels during the establishment of the hierarchy contribute toward their eventual rank. Later, after the hierarchy is settled and stable, rank is unrelated to plasma T (Baptista *et al.*, 1987; Buchanan *et al.*, 2010; Chase, 1982; Ramenofsky, 1984; Schlinger, 1987; Wiley *et al.*, 1999).

Given that engaging in aggressive behavior causes T to rise (Wingfield *et al.*, 1987), we must ask whether high T leads to the acquisition of a high rank, or vice versa. Some authors have reported that short-term treatment with exogenous T can alter stable hierarchies; the newly acquired ranks were main-tained in white-crowned sparrows (Baptista *et al.*, 1987) and domestic hens (Allee *et al.*, 1939) but not Japanese quail (*Coturnix coturnix japonica*; Selinger and Bermant, 1967). In the majority of such studies, T treatment affects eventual rank when it is done early during hierarchy establishment. After rank is stable, however, rank is usually unaffected by T administration (Buchanan *et al.*, 2010; Crook and Butterfield, 1968; Lumia, 1972; Mathewson, 1961; Rohwer and Rohwer, 1978; Wiley *et al.*, 1999). The lack of an effect of T treatment on established, stable ranks has been attributed to learning, or "social inertia" (Guhl, 1968; Wiley *et al.*, 1999), in groups of individuals that are familiar with each other. In house sparrows, which do not defend large breeding territories, dominance rank during the nonbreeding season is determined largely by rank at the end of the breeding season when plasma T level is falling (Buchanan et al., 2010). In other words, winners stay winners and losers stay losers (Chase, 1982). HPG activity during the breeding season may therefore contribute toward rank during the nonbreeding season. During both times of year, actual physical aggression is limited to periods of instability during which high-ranking positions up for grabs or otherwise in dispute. After the establishment of social relationships and boundaries, physical aggression is rare and plasma levels of T are much lower.

2. Territoriality in the nonbreeding season

Song sparrows in resident populations, despite being seasonal breeders that undergo gonadal regression in the fall, can maintain essentially the same territories year-round. Whereas the song sparrows studied by Wingfield in New York State (e.g., Wingfield, 1984a,b,c, 1985) abandon their territories and migrate at the conclusion of the breeding season, populations in Western Washington remain on the same territories, molt their feathers, and then, despite having completely regressed gonads and nondetectable levels of T, resume territorial defense (Wingfield and Hahn, 1994). Furthermore, young males just hatched the previous spring can establish new territories in the fall without an increase in T. STI during the fall induces the same behavioral responses as in spring, but without an accompanying HPG response (Soma and Wingfield, 2001; Wingfield and Hahn, 1994), and castration does not interfere with the ability to maintain a territory (Wingfield, 1994b). These results led to the hypothesis that territorial aggression and gonadal steroid secretion can become uncoupled in this and other species that defend territories outside the breeding season.

To address this question, Soma et al. (1999, 2000a,b) tested whether T action is necessary for autumnal aggression in a population of song sparrows in Western Washington. To block the effects of T, they administered androgen receptor antagonists and, because T can also act via conversion to estradiol (E2), blockers of that conversion. They found that blocking the action of aromatase, an enzyme that converts T to E2, reduced territorial aggression even in the fall when plasma levels of gonadal steroids are naturally very low. This result suggested that autumnal aggression is not in fact independent of steroid hormones. This reduction was reversed by treatment with E2 (Soma et al., 2000a).

Because gonads are not required for autumnal territorial defense (Wingfield, 1994b), the E2 that drives this behavior must come from a nongonadal source. One such source may be the brain itself, which contains high levels of aromatase. In song sparrows, aromatase mRNA is expressed in the brain at all times of year (Soma et al., 2003; Wacker et al., 2010) suggesting a possible source of E2 available to brain tissue year-round. Aromatase activity in the ventral telencephalon, which contains the putative avian homologue of the amygdala, is reduced during molt, at which time aggression is also low (Soma *et al.*, 2003). Brain-generated E2 may thus be an important regulator of territorial aggression in this species. To synthesize E2, the brain may make use of androgen precursors from regressed gonads or the adrenals (see Soma, 2006; Soma *et al.*, 2008 for reviews). Alternatively, some regions of the brain contain all of the enzymes necessary to synthesize E2 *de novo* from cholesterol substrate (reviewed by Soma *et al.*, 2008; Remage-Healey *et al.*, 2010), obviating the need for a peripheral steroid synthesis altogether. The regions of the brain important for aggressive behaviors, for example the nuclei that control singing behavior, are rich in these enzymes (Remage-Healey *et al.*, 2010). Aromatase expression is high in the ventromedial nucleus of the hypothalamus (VMH) during all times of year except during molt, when aggression is virtually absent (Wacker *et al.*, 2010). In mice, distinct populations of VMH neurons contribute to fighting and mating behavior (Lin *et al.*, 2011). In Section III, we will explore further the role of this region in aggression in songbirds.

Like androgen release from the gonad, E2 synthesis by the brain appears to be behaviorally regulated. Remage-Healey *et al.* (2008) recently showed that in zebra finches (*Taeniopygia guttata*), hearing conspecific song increases E2 concentrations in the auditory forebrain within minutes. Pradhan *et al.* (2010) showed subsequently that in nonbreeding, territorial song sparrows, exposure to STI rapidly increases activity of 3β-hydroxysteroid dehydrogenase, an enzyme necessary to synthesize E2 from androgen precursors. These findings demonstrate that the steroid environment within the brain is dynamic, sensitive to the social environment, and more independent of the gonad than previously thought. These discoveries of socially regulated, brain-generated steroid synthesis challenge the traditional view of hormone-mediated aggression and highlight the importance of songbird models to our understanding of how steroids regulate gene activity in the brain.

Forty years of research on free-living sparrows has shown that aggression depends on steroid hormones. Even when at first glance it appears that aggression and steroid hormones have come "uncoupled," for example, when resident song sparrows vigorously defend territories despite low plasma T, the evidence shows that low levels are necessary for the expression of territoriality and that the brain itself may produce sufficient amounts. The role of steroid hormones, particularly E2, in aggression seems very similar to their role in sexual receptivity (reviewed by Maney, 2010): plasma levels need not be high, and in fact seasonal peaks in plasma levels need not be associated in time with the behavior. A low level, however, is required for the behavior to be expressed. Because the frequency of aggressive behavior is clearly not always correlated directly with plasma levels of steroid hormones, it is possible that the hormones play a priming, or permissive, role and that other hormones or neurotransmitters are also important.

D. Evolution of aggression and life history strategies

Research on wild songbirds has demonstrated a robust, two-way relationship between aggression and HPG activity. What does that finding tell us about the evolution and genetic control of these behaviors? Because steroid hormones affect suites of behaviors, not just aggression, it is helpful to think about this issue in terms of behavioral strategies. As discussed in Section I above, investment in territory defense and mate-finding defines a strategy at one end of a behavioral continuum, with investment in survival and parenting at the other (Trivers, 1972). This trade-off appears to be mediated, at least in part, by HPG activity (Ketterson and Nolan, 1994; McGlothlin et al., 2007; Wingfield et al., 1990). In species with male parental care, T is high during territory establishment but falls during the parental phase (Fig. 5.1A, B). Exogenous administration of T during the parental phase inhibits parental behavior and increases territorial behavior (Hegner and Wingfield, 1987; Schoech et al., 1998; Silverin, 1980). In species without male parental care, for example polygynous species in which females do the bulk of the care, T remains high in males for the duration of the season (Fig. 5.1C). In a study by Wingfield (1984c), T treatment of male song sparrows not only reduced parental care but also induced polygyny in this normally monogamous species. These males spent much of their time singing and acquired huge territories, attracting multiple females. Their failure to provision the young, however, likely reduced their overall reproductive success (Hegner and Wingfield, 1987; Silverin, 1980). Alterations in HPG function could therefore result in large, cascading effects and make an important contribution to variation in social behavior and social strategies (Ketterson and Nolan, 1992; Sinervo and Svensson, 2002). When the effects of a hormone are antagonistic with respect to behavior, for example, in the case of T and parenting versus aggression, "antagonistic pleiotropy" can give rise to behavioral trade-offs (Finch and Rose, 1995). We hypothesize that suites of related genes, the expression of which is tightly governed by social cues, may act hierarchically to organize and regulate the hormonal and neural systems that promote territorial aggression and reduce parental behavior. In the next section, we explore the neural circuits that are involved and consider how evolution has shaped them in species with different behavioral strategies.

III. TRANSCRIPTIONAL ACTIVITY AND NEURAL MECHANISMS OF AGGRESSION IN BIRDS

A. Transcriptional traces of aggression reveal ubiquitous vertebrate themes

Although it has long been known that all vertebrates share some basic features in the organization of the amygdala, hypothalamic nuclei, and associated (limbic) areas of the basal forebrain and midbrain, the extent of these similarities has become much more clear in the past 10 years as investigators have combined genomic data with conventional neuroanatomical and functional approaches (e.g., using lesions and pharmacological manipulations). We now know that birds and rodents exhibit extraordinary similarities in the organization of limbic brain areas (Goodson, 2005; Newman, 1999), including distinct homologies at the subnuclear level (e.g., Goodson *et al.*, 2004a; Kingsbury *et al.*, 2011) and very specific similarities in the topographical patterns of transcriptional response to aggressive interactions (and other social interactions, as well), as established through the experimental induction of immediate early gene (IEG) transcripts and their protein products (Ball and Balthazart, 2001; Goodson, 2005). The IEGs most commonly used for such functional studies are *c-fos* and a gene variably known as *zif-268*, *egr-1*, *NGFI-A*, *krox-24*, or *Zenk* (the latter being a name often used in birds as an acronym for the other four names; Mello *et al.*, 1992).

Experimentally induced increases in IEG mRNA can be detected by *in situ* hybridization (ISH) within 15–30 min. For detection of IEG proteins by immunocytochemistry (ICC), most investigators harvest brain tissue 60–90 min after the experimental manipulation, such as an aggressive interaction. Because the induced IEG proteins are still elevated at 90 min and two half-lives of the protein have passed (Herdegen and Leah, 1998), it is possible to determine whether the experimental manipulation may have decreased IEG activity from control levels (e.g., Bharati and Goodson, 2006; Goodson and Wang, 2006). Most behavioral neuroscientists are primarily interested in IEGs not because of what they do inside the cell, but rather because they provide a proxy marker to indicate that a cell has responded in some way to a stimulus. That response may or may not be associated with action potentials and release of neurochemicals (Herdegen and Leah, 1998), but by labeling for IEG transcripts and proteins, investigators can gain a good idea about the functional properties of different brain areas or cell groups. The actual molecular functions of IEGs are varied, but typically include the regulation of other genes involved in experience-dependent neuroplasticity (Herdegen and Leah, 1998; Mello and Ribeiro, 1998), and thus IEGs probably have the ultimate effect of changing the way that cells function and behave during subsequent behavioral interactions.

In rodents, resident–intruder encounters induce IEG activity in a characteristic pattern that includes the medial amygdala (MeA), medial bed nucleus of the stria terminalis (BSTm, a component of the medial extended amygdala that shares many anatomical and functional properties with the MeA), anterior hypothalamus (AH), ventrolateral lateral septum (LS), ventrolateral subdivision of the ventromedial nucleus of the hypothalamus (VMHvl; or lateral VMH in hamsters), dorsal premammillary nucleus, and the dorsal midbrain periaqueductal gray (Kollack-Walker *et al.*, 1997; Motta *et al.*, 2009). Notably, along with the medial preoptic area, these same brain areas are central to the regulation of most other social behaviors, including communication behaviors, parental care, pair bonding, appetitive and consummatory sexual behavior, juvenile play, social recognition, and both same-sex and opposite-sex affiliation (Goodson, 2005; Newman, 1999). Despite the similarities, the relative amount of IEG activity across the different nodes of this "social behavior network" is distinctive for each social context, suggesting that functional relationships across the network nodes are dynamic and context-specific. Distinct patterns of correlated activity between network nodes in different social contexts have now been demonstrated in multiple vertebrate classes (Crews et al., 2006; Hoke et al., 2005; Yang and Wilczynski, 2007).

The areas comprising the social behavior network are readily identified in birds and are anatomically and functionally conserved across amniote vertebrates, and in fact, the basic features of this network are present even in fish (Goodson, 2005; Goodson and Bass, 2002). Consistent with this conservation, territorial songbirds housed in captivity exhibit a pattern of IEG activity after STI that is virtually identical to the pattern described for rodents after a resident–intruder encounter (Goodson and Evans, 2004; Goodson et al., 2005b). This work has been conducted in animals housed in their natural habitat, and data for catecholaminergic midbrain areas are even available from animals occupying natural territories (Maney and Ball, 2003). In addition, following exposure to same-sex conspecifics through a wire barrier in a quiet room (which elicits very little overt behavior), territorial finches exhibit relatively greater Fos and/or egr-1 responses than do gregarious species in a pattern similar to aggressive encounters (Goodson et al., 2005a). Hence, at least to an extent, the IEG activity of these brain areas reflects perceptual or motivational processes, not simply activation of aggression.

Although it is intuitive to interpret IEG induction as reflecting a positive relationship between a brain area and behavior, negative correlations between IEG cell counts and aggressive behavior are observed for multiple brain areas. These include the AH, both pallial and subpallial subdivisions of the LS, and the paraventricular nucleus of the hypothalamus (PVN; Fig. 5.2; Goodson et al., 2005b). This pattern of results suggests that aggression is under inhibitory control, at least by some areas. Consistent with this idea, lesions of the LS increase resident–intruder aggression in male field sparrows and pigeons (Columba livia; Goodson et al., 1999; Ramirez et al., 1988). Note, however, that such effects are not observed in some contexts, such as aggressive competition for mates in male zebra finches, a highly gregarious species (Goodson et al., 1999).

The PVN is heavily interconnected with the social behavior network and plays an important role in the regulation of autonomic and pituitary activity in relation to behavioral state. A subset of cells in the PVN produce arginine vasotocin (VT; homologue and evolutionary precursor of mammalian arginine vasopressin, VP), and the percentage of those cells that express Fos is also negatively correlated with aggressive response to an STI in male song sparrows

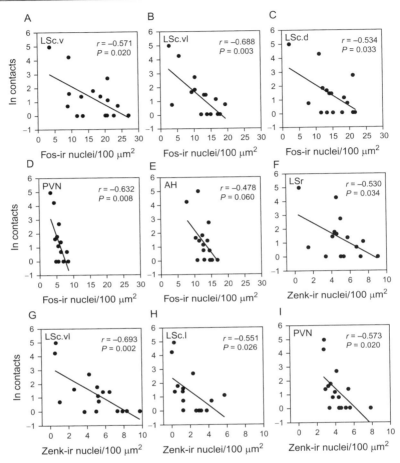

Figure 5.2. (A–E) Correlations between aggressive behavior and Fos-immunoreactive (-ir) cell counts in the subpallial (ventral, ventrolateral) and pallial (dorsal) zones of the caudal lateral septum (LSc.v, LSc.vl, and LSc.d; A–C, respectively), paraventricular hypothalamus (PVN; D), and anterior hypothalamus (AH; E) of male song sparrows exposed to STI ($n = 16$). The intruder's cage and a speaker broadcasting song were placed adjacent to the subject's cage. Subjects showed selective flights to the cage wall adjoining the intruder, providing a good measure of aggressive response. Data are shown as the natural log (ln) of the number of contacts with the wire barrier during a 10-min test. (F–I) Correlations between barrier contacts and Zenk-ir cell counts in the rostral LS (LSr; F), LS.vl (G), lateral zone of the LSc (LSc.l; H), and PVN (I). Cell counts are shown as the number of immunoreactive nuclei per 100 μm^2. Modified from Goodson *et al.* (2005b).

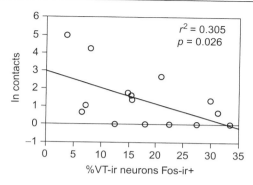

Figure 5.3. The percentage of arginine vasotocin (VT) neurons in the PVN that express Fos after a 10-min STI is negatively correlated with aggression (ln, the number of contacts with the cage wall adjoining the intruder's cage; see Fig. 5.2 caption; $n = 16$) in song sparrows. Modified from Goodson and Kabelik (2009).

(Fig. 5.3; Goodson and Kabelik, 2009). VT and VP are secretagogues for adrenocorticotropic hormone, and thus the lower IEG activity of the PVN VT cells in more aggressive males likely reflects a lower stress response to the encounter. Virtually identical results for VP–Fos colocalization are obtained in lab mice (Ho et al., 2010). As addressed in Section III.B below, this negative relationship between aggression and VT–Fos colocalization in the PVN accurately predicts pharmacological effects that vary in relation to the subject's dominance status (Goodson et al., 2009b).

B. Neurochemistry and major modulators

Although territorial aggression in birds has been the focus of hundreds of studies, including many that address proximate endocrine mechanisms (Goodson et al., 2005c; Konishi et al., 1989), a surprisingly small number of experiments have been conducted with the goal of delineating relevant neurochemical circuits in the brain. An early study of whole-brain neurochemistry shows that dopamine, norepinephrine, and acetylcholine are all associated with aggressive behavior in male Japanese quail (Edens, 1987), and catecholaminergic midbrain nuclei show increased IEG activity in response to STI in song sparrows (Maney and Ball, 2003).

Following aggressive competition for a potential mate, male zebra finches exhibit significant increases in the percentages of tyrosine hydroxylase-ir (TH-ir) cells expressing Fos within the "retrorubral" area (A8), substantial nigra (A9), ventral tegmental area (VTA; A10), and midbrain central gray (A11), but show a significant decrease in TH–Fos colocalization within the A12 neurons of the tuberal hypothalamus (Bharati and Goodson, 2006). TH is the rate-limiting enzyme for catecholamine synthesis, and all of the cell groups just listed are known to be dopaminergic. Despite these results, treatments with quinpirole, a dopamine D2 receptor agonist, significantly decrease aggression during mate competition. D1 and D4 agonists are without effect, although a modest inhibition is observed with the D3 agonist 7-OH-DPAT, which may reflect weak binding to the D2 receptor (Kabelik et al., 2010). These seemingly contradictory results likely reflect the fact that courtship is displayed at a high rate during the competition tests, and given that TH–Fos colocalization in the central gray and caudal VTA correlates positively with courtship singing (Goodson et al., 2009a), the increased colocalization of TH and Fos following mate competition is likely attributable to directed singing and not the display of aggression. The number of TH-ir cells in the central gray also correlates positively with the average number of songs that male zebra finches sing to females during courtship tests. Notably, territorial finch species exhibit fewer TH-ir cells in the caudal VTA than do gregarious species such as the zebra finch, although this may reflect a lower level of affiliation rather than a negative relationship between this cell group and aggression (Goodson et al., 2009a).

Dopaminergic mechanisms of song have also been examined in European starlings (Sturnus vulgaris), which sing in the context of breeding both to attract females and to repel other males. Antagonism of D1 receptors decreases song in breeding-condition males, whereas a dopamine reuptake inhibitor facilitates it (Schroeder and Riters, 2006), and aggressive song correlates negatively with D1 receptor density in numerous areas, including the LS, BSTm, medial preoptic area, and central gray (Heimovics et al., 2009).

The handful of other neurochemical manipulations that have been conducted in studies of avian aggression have focused on neuromodulators such as VT, vasoactive intestinal polypeptide (VIP), and endogenous opioids. Of these, the endogenous opioids have received the least attention, but are known to inhibit aggression in Japanese quail, at least partially via the delta receptor subtype (Kotegawa et al., 1997). These neuropeptides are each produced in multiple brain areas, and as suggested for VT, it may be the case that the different cell groups have divergent effects on behavior (a possibility that should also be considered in relation to the major neurotransmitters just discussed; Goodson and Kabelik, 2009).

Both VIP and VT exert complex effects on aggression that likely reflect the modulation of stress- and anxiety-related processes. For instance, in territorial male field sparrows housed in aviaries placed in their natural habitat, intraseptal infusions of VT decrease aggression in resident–intruder tests, but selectively facilitate the spontaneous use of an agonistic song type during the dawn song period (Goodson, 1998a). No effects are observed for the multipurpose song type that is used to attract females, and VIP tends to exert an opposite pattern of effects (Goodson, 1998a,b). Interestingly, the VT and VIP systems in the LS are both sensitive to sex steroids. Castration causes down- and upregulation of VT and VIP immunoreactivity, respectively, and T or E2 replacement reverses these effects (Aste et al., 1997; Panzica et al., 2001; Voorhuis et al., 1988; but see Wacker et al., 2008). The VT/VIP systems therefore represent a possible mechanism whereby gonadal steroids may modulate aggression.

In order to determine whether VT modulates stress-related processes, and whether it does so in a manner that is integrated with its effects on agonistic behavior, Goodson and Evans (2004) examined Zenk responses to nonsocial stress alone (capture in an outdoor flight cage and restraint for intraventricular infusions), or the same nonsocial stressors followed by STI. These manipulations were conducted in male song sparrows housed in flight cages placed in their natural habitat. Subjects were infused with either vehicle or a VT V_{1a} receptor antagonist and were sacrificed at the completion of testing for immunolabeling of Zenk and VT. In some brain areas, nonsocial and social stimuli induced Zenk within the same subset of cells, which was discernable because (1) in vehicle-treated animals, the nonsocial stressor induced a significant increase in Zenk-ir cell numbers, and subsequent exposure to the social challenge produced no further increase, but (2) blocking the Zenk response to handling with the V_{1a} antagonist revealed a sensitivity to the social challenge (i.e., by eliminating the ceiling effect). This "integrated" pattern of Zenk response was observed for numerous areas, including the AH, POA, lateral VMH, lateral BST, and most zones of the LS. Notably, in all of these cases, the antagonist exerted more pronounced effects in the subjects that were exposed to the nonsocial stress alone. However, in the BSTm and ventrolateral LS, Zenk responses to the social challenge were significantly greater than to the nonsocial stressor, even in vehicle-treated subjects, indicating that at least some cells in these areas are more selectively activated by social challenge. The BSTm showed particularly selective responses to the social challenge, which were completely blocked by the V_{1a} antagonist (Goodson and Evans, 2004).

Unfortunately, VT-ir cells of the BSTm were not detectable in this study (these neurons are weakly immunoreactive in most vertebrates and may store little peptide relative to the amount being released), but the VT-immunoreactive neurons of the PVN showed significant responses to social challenge and, most interestingly, the VT–Zenk colocalization was reduced by the V_{1a} antagonist only in the animals exposed to the STI (Goodson and Evans, 2004). As assessed in a later study with more robust immunolabeling of VT, VT–Fos colocalization in the BSTm of male song sparrows is not increased by social challenge, whereas colocalization in the PVN is negatively correlated with aggression (Fig. 5.2; Goodson and Kabelik, 2009).

In territorial estrildid finches, exposure to a same-sex conspecific through a wire barrier actually decreases VT–Fos colocalization in the BSTm, but the same manipulation increases VT–Fos colocalization in gregarious finch species, and the territorial birds do show large increases in VT–Fos colocalization if they are reunited with their pair-bond partner. Conversely, socially induced VT–Fos colocalization in the BSTm is blocked in the gregarious zebra finch if the subjects are intensely subjugated by a dominant bird (Goodson and Wang, 2006). Thus, the VT neurons of the BSTm exhibit an exquisite sensitivity to the valence of social stimuli, and more recent findings suggest that this valence sensitivity is not extended to nonsocial stimuli (Goodson *et al.*, 2009c).

The differential response profiles of the VT neurons in the BSTm and PVN may account for at least a portion of the context-specificity that is observed following central infusions of VT or V_1 receptor antagonists. For instance, in zebra finches, VT promotes aggression in the context of mate competition and a V_{1a} antagonist inhibits aggression (Goodson *et al.*, 2004b). These effects are consistent with the observation that VT–Fos colocalization is increased in the BSTm during mate competition, but not in the PVN. Conversely, resident–intruder aggression is inhibited by VT infusions in territorial species, consistent with the negative correlation between aggression and VT–Fos colocalization in the PVN (Goodson and Kabelik, 2009; Goodson and Wang, 2006). The different effects on mate competition and resident–intruder aggression (or nest defense in zebra finches) can be observed in the same species (Goodson *et al.*, 2009b; Kabelik *et al.*, 2009), as shown for the territorial violet-eared waxbill (*Uraeginthus granatina*) in Fig. 5.4A–B. However, in the violet-eared waxbill, males that are typically dominant do not show a behavioral response to the V_{1a} antagonist in standard resident–intruder tests whereas aggression is facilitated in subordinates (Fig. 5.4C; Goodson *et al.*, 2009b). Thus, perhaps only the subordinate males activate the PVN VT neurons during aggressive encounters and this activation inhibits aggression.

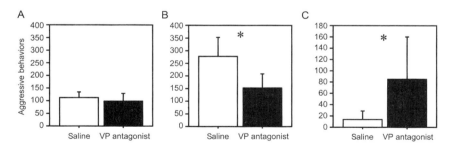

Figure 5.4. (A) Peripheral injections of a novel V1a antagonist that crosses the blood–brain barrier have no effect on resident–intruder aggression in male violet-eared waxbills that are aggressive and typically dominant, but aggression in the context of mate competition is significantly reduced by the antagonist in the same males (B). (C) In males that are typically subordinate, resident–intruder aggression is disinhibited by the same treatments. Modified from Goodson *et al.* (2009b).

The ability to label IEG products has provided immeasurable insight into the neural basis of aggression, allowing the identification and mapping of specific circuits that respond rapidly to social stimuli. Lacking until recently were powerful genomic methods necessary for a more complete understanding of the complex protein interactions involved in social responses. The sequencing of the zebra finch genome (Warren et al., 2010) has provided unprecedented insight into what happens inside the songbird brain during agonistic encounters. Using tools such as high throughput sequencing and microarray analysis, investigators can now look at the regulation of many hundreds of genes simultaneously. In a recent gene profiling study, Mukai et al. (2009) compared the expression of more than 11,500 different gene transcripts in free-living song sparrows responding either to STI or a control intrusion by a heterospecific. For behavioral manipulations conducted during the breeding season, 67 gene transcripts were differentially expressed in the hypothalamus following exposure to an STI compared to control. During the fall, when territorial aggression seems to be independent of gonadal steroid production (reviewed by Soma, 2006; Soma et al., 2008; see also Section II), 173 transcripts were affected (Mukai et al., 2009). There were significant interactions between season and STI for 88 transcripts (Mukai et al., 2009), which may in part reflect the differential regulation of the pituitary–gonadal axis across seasons. The expression of many of the gene transcripts was not, however, affected by season and therefore may be important for the regulation of aggressive behavior itself rather than endocrine responses to aggressive encounters. This study represents the early days of genomic analysis of social behavior in free-living, natural populations and sets the standard for many more sure to follow. In the next section, we consider a songbird species that because of a natural genetic anomaly is becoming a popular model for studying the genetic mechanisms underlying aggression.

IV. A NATURAL MODEL UNITING SOCIAL BEHAVIOR, HORMONES, AND GENETICS

A. The white-throated sparrow

The underlying genetic basis of variation in social behavior is of intense interest, yet only a handful of genes have been linked to specific social behaviors in vertebrates (reviewed by Robinson et al., 2005). Thus, there is an obvious need to identify populations, human or otherwise, in which there is clear linkage between genes and social behavior. A common wild songbird, the white-throated sparrow, offers such an opportunity. This species, in which socially monogamous pairs defend breeding territories, provision the young with food, and form flocks with stable dominance hierarchies in the winter, is a typical New World sparrow in nearly all respects. What sets it apart from other songbirds is that it exhibits alternative phenotypes, defined by a plumage polymorphism, that differ in their social behavior. Both males and females can be categorized into one of two plumage morphs that differ primarily in the color of the crown stripes (Lowther, 1961; Piper and Wiley, 1989; Watt, 1986; see Fig. 5.5). Behavioral studies conducted in the animals' natural habitat have established that individuals with a white medial stripe (WS) on the crown engage in a more aggressive strategy, whereas birds with a tan medial stripe (TS) are more parental.

The species represents a promising model in which to study the genetic basis of aggression because the plumage pattern segregates with the presence or absence of a structural rearrangement of chromosome 2. WS individuals are heterozygous for the rearranged chromosome (ZAL2^m), whereas those of the TS morph are homozygous for the wild-type chromosome (ZAL2; Thorneycroft, 1975). Once they molt into adult plumage the phenotype is fixed for the lifetime of the individual. Within a population, approximately half of the birds are WS (ZAL2/2^m), whereas the other half are TS (ZAL2/2; Lowther, 1961; Thorneycroft, 1975). This balanced polymorphism is maintained in the population by disassortative mating—WS and TS birds nearly always mate with individuals of the opposite morph (Knapton and Falls, 1983; Lowther, 1961; Thorneycroft, 1975; Tuttle, 1993). This mating pattern results in a virtual absence of birds homozygous for ZAL2^m, a genotype that Thorneycroft (1975) hypothesized may be less viable due to recessive deleterious mutations. Of more than 1000 individuals genotyped, only one was found to be homozygous for ZAL2^m (Maney et al. unpublished data; Romanov et al., 2009; Thorneycroft, 1975).

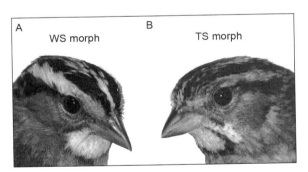

Figure 5.5. Plumage polymorphism in white-throated sparrows. (A) Individuals of the white-stripe (WS) morph have alternating black and white stripes on the crown, bright yellow lores, and a clear white throat patch. (B) Individuals of the tan-stripe (TS) morph have alternating brown and tan stripes on the crown, duller yellow lores, and dark bars within the white throat patch. Photos by Allison Reid. Reprinted from Maney (2008). (See Color Insert.)

The behavioral differences that segregate with the ZAL2^m chromosome have been well documented in field studies. Males and females of the WS morph are more aggressive, both in territorial defense and in mate-seeking, than their TS counterparts. WS males sing more in response to STI than TS males (Collins and Houtman, 1999; Kopachena and Falls, 1993a; Horton and Maney, unpublished observations) and are more likely to trespass onto the territories of other males (Tuttle, 2003). Whereas WS females sing and engage in active territorial defense independently of their mates, TS females do so only rarely (Kopachena and Falls, 1993a; Horton and Maney, unpublished observations). TS birds of both sexes feed young more often during the parental phase of the breeding season than do WS birds (Knapton and Falls, 1983; Kopachena and Falls, 1993b). The relative strategies employed by the different morphs of this species therefore fall onto different ends of the behavioral continuum between territoriality and parenting (Trivers, 1972).

B. Endocrine and neuroendocrine correlates of behavioral polymorphism

Because the behavioral trade-off between territorial defense and parenting is clearly mediated at least in part by HPG activity in songbirds (Ketterson and Nolan, 1994; McGlothlin et al., 2007; Wingfield et al., 1990), we should immediately suspect that, in white-throated sparrows, HPG function may vary according to morph. Spinney et al. (2006) found that in free-living birds in breeding condition, WS males do have larger testes and higher levels of circulating T than TS males. This phenomenon has also been demonstrated in captive populations (Maney, 2008; Swett and Breuner, 2009). In both the field and the lab, however, the difference in plasma T disappears when birds are not in breeding condition (Maney, 2008; Spinney et al., 2006). Interestingly, morph differences in aggression appear only during the breeding season, mirroring the morph difference in circulating T. In winter flocks and in laboratory-housed birds held on short days, morph is not related to dominance rank or to aggression (Dearborn and Wiley, 1993; Harrington, 1973; Piper and Wiley, 1989; Schlinger, 1987; Schwabl et al., 1988; Watt et al., 1984; Wiley et al., 1999). In contrast, when birds are held on long days and undergo gonadal recrudescence, WS birds engage in significantly more aggression than their TS cage-mates and tend to outrank them (Fig. 5.6; see also Watt et al., 1984). Morph differences in dominance and aggression may therefore depend on season and thus perhaps on differences in HPG function.

Because levels of gonadal steroids differ between the morphs, the behavioral polymorphism may be driven by the effects of these steroids on the brain. To evaluate this hypothesis, Maney et al. (2005) compared the morphs with respect to the VT and VIP neuropeptide systems, which are highly steroid dependent (Aste et al., 1997; Panzica et al., 2001; Voorhuis et al., 1988). WS birds

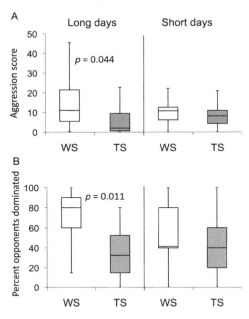

Figure 5.6. Medians, IQR, and ranges for (A) aggression scores (number of aggressive acts initiated per hour) and (B) individual ranks (as percent opponents dominated) within social groups. Males were introduced in single-sex groups of six birds (three WS and three TS per group) in indoor aviaries. Aggression scores and ranks were determined 10–14 days later by observing interactions and constructing dominance matrices. During spring-like day lengths (16L:8D), WS males were (A) more aggressive and (B) outranked TS males. Rank was unrelated to morph on short days (8L:16D). The long- and short-day experiments were conducted on different individuals. Data from Horton and Maney, unpublished.

had higher levels of VT-immunoreactivity in the BSTm and ventrolateral LS than TS birds. Since T is higher in WS males, this result is consistent with the idea that T may be engaging the VT system in the LS. Central administration of VT in the closely related white-throated sparrow induces agonistic song (Maney *et al.*, 1997), suggesting that engagement of this system may be directly related to aggressive behaviors. Central administration of VIP, in contrast, reduces agonistic song in field sparrows (Goodson, 1998a); immunoreactivity for this peptide was higher in the ventrolateral LS of the TS (less aggressive) morph. VIP immunoreactivity in this region is inversely proportional to T levels (Aste *et al.*, 1997) again supporting a possible role for gonadal sex steroids in the control of aggression in this species.

C. Causality and "phenotypic engineering"

Looking for morph differences in endocrine variables in unmanipulated individuals is an important endeavor, in that significant correlations can help illuminate possible physiological causes of aggression. Such correlations alone, however, can provide only limited information on causal mechanisms. The morph difference in plasma T, for example, could be a consequence, rather than a cause, of polymorphic behavior. Either scenario would explain the observed correlations between T and social behavior (Spinney et al., 2006). As discussed in Section II above, experiments involving manipulation of T or of social contexts in songbirds have revealed causal effects in both directions. Recall that in other songbird species, free-living males treated with T defend larger territories, engage in more aggression, acquire more mates, and provide less parental care than untreated males (Hegner and Wingfield, 1987; Schoech et al., 1998; Silverin, 1980; Wingfield, 1984b,c). Territorial intrusion or the presence of receptive females, however, cause release of endogenous T (Dufty and Wingfield, 1986; Moore, 1983; Wingfield and Hahn, 1994; Wingfield and Monk, 1994). A one-way causal effect of T on aggression and parenting may not completely explain polymorphic behavior in white-throated sparrows.

Some authors have suggested that the role of hormones in alternative phenotypes is best studied by performing hormonal manipulations, or "phenotypic engineering" (Ketterson and Nolan, 1992; Miles et al., 2007; Zera and Harshman, 2001). To test whether morph-dependent variation in territorial behavior in male white-throated sparrows is attributable entirely to variation in T, Maney et al. (2009) eliminated morph differences in T and then compared WS and TS responses to STI in the lab. Males in nonbreeding condition received silastic implants containing T, so that plasma levels in the WS and TS groups were high and equal. When presented with audio playback of conspecific male song, WS males sang significantly more often than TS males. This result suggests that WS males respond more aggressively to a territorial challenge than TS males, even when T levels are experimentally equalized between the morphs.

If morph differences in social behavior are not caused simply by differences in plasma T, then our search for causal factors should turn to other aspects of HPG function, for example, steroid binding or metabolism. The list of such factors is long and includes a large number of receptors, enzymes, and binding globulins. Comparative genomic approaches are required to conduct large-scale comparisons of gene expression as well as detailed analysis of the genetic differences between the morphs.

D. Mapping the ZAL2^m

The early genetic work in the white-throated sparrow, done more than 35 years ago (Thorneycroft, 1975), showed definitively that morph differences are associated with a clear, tractable chromosomal rearrangement. The ZAL2^m

chromosome thus offers a powerful starting point for understanding the mechanisms underlying aggressive behavior in birds and other vertebrates. In essence, nature has created a genetic manipulation that allows us to identify genes that are affected by the rearrangement and therefore potentially causal for heightened aggression in the WS individuals.

The identification of such genes first requires mapping of the $ZAL2^m$ rearrangement; genes that map within it can then be evaluated as likely candidates. Using a comparative genomic approach, Thomas *et al.* (2008) (see also Davis *et al.*, 2011; Huynh *et al.*, 2010a,b) began the initial modern genomic characterization of the $ZAL2^m$ rearrangement. By taking advantage of the genomic resources available for two other avian species, the chicken and the zebra finch, they established a comparative map of $ZAL2^m$ and found that the chromosome contains a complex rearrangement involving not one but at least two inversions around the centromere (Fig. 5.7). The two inversions may have occurred in succession, with the second completely contained within the first. Alternatively, ZAL2 and $ZAL2^m$ may each represent rearranged versions of an ancestral chromosome 2, having undergone rearrangement at different times. The rearrangement now spans the majority of the chromosome and could contain as many as 1000 protein-coding genes (Davis *et al.*, 2011; Thomas *et al.*, 2008). Thus, although the region containing the rearrangement is large, this work has delineated a finite set of genes linked to the behavioral and plumage polymorphisms in this species.

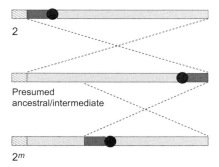

Figure 5.7. Model for the $ZAL2^m$ rearrangement. A minimum of two pericentric inversions, represented by the pairs of dashed lines, are hypothesized to have led to the $ZAL2/2^m$ polymorphism. ZAL2 (top) and $ZAL2^m$ (bottom) are shown along with a hypothetical chromosomal arrangement (middle) that could be either ancestral to both the ZAL2 and $ZAL2^m$ or an intermediate arrangement. Centromeres are represented by filled circles. Dark and light boxes represent segments originating on the short and long arms of the presumed ancestral chromosome, respectively. Free recombination between the ZAL2 and $ZAL2^m$ is limited to the tip of the short arm (hatched boxes).

Exactly how does the architecture of $ZAL2^m$ affect the expression of the genes inside the rearrangement and the proteins they encode? Inversions are hypothesized to affect gene and protein function in two main ways. First, genes at or near the breakpoints may be physically disrupted or otherwise directly affected by the breakage and subsequent change in position. So far, sequencing efforts have identified no genes physically disrupted by the $ZAL2^m$ breakpoints (Davis et al., 2011); however, position effects may have led to functionally distinct alleles for those nearby. For example, a cluster of genes encoding bitter taste receptors has been separated by one of the breakpoints and now maps to different arms of the chromosome (Davis et al., 2010). This separation, which appears to have led to nonsynonymous variants detected between the ZAL2 and $ZAL2^m$, may have implications for diet and habitat selection but is unlikely to explain morph differences in aggression and parenting.

The behavioral polymorphisms in social behavior in this species are more likely related to a second important consequence of pericentric inversions, which is the suppression of recombination and subsequent genetic differentiation of the inverted region. Thorneycroft (1975) observed that pairing in the $ZAL2/2^m$ bivalent during meiosis was limited to one arm of each chromosome. Cytogenetic mapping efforts (Davis et al., 2011; Thomas et al., 2008) suggest that single recombination events elsewhere in the chromosome would give rise to gametes with large duplications and deletions, thereby effectively preventing the inheritance of the recombined chromosomes. $ZAL2^m$ may therefore be largely isolated from ZAL2. Population genetics studies have confirmed differentiation between the ZAL2 and $ZAL2^m$ over the entire rearranged region as a result of suppression of recombination (Huynh et al., 2010a; Thomas et al., 2008). Alleles are shared between the haplotypes only at the tip of the short arm of $ZAL2/2^m$, which is outside the rearrangement (Fig. 5.7). The rearrangement itself contains a unique set of alleles that are not shared with ZAL2—an estimated 3000 fixed differences (Davis et al., 2011), and this set is inherited together. Thus, the lack of gene flow between the ZAL2 and $ZAL2^m$ has provided opportunity for the evolution of functionally distinct alleles that are restricted to one arrangement or the other. In the continuing analysis of the rearrangement, we should expect to find a series of genes, inherited together as a unit in WS birds, that are functionally distinct from the ZAL2 alleles with regard to either their protein products or patterns of expression. Given the important role of the HPG axis in aggression in this and other species, the strongest candidates will be closely related to HPG function.

V. FUTURE DIRECTIONS

Genetic research with comparative models will ultimately show how key genes, the molecular functions of which are conserved across evolutionary divergence, relate to complex and highly derived social behaviors such as aggression.

The mechanisms that underlie social behaviors in less accessible species, such as humans, are best studied in species that live in societies, particularly those that can be studied in their natural habitats or under naturalistic conditions (Insel and Fernald, 2004). The existing database on avian social behavior is unparalleled—for example, for over 4000 species, we know whether they are territorial or colonial, socially monogamous or polygynous, migratory, or sedentary. We have high-quality recordings of their vocalizations. No other group of animals, invertebrate or vertebrate, has been studied with such passion and intensity. This collective database, although it could provide profound insight into the neuroendocrine basis of diverse social behaviors, is underutilized by neuroscientists because the availability of genomic tools has, until recently, been limited. The recent sequencing of the zebra finch genome (Warren et al., 2010) now makes possible unprecedented advances in our understanding of social behavior because the resulting tools are applicable to all songbirds.

Advances in genomic technology, together with conservation of underlying mechanisms, will make it more and more feasible to bridge from well-characterized data-rich lab organisms, such as mice, to phenomena-rich wild species. These species, which include fish, lizards, songbirds, and voles, are proving to be rich resources for the analysis of social behavior and for the development of general principles (Robinson et al., 2005). Studies with these model organisms have demonstrated the power of a comparative approach—looking for neural or genetic differences among individuals with known behavioral differences (Bullock, 1984; Robinson, et al., 2005). The songbird model is preferable to more typical laboratory species in the study of social behavior because of the greater parallels with humans regarding societal structures and hormonal bases of behavioral strategies, as well as the potential to study free-living populations under natural conditions. Our work and that of others is making these natural and powerful models of vertebrate behavior feasible for genomic and neuroendocrine analysis.

Acknowledgments

The authors thank Jim Thomas for his contributions to the work described in Section IV. The research from the authors' labs was supported by NIH MH062656 to J. L. G., by NIH 1R01MH082833-01, NIH 5R21MH082046-02, and NSF IOS-0723805 to D. L. M. and the Center for Behavioral Neuroscience.

References

Adkins-Regan, E. (2005). Hormones and Animal Social Behavior. Princeton University Press, Princeton.
Allee, W. C. (1936). Analytical studies of group behavior in birds. Wilson Bull. 48, 145–151.
Allee, W. C. (1942). Group organization among vertebrates. Science 95, 289–293.

Allee, W. C., Collias, N. E., and Lutherman, C. Z. (1939). Modification of the social order in flocks of hens by the injection of testosterone propionate. *Physiol. Zool.* **12,** 412–440.

Arcese, P., Sogge, M. K., Marr, A. B., and Patten, M. A. (2002). Song Sparrow (*Melospiza melodia*). *In* "The Birds of North America Online" (A. Poole, ed.). Cornell Lab of Ornithology, Ithaca. doi: 10.2173/bna.704.

Archawaranon, M., and Wiley, R. H. (1988). Control of aggression and dominance in white-throated sparrows by testosterone and its metabolites. *Horm. Behav.* **22,** 497–517.

Archawaranon, M., Dove, L., and Wiley, R. H. (1991). Social inertia and hormonal control of aggression and dominance in white-throated sparrows. *Behaviour* **118,** 42–65.

Archer, J. (2006). Testosterone and human aggression: An evaluation of the challenge hypothesis. *Neurosci. Biobehav. Rev.* **30,** 319–345.

Aromaki, A. S., Lindman, R. E., and Eriksson, C. J. P. (1999). Testosterone, aggressiveness, and antisocial personality. *Aggress. Behav.* **25,** 113–123.

Aste, N., Viglietti-Panzica, C., Balthazart, J., and Panzica, G. C. (1997). Testosterone modulation of peptidergic pathways in the septo-preoptic region of male Japanese quail. *Poult. Avian Biol. Rev.* **8,** 77–93.

Ball, G. F., and Balthazart, J. (2001). Ethological concepts revisited: Immediate early gene induction in response to sexual stimuli in birds. *Brain Behav. Evol.* **57,** 252–270.

Baptista, L. F., DeWolfe, B. B., and Avery-Beausoleil, L. (1987). Testosterone, aggression, and dominance in Gambel's white-crowned sparrows. *Wilson Bull.* **99,** 86–91.

Baptista, L. F., Trail, P. W., DeWolfe, B. B., and Morton, M. L. (1993). Singing and its functions in female white-crowned sparrows. *Anim. Behav.* **46,** 511–524.

Berg, S., and Wynne-Edwards, K. E. (2001). Changes in testosterone, cortisol, and estradiol levels in men becoming fathers. *Mayo Clin. Proc.* **76,** 582–592.

Bernhardt, P. C., Dabbs, J. M., Fielden, J. A., and Lutter, C. D. (1998). Testosterone changes during vicarious experiences of winning and losing among fans at sporting events. *Physiol. Behav.* **65,** 59–62.

Bharati, I. S., and Goodson, J. L. (2006). Fos responses of dopamine neurons to sociosexual stimuli in male zebra finches. *Neuroscience* **143,** 661–670.

Blanchard, D. C., and Blanchard, R. J. (2003). What can animal aggression research tell us about human aggression? *Horm. Behav.* **44,** 171–177.

Blanchard, R. J., and Blanchard, D. C. (2005). Some suggestions for revitalizing aggression research. *Novartis Found. Symp.* **268,** 4–12.

Book, A. S., Starzyk, K. B., and Quinsey, V. L. (2001). The relationship between testosterone and aggression: A meta-analysis. *Aggr. Viol. Behav.* **6,** 579–599.

Booth, A., Shelley, G., Mazur, A., Tharp, G., and Kittok, R. (1989). Testosterone, and winning and losing in human competition. *Horm. Behav.* **23,** 556–571.

Buchanan, K. L., Evans, M. R., Roberts, M. L., Rowe, L., and Goldsmith, A. R. (2010). Does testosterone determine dominance in the house sparrow *Passer domesticus*? An experimental test. *J. Avian Biol.* **41,** 445–451.

Bullock, T. H. (1984). Comparative neuroscience holds promise for quiet revolutions. *Science* **225,** 473–478.

Carey, M., Burhans, D. E., and Nelson, D. A. (2008). Field Sparrow (*Spizella pusilla*). *In* "The Birds of North America Online" (A. Poole, ed.). Cornell Lab of Ornithology, Ithaca. doi: 10.2173/bna.103.

Chase, I. D. (1982). Dynamics of hierarchy formation: The sequential development of dominance relationships. *Behaviour* **80,** 218–240.

Chilton, G., Baker, M. C., Barrentine, C. D., and Cunningham, M. A. (1995). White-Crowned-Sparrow (*Zonotrichia leucophrys*). *In* "The Birds of North America Online" (A. Poole, ed.). Cornell Lab of Ornithology, Ithaca. doi: 10.2173/bna.183.

Collins, C. E., and Houtman, A. M. (1999). Tan and white color morphs of white-throated sparrows differ in their non-song vocal responses to territorial intrusion. *Condor* **101**, 842–845.

Crews, D., Lou, W., Fleming, A., and Ogawa, S. (2006). From gene networks underlying sex determination and gonadal differentiation to the development of neural networks regulating sociosexual behavior. *Brain Res.* **1126**, 109–121.

Crook, J. H., and Butterfield, P. A. (1968). Effects of testosterone propionate and luteinizing hormone on agonistic and nest building behavior of *Quelea quelea*. *Anim. Behav.* **16**, 370–384.

Dabbs, J. M., Jr., and Morris, R. (1990). Testosterone, social class, and antisocial behavior in a sample of 4,462 men. *Psychol. Sci.* **1**, 209–211.

Dabbs, J. M., Jr., Strong, R., and Milun, R. (1997). Exploring the mind of testosterone: A bleeper study. *J. Res. Pers.* **31**, 577–587.

Daitzman, R., and Zuckerman, M. (1980). Disinhibitory sensation seeking, personality and gonadal hormones. *Pers. Individ. Diff.* **1**, 103–110.

Davis, J. K., Lowman, J. J., Thomas, P. J., ten Hallers, B. F. H., Koriabine, M., Huynh, L. Y., Maney, D. L., de Jong, P. J., Martin, C. L., and NISC Comparative Sequencing Program Thomas, J. W. (2010). Evolution of a bitter taste receptor gene cluster in a New World sparrow. *Genome Biol. Evol.* **2**, 358–370.

Davis, J. K., Mittel, L. X., Lowman, J. J., Thomas, P. J., Maney, D. L., Martin, C. L., and Thomas, J. W. NISC Comparative Sequencing Program (2011). Haplotype-based genetic sequencing of a chromosomal polymorphism in the white-throated sparrow (*Zonotrichia albicollis*). *J. Heredity (accepted with minor revision).*

Dawson, A., Kin, V. M., Bentley, G. E., and Ball, G. F. (2001). Photoperiodic control of seasonality in birds. *J. Biol. Rhythms* **16**, 365–380.

Dearborn, D. C., and Wiley, R. H. (1993). Prior residence has a gradual influence on the dominance in captive white-throated sparrows. *Anim. Behav.* **46**, 39–46.

Dufty, A. M., and Wingfield, J. C. (1986). The influence of social cues on the reproductive endocrinology of male brown-headed cowbirds: Field and laboratory studies. *Horm. Behav.* **20**, 222–234.

Edens, F. W. (1987). Agonistic behavior and neurochemistry in grouped Japanese quail. *Comp. Biochem. Physiol. A Comp. Physiol.* **86**, 473–479.

Elekonich, M. M. (2000). Female song sparrow, *Melospiza melodia*, response to simulated conspecific and heterospecific intrusion across three seasons. *Anim. Behav.* **59**, 551–557.

Falls, J. B. (1969). Functions of territorial song in the white-throated sparrow. *In* "Bird Vocalizations" (R. A. Hinde, ed.), pp. 207–232. Cambridge University Press, Cambridge.

Falls, J. B. (1988). Does song deter territorial intrusion in white-throated sparrows (*Zonotrichia albicollis*)? *Can. J. Zool.* **66**, 206–211.

Falls, J. B., and Kopachena, J. G. (2010). *In* "White-Throated Sparrow (*Zonotrichia albicollis*), The Birds of North America Online" (A. Poole, ed.). Cornell Lab of Ornithology, Ithaca. doi: 10.2173/bna.128.

Ficken, R. W., Ficken, M. S., and Hailman, J. P. (1978). Differential aggression in genetically different morphs of the white-throated sparrow (*Zonotrichia albicollis*). *Z. Tierpsychol.* **46**, 43–57.

Finch, C. E., and Rose, M. R. (1995). Hormones and the physiological architecture of life history evolution. *Quart. Rev. Biol.* **70**, 1–52.

Fleming, A. S., Corter, C., Stallings, J., and Steiner, M. (2002). Testosterone and prolactin are associated with emotional responses to infant cries in new fathers. *Horm. Behav.* **42**, 399–413.

Gammie, S. C. (2005). Current models and future directions for understanding the neural circuitries of maternal behaviors in rodents. *Behav. Cogn. Neurosci. Rev.* **4**, 119–135.

Goldman, P. (1973). Song recognition by field sparrows. *Auk* **90**, 103–113.

Goodson, J. L. (1998a). Territorial aggression and dawn song are modulated by septal vasotocin and vasoactive intestinal polypeptide in male field sparrows (*Spizella pusilla*). *Horm. Behav.* **34**, 67–77.

Goodson, J. L. (1998b). Vasotocin and vasoactive intestinal polypeptide modulate aggression in a territorial songbird, the violet-eared waxbill (Estrildidae: *Uraeginthus granatina*). *Gen. Comp. Endocrinol.* **111,** 233–244.

Goodson, J. L. (2005). The vertebrate social behavior network: Evolutionary themes and variations. *Horm. Behav.* **48,** 11–22.

Goodson, J. L., and Bass, A. H. (2002). Vocal-acoustic circuitry and descending vocal pathways in teleost fish: Convergence with terrestrial vertebrates reveals conserved traits. *J. Comp. Neurol.* **448,** 298–322.

Goodson, J. L., and Evans, A. K. (2004). Neural responses to territorial challenge and nonsocial stress in male song sparrows: Segregation, integration, and modulation by a vasopressin V1 antagonist. *Horm. Behav.* **46,** 371–381.

Goodson, J. L., and Kabelik, D. (2009). Dynamic limbic networks and social diversity in vertebrates: From neural context to neuromodulatory patterning. *Front. Neuroendocrinol.* **30,** 429–441.

Goodson, J. L., and Wang, Y. (2006). Valence-sensitive neurons exhibit divergent functional profiles in gregarious and asocial species. *Proc. Natl. Acad. Sci. USA* **103,** 17013–17017.

Goodson, J. L., Eibach, R., Sakata, J., and Adkins-Regan, E. (1999). Effect of septal lesions on male song and aggression in the colonial zebra finch (*Taeniopygia guttata*) and the territorial field sparrow (*Spizella pusilla*). *Behav. Brain Res.* **98,** 167–180.

Goodson, J. L., Evans, A. K., and Lindberg, L. (2004a). Chemoarchitectonic subdivisions of the songbird septum and a comparative overview of septum chemical anatomy in jawed vertebrates. *J. Comp. Neurol.* **473,** 293–314.

Goodson, J. L., Lindberg, L., and Johnson, P. (2004b). Effects of central vasotocin and mesotocin manipulations on social behavior in male and female zebra finches. *Horm. Behav.* **45,** 136–143.

Goodson, J. L., Saldanha, C. J., Hahn, T. P., and Soma, K. K. (2005a). Recent advances in behavioral neuroendocrinology: Insights from studies on birds. *Horm. Behav.* **48,** 461–473.

Goodson, J. L., Evans, A. K., Lindberg, L., and Allen, C. D. (2005b). Neuro-evolutionary patterning of sociality. *Proc. R. Soc. Lond. B* **272,** 227–235.

Goodson, J. L., Evans, A. K., and Soma, K. K. (2005c). Neural responses to aggressive challenge correlate with behavior in nonbreeding sparrows. *Neuroreport* **16,** 1719–1723.

Goodson, J. L., Kabelik, D., Kelly, A. M., Rinaldi, J., and Klatt, J. D. (2009a). Midbrain dopamine neurons reflect affiliation phenotypes in finches and are tightly coupled to courtship. *Proc. Natl. Acad. Sci. USA* **106,** 8737–8742.

Goodson, J. L., Kabelik, D., and Schrock, S. E. (2009b). Dynamic neuromodulation of aggression by vasotocin: Influence of social context and social phenotype in territorial songbirds. *Biol. Lett.* **5,** 554–556.

Goodson, J. L., Rinaldi, J., and Kelly, A. M. (2009c). Vasotocin neurons in the bed nucleus of the stria terminalis preferentially process social information and exhibit properties that dichotomize courting and non-courting phenotypes. *Horm. Behav.* **55,** 197–202.

Gray, P. B. (2003). Marriage, parenting, and testosterone variation among Kenyan Swahili men. *Am. J. Phys. Anthropol.* **122,** 279–286.

Gray, P. B., Kahlenberg, S. M., Barret, E. S., Lipson, S. F., and Ellison, P. T. (2002). Marriage and fatherhood are associated with low testosterone in males. *Evol. Hum. Behav.* **23,** 193–201.

Guhl, A. M. (1968). Social inertia and social stability in chickens. *Anim. Behav.* **16,** 219–232.

Hamilton, W. D. (1971). Geometry for the selfish herd. *J. Theoret. Biol.* **31,** 295–311.

Harrington, B. A. (1973). Aggression in winter resident and spring migrant white-throated sparrows in Massachusetts. *Bird Banding* **44,** 314–315.

Hau, M. (2007). Regulation of male traits by testosterone: Implications for the evolution of vertebrate life histories. *Bioessays* **292,** 133–144.

Hegner, R. E., and Wingfield, J. C. (1986). Behavioral and endocrine correlates of multiple brooding in the semi-colonial house sparrow *Passer domesticus* I. Males. *Horm. Behav.* **20,** 294–312.

Hegner, R. E., and Wingfield, J. C. (1987). Effects of experimental manipulation of testosterone levels on parental investment and breeding success in male house sparrows. *Auk* **104**, 462–469.

Heimovics, S. A., Cornil, C. A., Ball, G. F., and Riters, L. V. (2009). D1-like dopamine receptor density in nuclei involved in social behavior correlates with song in a context-dependent fashion in male European starlings. *Neuroscience* **159**, 962–973.

Herdegen, T., and Leah, J. D. (1998). Inducible and constitutive transcription factors in the mammalian nervous system: Control of gene expression by Jun, Fos and Krox, and CREB/ATF proteins. *Brain Res. Rev.* **28**, 370–490.

Ho, J. M., Murray, J. H., Demas, G. E., and Goodson, J. L. (2010). Vasopressin cell groups exhibit strongly divergent responses to copulation and male-male interactions in mice. *Horm. Behav.* **58**, 368–377.

Hoke, K. L., Ryan, M. J., and Wilczynski, W. (2005). Social cues shift functional connectivity in the hypothalamus. *Proc. Natl. Acad. Sci. USA* **102**, 10712–10717.

Huynh, L. Y., Maney, D. L., and Thomas, J. W. (2010a). Chromosome-wide linkage disequilibrium caused by an inversion polymorphism in the white-throated sparrow. *Heredity (in press)*.

Huynh, L. Y., Maney, D. L., and Thomas, J. W. (2010b). Contrasting population genetic patterns within the white-throated sparrow genome (*Zonotrichia albicollis*). *BMC Genomics* **11**, 96.

Insel, T. R., and Fernald, R. D. (2004). How the brain processes social information: Searching for the social brain. *Annu. Rev. Neurosci.* **27**, 697–722.

Julian, T., and McHenry, P. C. (1989). Relationship of testosterone to men's family functioning at mid-life: A research note. *Aggress. Behav.* **15**, 281–289.

Kabelik, D., Klatt, J. D., Kingsbury, M. A., and Goodson, J. L. (2009). Endogenous vasotocin exerts context-dependent behavioral effects in a semi-naturalistic colony environment. *Horm. Behav.* **56**, 101–107.

Kabelik, D., Kelly, A. M., and Goodson, J. L. (2010). Dopaminergic regulation of mate competition aggression and aromatase-Fos colocalization in vasotocin neurons. *Neuropharmacology* **58**, 117–125.

Ketterson, E. D., and Nolan, V. (1992). Hormones and life histories: An integrative approach. *Am. Nat.* **140**, S33–S62.

Ketterson, E. D., and Nolan, V. (1994). Male parental behavior in birds. *Annu. Rev. Ecol. Syst.* **25**, 601–628.

Keys, G. C., and Rothstein, S. I. (1991). Benefits and costs of dominance and subordinance in white-crowned sparrows and the paradox of status signalling. *Anim. Behav.* **42**, 899–912.

Kingsbury, M. A., Kelly, A. M., Schrock, S. E., and Goodson, J. L. (2011). Mammal-like organization of the avian midbrain central gray and a reappraisal of the intercollicular nucleus. *PLoS One*.

Knapton, R. W., and Falls, J. B. (1983). Differences in parental contribution among pair types in the polymorphic white-throated sparrow. *Can. J. Zool.* **61**, 1288–1292.

Kollack-Walker, S., Watson, S. J., and Akil, H. (1997). Social stress in hamsters: Defeat activates specific neurocircuits within the brain. *J. Neurosci.* **17**, 8842–8855.

Konishi, M., Emlen, S. T., Ricklefs, R. E., and Wingfield, J. C. (1989). Contributions of bird studies to biology. *Science* **246**, 465–472.

Kopachena, J. G., and Falls, J. B. (1993a). Aggressive performance as a behavioral correlate of plumage polymorphism in the white-throated sparrow (*Zonotrichia albicollis*). *Behaviour* **124**, 249–266.

Kopachena, J. G., and Falls, J. B. (1993b). Re-evaluation of morph-specific variations in parental behavior of the white-throated sparrow. *Wilson Bull.* **105**, 48–59.

Kotegawa, T., Abe, T., and Tsutsui, K. (1997). Inhibitory role of opioid peptides in the regulation of aggressive and sexual behaviors in male Japanese quails. *J. Exp. Zool.* **277**, 146–154.

Krebs, J. R. (1971). Territory and breeding density in the great tit, *Parus major. Ecology* **52**, 2–22.

Krebs, J. (1976). Bird song and territorial defence. *New Sci.* **70**, 534–536.

Kroodsma, D. E. (1976). The effect of large song repertoires on neighbor "recognition" in male song sparrows. *Condor* **78**, 97–99.

Kroodsma, D. E., Bereson, R. C., Byers, B. E., and Minear, E. (1989). Use of song types by the chestnut-sided warbler: Evidence for both intra- and inter-sexual functions. *Can. J. Zool.* **67**, 447–456.

Lein, M. R. (1978). Song variation in a population of chestnut-sided warblers (*Dendroica pensylvanica*): Its nature and suggested significance. *Can. J. Zool.* **56**, 1266–1283.

Lin, D., Boyle, M. P., Dollar, P., Lee, H., Lein, E. S., Perona, P., and Anderson, D. J. (2011). Functional identification of an aggression locus in the mouse hypothalamus. *Nature* **470**, 221–226.

Lowther, J. K. (1961). Polymorphism in the white-throated sparrow, *Zonotrichia albicollis* (Gmelin). *Can. J. Zool.* **39**, 281–292.

Lumia, A. R. (1972). The relationships among testosterone, conditioned aggression, and dominance in male pigeons. *Horm. Behav.* **3**, 277–286.

Maney, D. L. (2008). Endocrine and genomic architecture of life history trade-offs in an avian model of social behavior. *Gen. Comp. Endocrinol.* **157**, 275–282.

Maney, D. L. (2010). Hormonal control of sexual behavior in female nonmammalian vertebrates. *In* "Encyclopedia of Animal Behavior" (M. D. Breed and J. Moore, eds.), Encyclopedia of Animal Behavior, Vol. 1, pp. 697–703. Elsevier, Oxford.

Maney, D. L., and Ball, G. F. (2003). Fos-like immunoreactivity in catecholaminergic brain nuclei after territorial behavior in free-living song sparrows. *J. Neurobiol.* **56**, 163–170.

Maney, D. L., Goode, C. T., and Wingfield, J. C. (1997). Intraventricular infusion of arginine vasotocin induces singing in a female songbird. *J. Neuroendocrinol.* **9**, 487–491.

Maney, D. L., Erwin, K. L., and Goode, C. T. (2005). Neuroendocrine correlates of behavioral polymorphism in white-throated sparrows. *Horm. Behav.* **48**, 196–206.

Maney, D. L., Lange, H. S., Raees, M. Q., and Sanford, S. E. (2009). Behavioral phenotypes persist after gonadal steroid manipulation in white-throated sparrows. *Horm. Behav.* **55**, 113–120.

Mathewson, S. F. (1961). Gonadotrophic hormones affect aggressive behavior in starlings. *Science* **134**, 1522–1523.

Maynard-Smith, J. (1977). Parental investment: A prospective analysis. *Anim. Behav.* **25**, 1–9.

McGlothlin, J. W., and Ketterson, E. D. (2007). Hormone-mediated suites as adaptations and evolutionary constraints. *Philos. Trans. R. Soc. Lond. B Biol. Sci.* **170**, 864–875.

McGlothlin, J. W., Jawor, J. M., and Ketterson, E. D. (2007). Natural variation in a testosterone-mediated trade-off between mating effort and parental effort. *Am. Nat.* **170**, 864–875.

Mello, C. V., and Ribeiro, S. (1998). ZENK protein regulation by song in the brain of songbirds. *J. Comp. Neurol.* **393**, 426–438.

Mello, C. V., Vicario, D. S., and Clayton, D. F. (1992). Song presentation induces gene expression in the songbird forebrain. *Proc. Natl. Acad. Sci. USA* **89**, 6818–6822.

Miles, D. B., Sinervo, B., Hazard, L. C., Svensson, E. I., and Costa, D. (2007). Relating endocrinology, physiology and behaviour using species with alternative mating strategies. *Funct. Ecol.* **21**, 653–665.

Møller, A. P. (1987). Variation in badge size in male house sparrows *Passer domesticus*: Evidence for status signalling. *Anim. Behav.* **35**, 1637–1644.

Moore, M. C. (1983). Effect of female displays on the endocrine physiology and behavior of male white-crowned sparrows, *Zonotrichia leucophrys*. *J. Zool.* **199**, 137–148.

Moore, M. C. (1991). Application of organization-activation theory to alternative male reproductive strategies: A review. *Horm. Behav.* **25**, 154–179.

Motta, S. C., Goto, M., Gouveia, F. V., Baldo, M. V. C., Canteras, N. S., and Swanson, L. W. (2009). Dissecting the brain's fear system reveals the hypothalamus is critical for responding in subordinate conspecific intruders. *Proc. Natl. Acad. Sci.* **106**, 4870–4875.

Moyer, K. E. (1968). Kinds of aggression and their physiological basis. *Commun. Behav. Biol.* **2**(A), 65–87.

Mukai, M., Replogle, K., Drnevich, J., Wang, G., Wacker, D., Band, M., Clayton, D. F., and Wingfield, J. C. (2009). Seasonal differences of gene expression profiles in song sparrow (*Melospiza melodia*) hypothalamus in relation to territorial aggression. *PLoS One* **4**, e8182.

Nelson, D. A., and Croner, L. J. (1991). Song categories and their functions in the field sparrow (*Spizella pusilla*). *Auk* **108**, 42–52.

Newman, S. W. (1999). The medial extended amygdala in male reproductive behavior: A node in the mammalian social behavior network. *Ann. N.Y. Acad. Sci.* **877**, 242–257.

Nice, M. M. (1943). Studies in the Life History of the Song Sparrow. II. The Behaviour of the Song Sparrow and Other Passerines. Dover, New York.

Nijhout, F. H. (2003). Development and evolution of adaptive polyphenisms. *Evol. Dev.* **5**, 9–18.

Nolan, Jr., V., Ketterson, E. D., Cristol, D. A., Rogers, C. M., Clotfelter, E. D., Titus, R. C., Schoech, S. J., and Snajdr, E. (2002). Dark-eyed Junco (*Junco hyemalis*). *In* "The Birds of North America Online" (A. Poole, ed.). Cornell Lab of Ornithology, Ithaca. doi:10.2173/bna.716.

Panzica, G. C., Aste, N., Castagna, C., Viglietti-Panzica, C., and Balthazart, J. (2001). Steroid-induced plasticity in the sexually dimorphic vasotocinergic innervation of the avian brain: Behavioral implications. *Brain Res. Rev.* **37**, 178–200.

Piper, W. H., and Wiley, R. H. (1989). Correlates of dominance in wintering white-throated sparrows: Age, sex and location. *Anim. Behav.* **37**, 298–310.

Pradhan, D. S., Newman, A. E., Wacker, D. W., Wingfield, J. C., Schlinger, B. A., and Soma, K. K. (2010). Aggressive interactions rapidly increase androgen synthesis in the brain during the non-breeding season. *Horm. Behav.* **57**, 381–389.

Ramenofsky, M. (1984). Agonistic behavior and endogenous plasma hormones in male Japanese quail. *Anim. Behav.* **32**, 698–708.

Ramirez, J. M., Salas, C., and Portavella, M. (1988). Offense and defense after lateral septal lesions in *Columba livia*. *Int. J. Neurosci.* **41**, 241–250.

Remage-Healey, L., Maidment, N. T., and Schlinger, B. A. (2008). Forebrain steroid levels fluctuate rapidly during social interactions. *Nat. Neurosci.* **11**, 1327–1334.

Remage-Healey, L., London, S. E., and Schlinger, B. A. (2010). Birdsong and the neural production of steroids. *J. Chem. Neuroanat.* **39**, 72–81.

Rhen, T., and Crews, D. (2002). Variation in reproductive behaviour within a sex: Neural systems and endocrine activation. *J. Neuroendocrinol.* **14**, 517–531.

Robinson, G. E., Grozinger, C. M., and Whitfield, C. W. (2005). Sociogenomics: Social life in molecular terms. *Nat. Rev. Genet.* **6**, 257–270.

Rohwer, S. (1975). The social significance of avian winter plumage variability. *Evolution* **29**, 593–610.

Rohwer, S., and Rohwer, F. C. (1978). Status signalling in Harris sparrows: Experimental deceptions achieved. *Anim. Behav.* **26**, 1012–1022.

Romanov, M. N., Tuttle, E. M., Houck, M. L., Modi, W. S., Chemnick, L. G., Korody, M. L., *et al.* (2009). The value of avian genomics to the conservation of wildlife. *BMC Genomics* **10**, S10.

Sabine, W. S. (1959). The winter society of the Oregon Junco: Intolerance, dominance, and the pecking order. *Condor* **61**, 110–135.

Schlinger, B. A. (1987). Plasma androgens and aggressiveness in captive winter white-throated sparrows (*Zonotrichia albicollis*). *Horm. Behav.* **21**, 203–210.

Schneider, K. C. J. (1984). Dominance, predation, and optimal foraging in white-throated sparrow flocks. *Ecology* **65**, 1820–1827.

Schoech, S. J., Ketterson, E. D., Nolan, V., Sharp, P. J., and Buntin, J. D. (1998). The effect of exogenous testosterone on parental behavior, plasma prolactin, and prolactin binding sites in the dark-eyed juncos. *Horm. Behav.* **34**, 1–10.

Schroeder, M. B., and Riters, L. V. (2006). Pharmacological manipulations of dopamine and opioids have differential effects on sexually motivated song in male European starlings. *Physiol. Behav.* **88,** 575–584.

Schwabl, H., Ramenofsky, M., Schwabl-Benzinger, I., Farner, D. S., and Wingfield, J. C. (1988). Social status, circulating levels of hormones, and competition for food in winter flocks of the white-throated sparrow. *Behaviour* **107,** 107–121.

Selinger, H. E., and Bermant, G. (1967). Hormonal control of aggressive behavior in Japanese quail (*Coturnix coturnix japonica*). *Behavior* **28,** 255–268.

Silverin, B. (1980). Effects of long-acting testosterone treatment on free living pied flycatchers, *Ficedula hypoleuca*, during the breeding season. *Anim. Behav.* **28,** 906–912.

Sinervo, B., and Svensson, E. (2002). Correlational selection and the evolution of genomic architecture. *Heredity* **89,** 329–338.

Soma, K. K. (2006). Testosterone and aggression: Berthold, birds and beyond. *J. Neuroendocrinol.* **18,** 543–551.

Soma, K. K., and Wingfield, J. C. (2001). Dehydroepiandrosterone in songbird plasma: Seasonal regulation and relationship to territorial aggression. *Gen. Comp. Endocrinol.* **123,** 144–155.

Soma, K. K., Sullivan, K., and Wingfield, J. C. (1999). Combined aromatase inhibitor and antiandrogen treatment decreases territorial aggression in a wild songbird during the nonbreeding season. *Gen. Comp. Endocrinol.* **115,** 442–453.

Soma, K. K., Tramontin, A. D., and Wingfield, J. C. (2000a). Oestrogen regulates male aggression in the non-breeding season. *Proc. R. Soc. Lond. B Biol. Sci.* **267,** 1089–1096.

Soma, K. K., Sullivan, K. A., Tramontin, A. D., Saldanha, C. J., Schlinger, B. A., and Wingfield, J. C. (2000b). Acute and chronic effects of an aromatase inhibitor on territorial aggression in breeding and nonbreeding male song sparrows. *J. Comp. Physiol. A* **186,** 759–769.

Soma, K. K., Schlinger, B. A., Wingfield, J. C., and Saldanha, C. J. (2003). Brain aromatase, 5-alpha reductase, and 5-beta reductase change seasonally in wild male song sparrows: Relationship to aggressive and sexual behavior. *J. Neurobiol.* **56,** 209–221.

Soma, K. K., Scotti, M. A., Newman, A. E., Charlier, T. D., and Demas, G. E. (2008). Novel mechanisms for neuroendocrine regulation of aggression. *Front. Neuroendocrinol.* **29,** 476–489.

Spinney, L. H., Bentley, G. E., and Hau, M. (2006). Endocrine correlates of alternative phenotypes in the white-throated sparrow (*Zonotrichia albicollis*). *Horm. Behav.* **50,** 762–771.

Storey, A. E., Walch, C. J., Quinton, R. L., and Wynne-Edwards, K. E. (2000). Hormonal correlates of paternal responsiveness in new and expectant fathers. *Evol. Hum. Behav.* **21,** 79–95.

Swett, M. B., and Breuner, C. W. (2009). Plasma testosterone correlates with morph type across breeding substages in male white-throated sparrows. *Physiol. Biochem. Zool.* **82,** 572–579.

Thomas, J. W., Cáceres, M., Lowman, J. J., Morehouse, C. B., Short, M. E., Baldwin, E. L., Maney, D. L., and Martin, C. L. (2008). The chromosomal polymorphism linked to variation in social behavior in the white-throated sparrow (*Zonotrichia albicollis*) is a complex rearrangement that suppresses recombination. *Genetics* **179,** 1455–1468.

Thorneycroft, H. B. (1975). A cytogenetic study of the white-throated sparrow, *Zonotrichia albicollis*. *Evolution* **29,** 611–621.

Tompkins, G. (1933). Individuality and territoriality as displayed in winter by three passerine species. *Condor* **35,** 98–106.

Trainor, B. C., Finy, M. S., and Nelson, R. J. (2008). Paternal aggression in a biparental mouse: Parallels with maternal aggression. *Horm. Behav.* **53,** 200–207.

Trivers, R. L. (1972). Parental investment and sexual selection. In "Sexual Selection and the Descent of Man" (B. Campbell, ed.), pp. 139–179. Aldine, Chicago.

Tuttle, E. M. (1993). Mate choice and stable polymorphism in the white-throated sparrow. Ph.D, Dissertation, State University of New York at Albany.

Tuttle, E. M. (2003). Alternative reproductive strategies in the white-throated sparrow: Behavioral and genetic evidence. *Behav. Ecol.* **14,** 425–432.

Udry, J. R. (1990). Biosocial models of adolescent problem behaviors. *Social Biol.* **37,** 1–10.

Veenema, A. H., Bredewold, R., and Neumann, I. D. (2007). Opposite effects of maternal separation on intermale and maternal aggression in C57BL/6 mice: Link to hypothalamic vasopressin and oxytocin immunoreactivity. *Psychoneuroendocrinology* **32,** 437–450.

Voorhuis, T. A. M., Kiss, J. Z., de Kloet, E. R., and de Wied, D. (1988). Testosterone-sensitive vasotocin-immunoreactive cells and fibers in the canary brain. *Brain Res.* **442,** 139–146.

Wacker, D. W., Schlinger, B. A., and Wingfield, J. C. (2008). Combined effects of DHEA and fadrozole on aggression and neural VIP immunoreactivity in the non-breeding male song sparrow. *Horm. Behav.* **53,** 287–294.

Wacker, D. W., Wingfield, J. C., Davis, J. E., and Meddle, S. L. (2010). Seasonal changes in aromatase and androgen receptor, but not estrogen receptor mRNA expression in the brain of the free-living male song sparrow, *Melospiza melodia morphna. J. Comp. Neurol.* **518,** 3819–3835.

Walker, B. G., Boersma, P. D., and Wingfield, J. C. (2005). Field endocrinology and conservation biology. *Integr. Comp. Biol.* **45,** 12–18.

Warren, W. C., Clayton, D. F., Ellegren, H., Arnold, A. P., Hillier, L. W., Künstner, A., *et al.* (2010). The genome of a songbird. *Nature* **464,** 757–762.

Watt, D. J. (1986). Plumage brightness index for white-throated sparrows. *J. Field Ornithol.* **57,** 105–113.

Watt, D. J., Ralph, C. J., and Atkinson, C. T. (1984). The role of plumage polymorphism in dominance relationships of the white-throated sparrow. *Auk* **101,** 110–120.

Wiley, R. H., Piper, W. H., Archawaranon, M., and Thompson, E. W. (1993). Singing in relation to social dominance and testosterone in white-throated sparrows. *Behaviour* **127,** 175–190.

Wiley, R. H., Steadman, L., Chawick, L., and Wollerman, L. (1999). Social inertia in white-throated sparrows results from recognition of opponents. *Anim. Behav.* **57,** 453–463.

Wingfield, J. C. (1984a). Environmental and endocrine control of reproduction in the song sparrow, Melospiza melodia I. Temporal organization of the breeding cycle. *Gen. Comp. Endocrinol.* **56,** 406–416.

Wingfield, J. C. (1984b). Environmental and endocrine control of reproduction in the song sparrow, Melospiza melodia II. Agonistic interactions as environmental information stimulating secretion of testosterone. *Gen. Comp. Endocrinol.* **56,** 417–424.

Wingfield, J. C. (1984c). Androgens and mating systems: Testosterone-induced polygyny in normally monogamous birds. *Auk* **101,** 665–671.

Wingfield, J. C. (1985). Short-term changes in plasma levels of hormones during establishment and defense of a breeding territory in male song sparrows, *Melospiza melodia. Horm. Behav.* **19,** 174–187.

Wingfield, J. C. (1994a). Control of territorial aggression in a changing environment. *Psychoneuroendocrinology* **19,** 709–721.

Wingfield, J. C. (1994b). Regulation of territorial behavior in the sedentary song sparrow, *Melospiza melodia morphna. Horm. Behav.* **28,** 1–15.

Wingfield, J. C., and Farner, D. S. (1976). Avian endocrinology—Field investigations and methods. *Condor* **78,** 570–573.

Wingfield, J. C., and Farner, D. S. (1978). The endocrinology of a naturally breeding population of white-crowned sparrow, *Zonotrichia leucophrys pugetensis. Physiol. Zool.* **51,** 188–205.

Wingfield, J. C., and Farner, D. S. (1993). Endocrinology of reproduction in wild species. *In* "Avian Biology" (D. S. Farner, J. King, and K. Parkes, eds.), pp. 163–327. Academic Press, San Diego.

Wingfield, J. C., and Goldsmith, A. R. (1990). Plasma levels of prolactin and gonadal steroids in relation to multiple-brooding and renesting in free-living populations of the song sparrow, *Melospiza melodia. Horm. Behav.* **24,** 89–103.

Wingfield, J. C., and Hahn, T. P. (1994). Testosterone and territorial behavior in sedentary and migratory sparrows. *Anim. Behav.* **47,** 77–89.

Wingfield, J. C., and Monk, D. (1994). Behavioral and hormonal responses of male song sparrows to estradiol-treated females during the non-breeding season. *Horm. Behav.* **28,** 146–154.

Wingfield, J. C., and Wada, M. (1989). Male-male interactions increase both luteinizing hormone and testosterone in the song sparrow, *Melospiza melodia*: Specificity, time course and possible neural pathways. *J. Comp. Physiol. A* **166,** 189–194.

Wingfield, J. C., Ball, G. F., Dufty, A. M., Jr., Hegner, R. E., and Ramenofsky, M. (1987). Testosterone and aggression in birds: Tests of the "challenge hypothesis". *Am. Sci.* **75,** 602–608.

Wingfield, J. C., Hegner, R. F., Dufty, A. M. J., and Ball, G. F. (1990). The challenge hypothesis: Theoretical implications for patterns of testosterone secretion, mating systems and breeding strategies. *Am. Nat.* **136,** 829–846.

Wingfield, J. C., Lynn, S. E., and Soma, K. K. (2001). Avoiding the 'costs' of testosterone: Ecological bases of hormone-behavior interactions. *Brain Behav. Evol.* **57,** 239–251.

Wynne-Edwards, K. E. (2001). Hormonal changes in mammalian fathers. *Horm. Behav.* **40,** 139–145.

Yang, E. J., and Wilczynski, W. (2007). Social experience organizes parallel networks in sensory and limbic forebrain. *Dev. Neurobiol.* **67,** 285–303.

Zera, A. J., and Harshman, L. G. (2001). The physiology of life history trade-offs in animals. *Annu. Rev. Ecol. Syst.* **32,** 95–126.

6

Genetics of Aggression in Voles

Kyle L. Gobrogge[1] and Zuoxin W. Wang

Department of Psychology and Program in Neuroscience, Florida State University, Tallahassee, Florida, USA

ABSTRACT

Prairie voles (*Microtus ochrogaster*) are socially monogamous rodents that form pair bonds—a behavior composed of several social interactions including attachment with a familiar mate and aggression toward conspecific strangers.

[1]Present address: Department of Neurobiology, Harvard Medical School, Boston, Massachusetts, USA

Advances in Genetics, Vol. 75
0065-2660/11 $35.00
DOI: 10.1016/B978-0-12-380858-5.00003-4

Therefore, this species has provided an excellent opportunity for the study of pair bonding behavior and its underlying neural mechanisms. In this chapter, we discuss the utility of this unique animal model in the study of aggression and review recent findings illustrating the neurochemical mechanisms underlying pair bonding-induced aggression. Implications of this research for our understanding of the neurobiology of human violence are also discussed.   © 2011, Elsevier Inc.

I. INTRODUCTION

Mating induces aggression in several organisms throughout the animal kingdom. Within species, patterns of inter- and intrasexual aggression vary as a function of monogamy, parental investment, and group structure. In the wild, the appropriate coordination of social behavior is necessary for survival and reproductive success. How organisms make decisions about which behavior to display in the natural environment remains an important area of biological investigation. To address these questions, previous work has relied on using traditional laboratory rodents. However, these animals do not readily display certain types of social behaviors and thus are not appropriate for some investigations. For example, laboratory rats and mice do not exhibit strong social bonds between mates, and males typically do not display paternal behavior or female-directed aggression. Because mating naturally induces pair bonding, aggression, and biparental behavior in the socially monogamous prairie vole (*Microtus ochrogaster*), this species represents a unique animal model to study the underlying neural mechanisms regulating social behavior associated with a monogamous life strategy.

In this chapter, we begin by describing the prairie vole model and reviewing the neural correlates of pair bonding behavior. We focus on the neuropeptides arginine vasopressin (AVP) and oxytocin; neurotransmitters dopamine (DA), gamma-aminobutyric acid (GABA), and glutamate; and steroid hormones testosterone and estrogen in the regulation of aggression. We highlight the molecular genetics underlying courtship and aggression associated with monogamous pair bonding in voles and humans. Finally, we speculate on the potential for translation, of aggression studies in prairie voles, for research examining the etiology of violence in human populations—with a particular emphasis on the interactions between drug abuse and social behavior.

II. THE PRAIRIE VOLE MODEL

Voles are microtine (*Microtus*) rodents that are taxonomically and genetically similar, yet show remarkable differences in their social behavior (Young and Wang, 2004; Young *et al.*, 2008, 2011a). These animals have provided an excellent opportunity for comparative studies examining social behaviors associated with different life strategies. For example, prairie (M. ochrogaster) and pine (M. pinetorum) voles are highly social and monogamous, whereas meadow (M. pennsylvanicus) and montane (M. montanus) voles are asocial and promiscuous (Dewsbury, 1987; Insel and Hulihan, 1995; Jannett, 1982). In the laboratory, prairie and pine voles are biparental, with both parents equally caring for their young, while meadow and montane voles are primarily maternal and males do not stay in the natal nest after female parturition (McGuire and Novak, 1984, 1986; Oliveras and Novak, 1986). Following mating, prairie voles develop pair bonds between mates (Fig. 6.1A; Young and Wang, 2004) and males even display aggression selectively toward conspecific strangers but not toward their familiar partner (Fig. 6.1B; Aragona et al., 2006; Gobrogge et al., 2007, 2009; Winslow et al., 1993)—behaviors that are not exhibited by promiscuous meadow or montane voles (Insel et al., 1995; Lim et al., 2004). Interestingly, these vole species do not differ in their nonsocial behavior (Tamarin, 1985), further indicating associations between species-specific social behavior and life strategy (Carter et al., 1995; Insel et al., 1998; Wang and Aragona, 2004; Young and Wang, 2004; Young et al., 1998). Therefore, prairie voles represent a unique model system to dissect the neural mechanisms underlying ethologically relevant social behavior.

One behavioral index of pair bonding is selective aggression, which is more prominent in male than in female prairie voles. It has been suggested that selective aggression is a behavioral trait associated with mate guarding that is important for pair bonding (Carter et al., 1995). Selective aggression is studied using a resident–intruder test (RIT) (Fig. 6.1C; Gobrogge et al., 2007, 2009; Winslow et al., 1993; Wang et al., 1997a). A conspecific intruder is introduced into the male resident cage and their behavioral interactions are videotaped for 5–10 min (Aragona et al., 2006; Gobrogge et al., 2007, 2009; Wang et al., 1997a; Winslow et al., 1993). Subject's behavioral interactions with the intruder are recorded and the frequency of aggressive behaviors including attacks, bites, chases, defensive/offensive upright postures, offensive sniffs, threats, and retaliatory attacks are calculated as a composite score (Gobrogge et al., 2007, 2009) as well as the duration of affiliative side-by-side physical contact (Gobrogge et al., 2007, 2009; Winslow et al., 1993). It is important to note that both offensive and defensive types of aggression are critical components of selective aggression in male prairie voles (Wang et al., 1997a; Winslow et al., 1993).

Selective aggression is associated with mating, as cohabitation in the absence of mating does not induce this behavior in male prairie voles (Insel et al., 1995; Wang et al., 1997a; Winslow et al., 1993). Selective aggression is also enduring (Aragona et al., 2006; Gobrogge et al., 2007, 2009; Insel et al., 1995) and lasts for at least 2 weeks after partner preference formation (Aragona et al., 2006; Gobrogge et al., 2007, 2009), even in the absence of continuous exposure

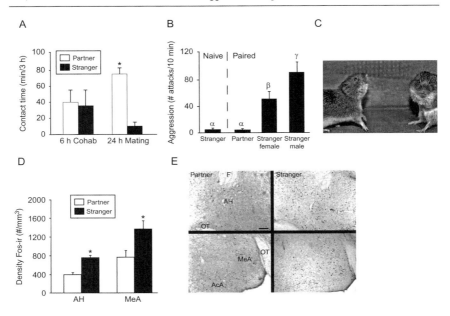

Figure 6.1. Neural correlates of selective aggression. (A) After 24 h, but not 6 h, of mating, male and female prairie voles display significantly more time in side-by-side physical contact with an opposite sex familiar partner than with a stranger. (B) Sexually naïve male prairie voles (Naïve) do not display aggression toward a stranger female, whereas 2 weeks of sexual and social experience (Paired) induces selective aggression toward both male and female strangers but not toward familiar female partners. (C) Photo depicts a pair bonded male prairie vole (left) preparing to attack a—sexually receptive—stranger female prairie vole (right). (D) Stereological estimates reveal a significantly higher density of Fos-ir cells in the anterior hypothalamus (AH) and medial amygdala (MeA) of pair bonded males displaying aggression toward a stranger female (Stranger) compared to males displaying affiliation toward their familiar female partner (Partner). (E) Photomicrographs showing Fos-ir (dark nuclear staining) in the AH and MeA from pair bonded males re-exposed to their familiar female partner (Partner) or to a stranger female (Stranger). F: fornix; OT: optic tract. Bars indicate means ± standard error of the mean. Bars with different Greek letters differ significantly from each other. *$p < 0.05$. Scale bar = 100 μm. Adapted from Aragona and Wang (2009), Gobrogge and Wang (2009), Gobrogge et al. (2007), Wang et al. (1997a), Winslow et al. (1993), and Young et al. (2011a).

to a partner (Insel et al., 1995). Importantly, males display aggression not only toward conspecific males (Fig. 6.1B; Aragona et al., 2006; Gobrogge et al., 2007, 2009; Insel et al., 1995; Wang et al., 1997a; Winslow et al., 1993) but also toward sexually receptive females (Fig. 6.1B and C; Gobrogge et al., 2007, 2009). This selective aggression functions to maintain monogamous pair bonds as males reject potential female mates (Fig. 6.1B and C; Aragona et al., 2006; Gobrogge *et al.*, 2007, 2009; Wang *et al.*, 1997a). Together, these reliably expressed and measurable agonistic behaviors make the prairie vole an excellent model for investigation of the neural mechanisms underlying naturally occurring aggression associated with monogamy. It should be mentioned that although female aggression has been less studied in voles, female prairie voles exhibit similar aggressive behavior as males and this behavior is influenced by a female's social and sexual experience (Bowler *et al.*, 2002).

III. NEURAL CORRELATES

With considerable overlap of brain areas involved in several forms of social (Newman, 1999) and agonistic (Table 6.1) behaviors, there is a significant amount of ambiguity regarding which brain areas may be involved in the regulation of selective aggression. Using a neuronal activation marker of an immediate early gene, *c-fos*, previous studies in voles have examined neuronal

Table 6.1. Summary of Brain Areas Implicated in Aggression

Brain area	Species	References
Anterior hypothalamus (AH)	Human	Sano *et al.* (1966), Ramamurthi (1988)
	Prairie vole	Gobrogge *et al.* (2007, 2009), Gobrogge and Wang (2009)
	Rat	Veening *et al.* (2005), Kruk (1991), Bermond *et al.* (1982), Kruk *et al.* (1984), Adams *et al.* (1993), Roeling *et al.* (1993), Haller *et al.* (1998), Veenema *et al.* (2006), Motta *et al.* (2009)
	Syrian hamster	Delville *et al.* (2000), Ferris and Potegal (1988), Caldwell and Albers (2004), Ferris *et al.* (1997, 1989), Albers *et al.* (2006), Harrison *et al.* (2000b), Jackson *et al.* (2005), Grimes *et al.* (2007), Ricci *et al.* (2009), Schwartzer *et al.* (2009), Schwartzer and Melloni (2010a,b)
Lateral septum (LS)	Rat	Veenema *et al.* (2010)
Medial amygdala (MeA)	Human	Ramamurthi (1988)
	Prairie vole	Wang *et al.* (1997a), Gobrogge and Wang (2009)
	Rat	Koolhaas *et al.* (1990)
Nucleus accumbens (NAcc)	Prairie vole	Aragona *et al.* (2006)
Ventromedial hypothalamus (VMH)	Mouse	Choi *et al.* (2005), Lin *et al.* (2011)

Brain structures involved in aggression, across species, with corresponding references to ground brain area acronyms used throughout the chapter.

activity associated with aggression (Wang et al., 1997a), maternal (Katz et al., 1999) and paternal behavior (Kirkpatrick et al., 1994), mating (Curtis and Wang, 2003; Lim and Young, 2004), anxiety (Stowe et al., 2005), spatial learning (Kuptsov et al., 2005), chemosensory processing (Hairston et al., 2003; Tubbiola and Wysocki, 1997), social experience (Cushing et al., 2003a; Kramer et al., 2006), or pharmacological challenges (Curtis and Wang, 2005b; Cushing et al., 2003b; Gingrich et al., 1997). Within these studies, however, typically only one type of behavior was investigated. Little focus was aimed at examining other forms of social behavior, including affiliation or general social olfactory processing. Consequently, there is considerable overlap in brain–behavior relationships among these studies leading to ambiguity as to which brain areas regulate selective aggression. Nevertheless, in an early study, male prairie voles displayed aggression toward a male intruder following 24 h of mating, but not following 24 h of cohabitation with a female without mating (Wang et al., 1997a). However, despite their differences in sociosexual experience and in aggressive behavior, both types of male exposure led to equal levels of Fos-ir (immunoreactivity) expression in some brain areas, such as the bed nucleus of the stria terminalis (BNST). Males that mated for 24 h and displayed high levels of aggression toward either a male or a female intruder showed increased levels of Fos-ir expression in the medial amygdala (MeA; Wang et al., 1997a), compared to males that cohabitated with a female without mating, implicating the MeA as a brain area associated with the display of mating-induced selective aggression (Fig. 6.1D and E).

In a more recent study, several brain areas including the BNST, medial preoptic area (MPOA), paraventricular nucleus (PVN), and lateral septum (LS) showed higher levels of Fos expression in pair bonded males that had experienced an RIT compared to pair bonded males not exposed to a social intruder (Gobrogge et al., 2007). However, no group differences in Fos expression across these brain areas were found among males that were exposed to different social stimuli or displaying different patterns of social behavior, including aggression or affiliation toward intruders. These data suggest that the increased neuronal activation in these brain regions is probably due to olfactory stimulation or general arousal associated with exposure to a conspecific, but such a response is nonselective. A unique pattern of Fos expression was found in the anterior hypothalamus (AH), in which exposure to a conspecific stranger, either male or female, induced a significant increase in AH-Fos over those reexposed to their familiar partner (Fig. 6.1D and E; Gobrogge et al., 2007). This increase in Fos staining may indicate a stimulus-specific response. The AH appears to be more responsive to chemosensory, tactile, and/or visual cues associated with conspecific strangers, but not familiar partners (Gobrogge et al., 2007). These data indicate that the increased neuronal activation in the AH may be involved in aggressive behavior displayed by pair bonded male prairie voles (Gobrogge et al.,

2007). This notion is corroborated by previous research documenting a critical role of the hypothalamus in regulating aggression across several mammalian species. For example, the ventromedial hypothalamus (VMH) in mice (Choi et al., 2005; Lin et al., 2011) and AH in rats (Veening et al., 2005) is responsive to conspecific chemosensory cues, which elicit aggressive behavior. Electrical stimulation applied directly to the AH induces attack toward conspecifics in rats (Kruk, 1991) and other animals (Albert and Walsh, 1984; Siegel et al., 1999). Interestingly, in humans, surgical lesioning of the AH reduces physical violence (Ramamurthi, 1988; Sano et al., 1966). In summary, data from vole studies demonstrate that activation of the MeA and AH is associated with the display of selective aggression (Gobrogge et al., 2007; Wang et al., 1997a).

IV. NEURAL CIRCUITRY

To directly evaluate the neural circuitry programming selective aggression, we performed a series of tract tracing experiments focusing on the AH and MeA. Intra-AH injections of an anterograde tracer, biotinylated dextran amine (BDA), resulted in BDA-ir staining in several brain regions. The AH projected to areas involved in processing chemosensory cues including the MeA; areas important for regulating social behavior including the LS, BNST, MPOA, VMH, and dorsal raphe (DR); and areas coordinating motor output, such as the periaqueductal gray (Gobrogge and Wang, 2009). The AH also projected to brain areas implicated in evaluating incentive salience including the medial prefrontal cortex (mPFC), nucleus accumbens (NAcc), and ventral pallidum (VP), as well as areas involved in memory formation and consolidation including hippocampal regions CA3 and the dentate gyrus (Gobrogge and Wang, 2009). Further, site-specific injections of a retrograde tracer, fluorogold (FG), into the LS, NAcc, or MeA resulted in FG-ir staining in the AH, indicating reciprocal connections between the AH and regions involved in motivation and chemo-sensory communication (Gobrogge and Wang, 2009).

Fos-ir staining was also used to assess neuronal activation, in this circuit, associated with the display of pair bonding behavior. Males displaying aggression toward an unfamiliar female showed a significantly higher density of Fos-ir in the AH and MeA relative to males displaying affiliation toward their female partner (Fig. 6.1D and E), replicating our previous findings (Gobrogge et al., 2007; Wang et al., 1997a). Interestingly, we identified a MeA-AH-LS circuit that was activated when males were displaying aggression and a DR-AH circuit that was recruited when males were displaying affiliation (Gobrogge and Wang, 2009). The identification of these two neural circuits indicates a specific neuronal framework associated with the choice between affiliation (DR-AH circuit) and aggression (MeA-AH-LS circuit) in pair bonded male prairie voles.

V. NEUROCHEMICAL REGULATION OF SELECTIVE AGGRESSION

Previous work has primarily focused on partner preference formation and documented a growing list of neurochemicals, including oxytocin (OT), AVP, corticotropin releasing hormone, DA, GABA, and glutamate, as well as their interactions in the regulation of pair bonding behavior (Aragona et al., 2003; Curtis and Wang, 2005a,b; Carter et al., 1995; DeVries et al., 1995; Gingrich et al., 2000; Lim and Young, 2004; Liu and Wang, 2003; Lim et al., 2004; Liu et al., 2001; Smeltzer et al., 2006; Wang et al., 1998, 1999; Williams et al., 1992, 1994; Winslow et al., 1993). Importantly, data from several studies indicate a select subset of neurochemicals in the regulation of selective aggression.

A. Neuropeptides

Comparative approaches have been utilized in studies examining neuroendocrine mechanisms regulating social behavior in voles (Insel, 2010). Studies have focused on examining the central AVP system, a nine amino acid neuropeptide with diverse forebrain projections, across monogamous and promiscuous vole species. AVP is an antidiuretic hormone and has been shown to stimulate three structurally distinct receptors V1a, V1b, and V2, each activating very specific second messenger systems (Michell et al., 1979). Classically, AVP was first described as a primary homeostatic factor controlling kidney water reabsorption, blood volume/pressure, and vasodilatation in the peripheral nervous system. AVP and its receptors have been shown to be widely expressed in the central nervous system (Thibonnier, 1992), within specific brain regions (Johnson et al., 1993).

The V1a AVP receptor subtype (V1aR), in particular, has been extensively studied in the regulation of social behavior (Insel et al., 1994) including aggression (Albers et al., 2006; Cooper et al., 2005; Ferris et al., 2006; Winslow et al., 1993). V1aRs are directly coupled to stimulatory (s) Gq-11 proteins (Thibonnier et al., 1993). Stimulation of these G-proteins leads to activation of adenylate cyclase, cAMP, protein kinase C, and phospholipases C, A_2, and D (Raggenbass et al., 1991; Thibonnier, 1992; Thibonnier et al., 1992, 1994) enhancing calcium influx through L-type calcium channels (Son and Brinton, 2001). Such activation enhances learning and memory in the aging brain (Deyo et al., 1989; Yamada et al., 1996) via direct effects on gene expression (Murphy et al., 1991).

Because AVP regulates species-specific social behaviors such as aggression (Ferris et al., 1984; Ryding et al., 2008), it was hypothesized that the organization of central AVP systems may differ between monogamous and promiscuous vole species (Bamshad et al., 1993b; Insel and Shapiro, 1992). To test this hypothesis, the distribution pattern of AVP cells, fibers, and receptors were mapped in the vole brain. In all vole species examined, AVP-ir neurons were found in several brain regions, including the PVN and SON (supraoptic nucleus) of the hypothalamus, the BNST, MeA, AH, and MPOA (Bamshad et al., 1993b; Gobrogge et al., 2007; Wang, 1995; Wang et al., 1996). AVP-ir fibers were localized in the LS, lateral habenular nucleus, diagonal band, BNST, MPOA, and MeA (Bamshad et al., 1993b; Wang et al., 1996). Overall, AVP distribution patterns were highly conserved between monogamous and promiscuous vole species (Wang, 1995; Wang et al., 1996). Dramatic species differences in the distribution patterns and regional densities of V1aRs, however, were observed between vole species exhibiting different life strategies (Hammock and Young, 2002). For example, prairie voles have higher densities of V1aRs in the BNST, VP, central and basolateral nuclei of the amygdala, and accessory olfactory bulb, whereas montane voles exhibit a higher density of V1aRs in the LS and mPFC (Insel et al., 1994; Lim et al., 2004; Smeltzer et al., 2006; Wang et al., 1997c; Young et al., 1997). Further, prairie and pine voles exhibit similar patterns of V1aR binding, which differ from that of promiscuous meadow and montane voles (Insel et al., 1994; Lim et al., 2004). Species differences in V1aR distribution are stable across the lifespan (Wang et al., 1997b,c) and are receptor-specific, as no species differences are found in either the benzodiazepine or opiate receptor systems (Insel and Shapiro, 1992). In addition, monogamous prairie and promiscuous meadow voles differ in central AVP activity during mating and reproduction (Bamshad et al., 1993a; Wang et al., 1994). Because of these anatomical and functional differences, central AVP was thought to underlie selective aggression in male prairie voles (Winslow et al., 1993).

Among the neuropeptides underlying aggression (Miczek et al., 2007; Siever, 2008), AVP, and its homolog vasotocin, have been found to regulate several forms of aggression across species (Caldwell et al., 2008; Riters and Panksepp, 1997) and diverse taxa (Backstrom and Winberg, 2009; Goodson, 2008). In humans, central AVP correlates with aggressive behavior (Coccaro et al., 1998) and mediates anger (Thompson et al., 2004). Thus, the central AVP system may have evolved to be primed by a wide variety of experiences to induce aggression, when appropriate, in social animals (Donaldson and Young, 2008). Because central AVP underlies territorial aggression in other rodents (Ferris et al., 1984), AVP was proposed to regulate mating-induced aggression in prairie voles. In a pharmacological study, injections of a V1aR antagonist (V1aR Ant) into the lateral ventricle of male prairie voles blocked selective aggression induced by mating whereas injections of AVP-induced aggression toward an intruder in the absence of mating (Winslow et al., 1993). These effects were neuropeptide specific, as intracerebroventricular (ICV) infusion of an OT receptor antagonist had no effect on mating-induced aggression (Winslow et al., 1993). Importantly, developmental exposure to either AVP in male prairie voles (Stribley and Carter, 1999) or OT in female prairie voles (Bales and

Carter, 2003) facilitates aggression in adulthood. Together, these data highlight a critical role of neuropeptide regulation of prairie vole aggression. However, the action site and release dynamics of neuropeptides in the regulation of selective aggression were unclear.

In a more recent study, it was found that the display of selective aggression was associated with increased neuronal activation in the AH, specifically in neurons expressing AVP (Fig. 6.2A and B; Gobrogge et al., 2007). In a previous study in Syrian hamsters (*Mesocricetus auratus*), aggression has also been shown to be associated with an increase in AVP-ir/Fos-ir double-labeled cells in the nucleus circularis (NC), medial SON (mSON), and surrounding areas ventral to the fornix in the AH (Delville et al., 2000). Our data, combined with previous results from other species, suggest that the AH may be a brain area in which AVP regulates selective aggression. Indeed, AH-AVP has been implicated in various forms of aggressive behavior. In Syrian hamsters, for example, blockade of V1aRs in the AH diminished offensive aggression (Caldwell and Albers, 2004; Ferris and Potegal, 1988) whereas intra-AH administration of a V1aR agonist enhanced aggressive behavior (Caldwell and Albers, 2004; Ferris et al., 1997). More recently, the density of V1aRs in the AH has been found to increase significantly in Syrian hamsters after their display of offensive aggression (Albers et al., 2006).

Because selective aggression was associated with neuronal activation in the AH, specifically in AVP-containing neurons (Fig. 6.2A and B; Gobrogge et al., 2007), we tested the hypothesis that selective aggression is associated with AH-AVP release. *In vivo* brain microdialysis, with ELISA, was performed on male prairie voles that were pair bonded for 2 weeks. Pair bonded males displayed significantly higher levels of aggression toward novel females but more side-by-side affiliation with their familiar female partner (Gobrogge et al., 2009). ELISA analysis indicated that exposure to a stranger female, compared to a familiar partner, increased AH-AVP release (Fig. 6.2C), which is further confirmed by correlation analyses indicating that AH-AVP release was coupled positively with aggression and negatively with affiliation (Gobrogge et al., 2009). Moreover, intra-AH administration of AVP at a high (500 ng/side), but not a low (5 ng/side), dose in sexually naïve males induced aggression toward a novel female, and this effect was mediated by V1aR as concurrent administration of a 10-fold higher dose of a V1aR Ant blocked AVP-induced aggression (Fig. 6.2D; Gobrogge et al., 2009). Further, intra-AH infusions of a V1aR Ant (5 μg/side), in pair bonded males, blocked aggression and facilitated social affiliation toward novel females (Fig. 6.2D; Gobrogge et al., 2009). Thus, AH-AVP is both necessary and sufficient to regulate mating-induced selective aggression in male prairie voles.

Prior research has shown that the social environment has a significant impact on signaling and structural components of AVP systems in the brain. In marmoset monkeys, for example, prefrontal V1aR increases during fatherhood (Kozorovitskiy et al., 2006). In hamsters, several social and drug paradigms have

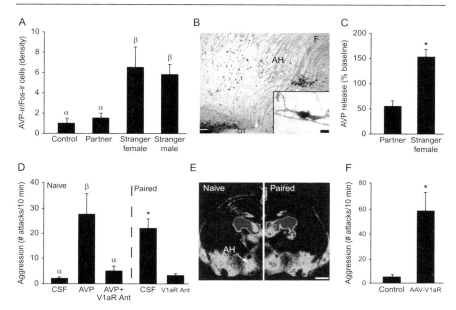

Figure 6.2. Vasopressin (AVP) regulation of selective aggression. (A) Pair bonded male prairie voles that display aggression toward female or male strangers exhibit a significantly higher density of AVP-ir/Fos-ir double-labeled cells in the AH relative to pair bonded males re-exposed to their female partner or to males not exposed to any social stimulus (Control). (B) Photomicrograph of AVP-ir cell bodies and fibers (brown cytoplasmic staining), Fos-ir (dark nuclear staining), or AVP-ir/Fos-ir double labeled cells (insert) in the anterior hypothalamus (AH). (C) *In vivo* brain microdialysis reveals a significant increase in AH-AVP release in pair bonded male prairie voles displaying aggression toward a stranger female compared to males reexposed to their female partner. (D) Intra-AH AVP microinfusion, in sexually naïve males (Naïve), is sufficient to induce aggression toward a stranger female compared to control males infused with cerebral spinal fluid (CSF). AH-AVP-induced aggression is blocked with concurrent administration of AVP with a V1aR Ant. Pair bonded males (Paired) display aggression toward stranger females, which is blocked by intra-AH infusion of a V1aR Ant. (E) Pair bonded males (Paired) exhibit a significantly higher density of AVP-V1a receptor (V1aR) binding in the AH relative to sexually naïve males (Naïve). (F) Sexually naïve males infused with an adeno-associated virus expressing the *V1aR* gene (AAV-*V1aR*) in the AH exhibit enhanced aggression toward stranger females relative to males infused with the *LacZ*-gene (Control). F, fornix; OT, optic tract. Bars indicate means ± standard error of the mean. Bars with different Greek letters differ significantly from each other. *$p < 0.05$. Scale bars = 100 µm, insert scale bars = 10 µm. Adapted from Aragona *et al.* (2006), Gobrogge *et al.* (2007, 2009), and Young *et al.* (2011a). (See Color Insert.)

been shown to directly alter the AH-AVP system to regulate offensive aggression (Ferris *et al.*, 1989; Grimes *et al.*, 2007; Harrison *et al.*, 2000b; Jackson *et al.*, 2005). In a previous study in male prairie voles, cohabitation with a female significantly increased the number of AVP mRNA labeled cells in the BNST (Wang *et al.*, 1994). Because social isolation increases the density of V1aRs in the AH to regulate offensive aggression in golden hamsters (Albers *et al.*, 2006), we tested the hypothesis that the density of AH-V1aR changes with pair bonding experience to engage selective aggression. Pair bonded males showed higher densities of V1aR binding, site specifically, in the AH (Fig. 6.2E), with no change in OT receptor binding, demonstrating that pair bonding experience induces a neural plastic reorganization of V1aRs in a region- and receptor-specific manner (Gobrogge *et al.*, 2009). To determine whether this increase in V1aR density in the AH, following pair bonding, was directly related to the emergence of aggression toward novel females, we used viral vector mediated gene transfer to artificially elevate V1aR density in the AH. Males that received intra-AH infusions of the AAV-V1aR displayed higher levels of aggression toward a novel female compared to control males that received infusions of the *LacZ*-gene (Fig. 6.2F; Gobrogge *et al.*, 2009). Similar viral vector mediated increases in V1aR expression in the VP in voles (Lim *et al.*, 2004; Pitkow *et al.*, 2001) and LS in mice (Bielsky *et al.*, 2005) enhanced affiliation. In male rats, intermale aggression correlates with AVP release in the LS while AVP release in the BNST is inversely related to aggression levels (Veenema *et al.*, 2010). Together, these data support the notion that region-specific AVP functioning regulates specific types of social behaviors, and multiple brain regions serve as a circuit in which AVP coordinates a range of adaptive behaviors important for reproductive success.

B. Dopamine

Anatomically, central DA is divided into three distinct pathways: nigrostriatal, incertohypothalamic, and mesocorticolimbic. DA cell bodies projecting from the substantia nigra synapse in the dorsal striatum and comprise the nigrostriatal path (Swanson, 1982). Incertohypothalamic paths extend from DA cell bodies of the A12–14 cell groups and project to the MPOA and PVN (Cheung *et al.*, 1998). The mesocorticolimbic path represents DA cell bodies originating in the ventral tegmental area (VTA; Fig. 6.3C) projecting to the mPFC and NAcc (Fig. 6.3C; Carr and Sesack, 2000; Swanson, 1982). In addition, DA cells in the AH (Fig. 6.3B) project to forebrain areas including the striatum, LS, NAcc, and mPFC (Alcaro *et al.*, 2007; Lindvall and Stenevi, 1978; Maeda and Mogenson, 1980).

DA preferentially binds to two families of receptors: D1-like and D2-like. Both types of DA receptors are found in the mPFC, NAcc, LS, and MeA (Boyson *et al.*, 1986). D1-like receptors are directly coupled to both stimulatory (s) $G\alpha$ and $G\alpha_{olf}$ proteins (Neve *et al.*, 2004). Stimulation of these G-proteins leads to activation of adenylate cyclase, cAMP, and protein phosphatase-1

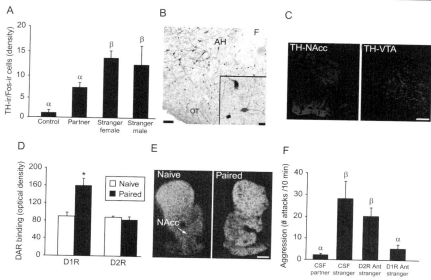

Figure 6.3. Dopamine (DA) regulation of selective aggression. (A) Pair bonded male prairie voles that display aggression toward female or male strangers exhibit a significantly higher density of TH-ir/Fos-ir double-labeled cells in the AH relative to pair bonded males re-exposed to their female partner or to males not exposed to any social stimulus (Control). (B) Photomicrograph of TH-ir cell bodies and fibers (brown cytoplasmic staining), Fos-ir (dark nuclear staining), or TH-ir/Fos-ir double labeled cells (insert) in the anterior hypothalamus (AH). (C) Photomicrograph of TH-ir fibers in the nucleus accumbens (NAcc) and neurons in the ventral tegmental area (VTA). (D, E) Pair bonded males (Paired) exhibit a significantly higher density of DA D1-like (D1R), but not D2-like (D2R), receptor binding in the NAcc compared to sexually naïve males (Naïve). (F) Pair bonded males, infused with CSF in the NAcc, display aggression toward a stranger female but not toward their female partner. Concurrent infusion of CSF with a DA D1R antagonist (D1R Ant), but not D2R antagonist (D2R Ant), in the NAcc is sufficient to attenuate pair bonding-induced aggression toward a stranger female. F, fornix; OT, optic tract. Bars indicate means ± standard error of the mean. Bars with different Greek letters differ significantly from each other. $*p < 0.05$. Scale bars = 100 μm, insert scale bars = 10 μm. Adapted from Aragona et al. (2006), Gobrogge et al. (2007, 2009), and Young et al. (2011a).

inhibitor DARP-32 (Neve et al., 2004). Conversely, D2-like receptors couple to inhibitory (i) Gα proteins and, when activated, downregulate adenylate cyclase, cAMP, and protein phosphatase-1 inhibitor DARP-32 (Neve et al., 2004). D1-receptor stimulation plays a critical role in calcium influx, via L-type calcium channels, which is important for cellular long-term potentiation facilitating learning and memory in the aging brain (Deyo et al., 1989; Yamada et al., 1996) through direct influences on gene expression (Murphy et al., 1991).

Because mesocorticolimbic DA underlies partner preference formation (Aragona and Wang, 2009; Aragona *et al.*, 2003; Gingrich *et al.*, 2000; Gobrogge *et al.*, 2008; Wang *et al.*, 1999), studies focused on examining the potential role of central DA regulating selective aggression (Gobrogge *et al.*, 2008). Pair bonded male prairie voles have significantly higher levels of DA D1-type receptors (D1Rs), but not D2-type receptors (D2Rs), in the NAcc (Fig. 6.3D and E; Aragona *et al.*, 2006). Males that cohabitated with a female for 24 h, with or without mating, did not exhibit an increase in D1Rs in the NAcc, supporting the idea that upregulation of NAcc D1Rs, after pair bonding, may directly regulate selective aggression. To test this notion, pair bonded male prairie voles were injected with a D1R antagonist (D1R Ant) into the NAcc. NAcc-D1R, not D2R (D2R Ant), antagonism was sufficient to block selective aggression in pair bonded male prairie voles (Fig. 6.3F; Aragona *et al.*, 2006). In other work, both brief and extended cohabitation with unfamiliar conspecifics in female prairie voles significantly increased the number of DA-ergic cells in the BNST and MeA (Cavanaugh and Lonstein, 2010) and blocking D2Rs, during development, decreased aggression-related behavior including infanticide in adult female, but not male, prairie voles (Hostetler *et al.*, 2010). Further, pair bonded male prairie voles—displaying aggression toward either male or female intruders, had a significantly higher density of cells in the AH that were double-labeled for tyrosine hydroxylase-ir (TH—rate-limiting enzyme in DA biosynthesis) and Fos-ir than males not exposed to a social stimulus or males that were re-exposed to their familiar female partner (Fig. 6.3A and B), implicating AH-DA involvement in selective aggression (Gobrogge *et al.*, 2007).

C. Steroid hormones

Physical aggression is significantly more common in males than females and these behavioral sex differences have been observed across many species (Gatewood *et al.*, 2006). Research describing biological contributions underlying these sex differences has focused primarily on steroid hormones (Gatewood *et al.*, 2006). Several studies have examined the role of androgens in the development of aggressive behavior, both organizationally (e.g., treatment with prenatal testosterone) and activationally (e.g., treatment with postnatal testosterone). Previous research has found that prenatal androgen exposure increases the behavioral expression of adult aggression (Michard-Vanhee, 1988; Vale *et al.*, 1972). Although organizational and activational influences of androgen on aggression have been noted, some inconsistent results have been reported. For example, castration in male rats (Koolhaas *et al.*, 1990) and male prairie voles (Demas *et al.*, 1999) has no affect on aggression. Thus, circulating testosterone, alone, cannot solely contribute to the expression of aggressive behavior in all rodent species. However, these findings do not rule out the possible effects of testosterone having organizational influences on other neurochemical systems in the brain—which, together, may regulate aggression in adulthood. Thus, neurochemical–steroid hormone interactions—underlying aggression—have also been examined. For example, AVP administered directly into the MeA facilitates territorial aggression in male rats and is sufficient to block the effects of castration on reducing aggression (Koolhaas et al., 1990). Further, castration (Bermond et al., 1982), but not ovariectomy (Kruk et al., 1984), decreases the excitability of neurons in the AH—blocking electrically induced aggression, which in castrates can be reversed by testosterone treatment (Bermond et al., 1982). Therefore, circulating testosterone may be acting as a potent neuromodulator—interacting with neurochemicals, like AVP, to regulate aggression.

Additional evidence demonstrating steroid hormone–neurotransmitter interactions exists in research investigating central DA. For example, 75% of TH-ir expressing cells in the hamster posterior MeA contain androgen receptors, are DA-ergic (i.e., they do not co-label with DA beta hydroxylase), and are highly influenced by gonadal hormones compared to TH-ir cells found in the anterior MeA (Asmus and Newman, 1993; Asmus et al., 1992). Interestingly, this same group of TH-ir cells is found in the posterior MeA and BNST in male prairie voles, which appears to be influenced by testosterone (Northcutt et al., 2007) and activated after mating and social experience (Northcutt and Lonstein, 2009). Together, these data suggest interactions between steroid hormones and central DA, in areas such as the MeA, in processing chemosensory cues related to social communication.

D. Classical neurotransmitters

In addition to the effects of neurotransmitters and hormones on aggression, neuromodulators such as GABA and glutamate have also been shown to be involved in the display of agonistic behavior. For example, microinjection of a GABA antagonist (Adams et al., 1993; Roeling et al., 1993) with concurrent treatment of a glutamate agonist (Haller et al., 1998) in the AH facilitates attack behavior in rodents—with higher doses having greater behavioral effects. Further, reverse microdialysis infusion with a glutamate agonist and a GABA-A antagonist into the AH of rats, recently having experienced an agonistic encounter, also facilitates aggression (Haller et al., 1998). Finally, it is worth mentioning that neuromodulators in the VTA, which provides the major source of DA projections to the NAcc (Fig. 6.3C) and mPFC, are also involved in pair bonding behavior. Glutamate and GABA receptor blockade in the VTA, which alters DA activity in the NAcc, induces partner preference formation in the absence of mating in male prairie voles (Curtis and Wang, 2005b).

VI. MOLECULAR GENETICS OF SELECTIVE AGGRESSION

Monogamous male prairie voles carry several repetitive microsatellite DNA sequences in the promoter region of the V1aR gene that are not found in promiscuous male voles (Hammock and Young, 2002, 2004; Young, 1999). These genetic differences directly contribute to species differences in social organization (Landgraf et al., 2003; Lim et al., 2004; Pitkow et al., 2001). Further, mice carrying a transgene coding for the prairie vole V1aR exhibit central V1aR patterns similar to prairie voles and, when injected with AVP, display enhanced social affiliation (Young et al., 1999). Male voles injected in the VP, with a virus expressing V1aR, display enhanced partner preference in the absence of mating (Lim et al., 2004; Pitkow et al., 2001). Interestingly, individual variability in the genetic sequences coding V1aRs has revealed remarkable within species differences in the strength of monogamous pair bonds in prairie voles (Hammock and Young, 2005; Ophir et al., 2008).

Because prairie voles carry varying lengths of DNA to code V1aRs (Hammock and Young, 2005), this genomic predisposition enables their brain to dynamically express V1aRs after sociosexual experience. This genetic loading distinguishes prairie voles from traditional laboratory rodents that lack this genetic makeup. Future work would benefit from comparing aggression levels between male prairie voles carrying long versus short versions of the promoter region encoding the V1aR gene. In humans, polymorphisms in the promoter region encoding V1aR are associated with differences in sociosexual bonding behaviors (Prichard et al., 2007; Walum et al., 2008) including altruism (Israel et al., 2008) and deficits in social communication observed in individuals with autistic spectrum disorders (Israel et al., 2008; Kim et al., 2002; Meyer-Lindenberg et al., 2008). To date, no study has examined potential associations between the V1aR gene and patterns of aggression in humans. Thus, it would be interesting to examine polymorphisms in the V1aR gene in patients with a lifetime history of pathological violence, as the V1aR system may harbor susceptibility genes underlying extreme forms of aggression, increasing the prevalence of homicide and suicide in human populations.

VII. DRUG-INDUCED AGGRESSION

The prairie vole has been established as an animal model for depression related to social separation (Bosch et al., 2009; Grippo et al., 2007). Further, chronic metal ingestion—a potential model for autism—produces social avoidance in male, but not female, prairie voles exposed to unfamiliar same-sex conspecific strangers (Curtis et al., 2010), suggesting a developmental mechanism underlying the social deficits associated with autistic spectrum disorders in humans. Recently, prairie voles have also been utilized to examine the effects of drugs of abuse on pair bonding behavior (Gobrogge et al., 2009; Liu et al., 2010; Young et al., 2011b).

Drug addiction is a significant problem for many humans because drug abuse has such a powerful control over social behavior essential for survival (Kelley and Berridge, 2002; Nesse and Berridge, 1997; Panksepp et al., 2002). In humans, substance abuse has been associated with weapon-related violence and homicide (Hagelstam and Hakkanen, 2006; Madan et al., 2001; Spunt et al., 1998), intimate partner aggression, including partner-directed physical and psychological aggression (Chermack et al., 2008; O'Farrell and Fals-Stewart, 2000), sexual (El-Bassel et al., 2001), and child abuse (Haapasalo and Hamalainen, 1996; Mokuau, 2002; Walsh et al., 2003). Collectively, drug-related violence leads to family system dysfunction and incarceration (Krug et al., 2002), creating significant societal concerns. While aggression research in humans has provided valuable information regarding relationships between drug abuse and violence, animal models have been used to examine neural mechanisms underlying drug-induced aggression.

Drug use can override neurobiological programs to activate maladaptive forms of agonistic behavior, engaging inappropriate types of physical aggression (Swartz et al., 1998) such as domestic violence (Moore et al., 2008) and intimate partner homicide (Farooque et al., 2005). As a result, chronic drug abuse can cause permanent neural reorganization (Nestler and Aghajanian, 1997; White and Kalivas, 1998), impairing the adaptive—social brain (Panksepp et al., 2002), leading to the display of maladaptive social behavior (Wise, 2002). Multiple studies have demonstrated that aggression may be altered shortly after drug exposure and that the directionality of these effects depends on drug, dose, and individual differences between subjects.

Repeated exposure to several drugs of abuse, during adolescence or adulthood, persistently enhances agonistic behaviors, specifically those associated with offensive aggression. For example, Syrian hamsters treated during adolescence with cocaine (DeLeon et al., 2002a; Harrison et al., 2000a; Jackson et al., 2005; Knyshevski et al., 2005a,b; Melloni et al., 2001) or anabolic-androgenic steroids (AASs) (DeLeon et al., 2002b; Harrison et al., 2000b; Melloni and Ferris, 1996; Melloni et al., 1997) display enhanced offensive aggression in adulthood. Interestingly, these drug experiences reorganize AVP (Grimes et al., 2007; Harrison et al., 2000b; Jackson et al., 2005), DA (Ricci et al., 2009; Schwartzer et al., 2009), and GABA (Schwartzer et al., 2009) signaling in the AH.

For example, when compared with nonaggressive sesame oil-treated control males, aggressive AAS-treated males exhibit significant neuroplastic changes in the AH including increased AVP-ir fiber density and AVP content (Harrison et al., 2000b), an increase in TH-ir cell and fiber density (Ricci et al.,

2009), enhanced DA-D2R expression (Schwartzer *et al.*, 2009), a higher number of GAD_{67}-ir cells (Schwartzer *et al.*, 2009), and decreased $GABA_A$ receptor expression (Schwartzer *et al.*, 2009). Further, pharmacological blockade of D2 (Schwartzer and Melloni, 2010a,b), but not D5 (Schwartzer and Melloni, 2010b), DA receptors in the AH abolishes these effects. Together, results from this work suggests that AVP and DA signaling facilitates aggression by GABA inhibition in the AH of AAS-treated male Syrian hamsters.

As noted above, previous work has shown that exposure to drugs of abuse, such as cocaine, enhances male–male aggression by reorganizing the AH-AVP system in hamsters (Jackson *et al.*, 2005). Therefore, we tested the hypothesis that amphetamine (AMPH), another commonly abused psychostimulant, would act in a similar fashion in affecting male-to-female aggression in prairie voles. Because our recent data revealed that repeated AMPH treatment—(1 mg/kg) for 3 consecutive days—induces a conditioned place preference (Aragona *et al.*, 2007) and blocks mating-induced partner preference (Fig. 6.4A; Liu *et al.*, 2010), this treatment regimen was used. To examine the selectivity of AMPH-induced aggression, males treated with saline or AMPH were tested for aggression toward an unfamiliar female or a familiar female (that cohabitated with a male across a wire mesh screen for 24 h without mating). Compared with saline-treated controls, AMPH-treated males displayed significantly higher levels of aggression toward either familiar or unfamiliar females (Fig. 6.4B), indicating that AMPH exposure induces generalized aggression, rather than being selective to novel females. This AMPH treatment also induced an increase in the density of AVP-V1aR binding in the AH, but not MPOA, relative to saline control males (Fig. 6.4C). Further, intra-AH infusions of CSF containing the V1aR Ant, but not CSF alone, diminished AMPH-induced aggression toward novel females (Fig. 6.4D). These data suggest that repeated exposure to AMPH can induce female-directed aggression and that this behavior is mediated by AH-AVP. Interestingly, these behavioral effects coincide with upregulation of D1Rs in the NAcc (Liu *et al.*, 2010) and V1aRs in the AH (Gobrogge *et al.*, 2009)— which both facilitate aggression toward novel females (Aragona *et al.*, 2006; Gobrogge *et al.*, 2009); indicating that drugs of abuse can hijack neuroplasticity evolved to maintain monogamous pair bonds.

VIII. CONCLUSIONS AND FUTURE DIRECTIONS

Prairie voles have provided an excellent model system to study the neurobiology of ethologically meaningful aggression associated with monogamous pair bonds. Aggression can be easily manipulated under laboratory conditions and reliably expressed following mating and social cohabitation. Several brain areas: MeA, AH, and NAcc, work in a neural circuit to regulate selective aggression via AVP

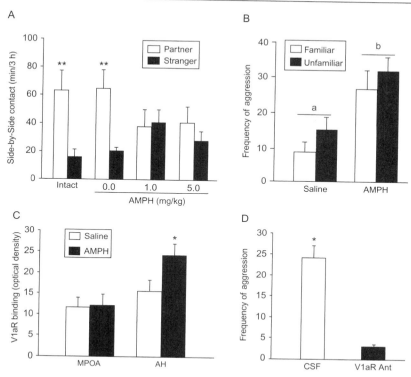

Figure 6.4. Drug experience impairs pair bonding behavior. (A) Pair bonded (Intact) and saline treated (0.0) male prairie voles, receiving 3-day—once daily; repeated injections, spend significantly more time in physical side-by-side contact with their familiar female partner than with an unfamiliar stranger female. Pair bonded males injected (i.p., intraperitoneal) with 1.0 or 5.0 mg/kg amphetamine (AMPH) spent equal amounts of time in physical side-by-side contact with their female partner as with a stranger female. (B, C) Repeated AMPH exposure, in sexually naïve males, increases aggression toward both familiar and unfamiliar females and enhances the density of vasopressin (AVP) receptor (V1aR) expression in the anterior hypothalamus (AH) but not in the medial preoptic area (MPOA). (D) Site-specific microinfusion of an AVP-V1aR antagonist (V1aR Ant) into the AH, of males receiving 3-day repeated AMPH exposure (i.p.), significantly decreases AMPH-induced aggression toward novel females relative to males receiving intra-AH infusions of cerebral spinal fluid (CSF). Bars indicate means ± standard error of the mean. Bars with different Greek letters differ significantly from each other. $*p < 0.05$; $**p < 0.01$. Adapted from Gobrogge et al. (2009), Liu et al. (2010), and Young et al. (2011a).

and DA. Commonly abused drugs, like AMPH, can usurp these neurochemical circuits to engage maladaptive social behavior impairing monogamous pair bonds.

Although male-to-male aggression has been studied in a variety of mammals, we know surprisingly little about male-to-female aggression and its underlying neuromechanisms. Interestingly, pair bonded male prairie voles naturally display aggression toward conspecific females but not toward their female partner and, therefore, selective aggression allows for investigation of the neurobiology of male-to-female aggression. Data have demonstrated that this form of selective aggression is mediated by elevated AVP release and increased V1aR expression in the AH—priming male prairie voles to respond aggressively to novel females. In addition, data have also shown that the same AH-AVP system mediates generalized, female-directed aggression induced by AMPH. Together with previous research from other animals (Ferris *et al.*, 1989, 1997; Grimes *et al.*, 2007; Harrison *et al.*, 2000b; Jackson *et al.*, 2005; Veenema *et al.*, 2006), these data demonstrate a unique point of convergence in the mammalian brain (Choi *et al.*, 2005; Motta *et al.*, 2009). The AH-AVP system is highly conserved and functions to control different forms of aggression to maintain a wide range of resources important for reproductive success. These highly evolved neuropeptide systems appear to be extremely vulnerable to drugs of abuse, as our data show that hypothalamic AVP controls both naturally occurring as well as drug-facilitated female-directed aggression, suggesting that psychostimulant drugs, like AMPH, are capable of switching adaptive (functional) forms of aggression (e.g., mate guarding) to aberrant (dysfunctional) forms of violent behavior (e.g., partner-directed aggression). Together, these data demonstrate the utility of the prairie vole model for evaluation of the effects of drug abuse on neural systems controlling adaptive forms of aggression—such as mate guarding.

Finally, because other neurochemicals, such as DA (Aragona *et al.*, 2006) and serotonin (Villalba *et al.*, 1997), also regulate selective aggression in prairie voles, offensive aggression related to drug experience (Tidey and Miczek, 1992) and AVP/5-HT interactions mediate aggression in other rodents (Ferris *et al.*, 1997; Veenema *et al.*, 2006), future studies should examine potential neurochemical interactions in the regulation of selective aggression. By understanding the basic neuroendocrinology of pair bonding in prairie voles, we may eventually be able to better clarify the neural chemistry of mental health deficits associated with aberrations in social behavior in patients suffering from drug addiction or pathological violence.

Acknowledgments

The authors would like to thank Benjamin William Tyson for critically reading this manuscript. The work reviewed in this chapter was supported by National Institutes of Health grants MHF31-79600 to K. L. G., MHR01-58616, MHR01-89852, DAR01-19627, and DAK02-23048 to Z. X. W., and NIH Program Training Grant T32 NS-07437.

References

Adams, D. B., Boudreau, W., Cowan, C. W., Kokonowski, C., Oberteuffer, K., and Yohay, K. (1993). Offense produced by chemical stimulation of the anterior hypothalamus of the rat. *Physiol. Behav.* **53,** 1127–1132.

Albers, E. H., Dean, A., Karom, M. C., Smith, D., and Huhman, K. L. (2006). Role of V1a vasopressin receptors in the control of aggression in Syrian hamsters. *Brain Res.* **1073–1074,** 425–430.

Albert, D. J., and Walsh, M. L. (1984). Neural systems and the inhibitory modulation of agonistic behavior: A comparison of mammalian species. *Neurosci. Biobehav. Rev.* **8,** 5–24.

Alcaro, A., Huber, R., and Panksepp, J. (2007). Behavioral functions of the mesolimbic dopaminergic system: An affective neuroethological perspective. *Brain Res. Rev.* **56,** 283–321.

Aragona, B. J., and Wang, Z. (2009). Dopamine regulation of social choice in a monogamous rodent species. *Front. Behav. Neurosci.* **3,** 15.

Aragona, B. J., Liu, Y., Curtis, J. T., Stephan, F. K., and Wang, Z. (2003). A critical role for nucleus accumbens dopamine in partner-preference formation in male prairie voles. *J. Neurosci.* **23,** 3483–3490.

Aragona, B. J., Liu, Y., Yu, Y. J., Curtis, J. T., Detwiler, J. M., Insel, T. R., and Wang, Z. (2006). Nucleus accumbens dopamine differentially mediates the formation and maintenance of monogamous pair bonds. *Nat. Neurosci.* **9,** 133–139.

Aragona, B. J., Detwiler, J. M., and Wang, Z. (2007). Amphetamine reward in the monogamous prairie vole. *Neurosci. Lett.* **418,** 190–194.

Asmus, S. E., and Newman, S. W. (1993). Tyrosine hydroxylase neurons in the male hamster chemosensory pathway contain androgen receptors and are influenced by gonadal hormones. *J. Comp. Neurol.* **331,** 445–457.

Asmus, S. E., Kincaid, A. E., and Newman, S. W. (1992). A species-specific population of tyrosine hydroxylase-immunoreactive neurons in the medial amygdaloid nucleus of the Syrian hamster. *Brain Res.* **575,** 199–207.

Backstrom, T., and Winberg, S. (2009). Arginine-vasotocin influence on aggressive behavior and dominance in rainbow trout. *Physiol. Behav.* **96,** 470–475.

Bales, K. L., and Carter, C. S. (2003). Sex differences and developmental effects of oxytocin on aggression and social behavior in prairie voles (Microtus ochrogaster). *Horm. Behav.* **44,** 178–184.

Bamshad, M., Novak, M. A., and De Vries, G. J. (1993a). Sex and species differences in the vasopressin innervation of sexually naive and parental prairie voles, Microtus ochrogaster and meadow voles, Microtus pennsylvanicus. *J. Neuroendocrinol.* **5,** 247–255.

Bamshad, M., Novak, M. A., and deVries, A. C. (1993b). Sex and species differences in the vasopressin innervation of sexually naive and parental prairie voles (Microtus ochrogaster) and meadow voles (Microtus pennsylvanicus). *J. Neuroendocrinol.* **5,** 247–255.

Bermond, B., Mos, J., Meelis, W., van der Poel, A. M., and Kruk, M. R. (1982). Aggression induced by stimulation of the hypothalamus: Effects of androgens. *Pharmacol. Biochem. Behav.* **16,** 41–45.

Bielsky, I. F., Hu, S. B., Ren, X., Terwilliger, E. F., and Young, L. J. (2005). The V1a vasopressin receptor is necessary and sufficient for normal social recognition: A gene replacement study. *Neuron* **47,** 503–513.

Bosch, O. J., Nair, H. P., Ahern, T. H., Neumann, I. D., and Young, L. J. (2009). The CRF system mediates increased passive stress-coping behavior following the loss of a bonded partner in a monogamous rodent. *Neuropsychopharmacology* **34,** 1406–1415.

Bowler, C. M., Cushing, B. S., and Carter, C. S. (2002). Social factors regulate female-female aggression and affiliation in prairie voles. *Physiol. Behav.* **76,** 559–566.

Boyson, S. J., McGonigle, P., and Molinoff, P. B. (1986). Quantitative autoradiographic localization of the D1 and D2 subtypes of dopamine receptors in rat brain. *J. Neurosci.* **6,** 3177–3188.

Caldwell, H. K., and Albers, H. E. (2004). Effect of photoperiod on vasopressin-induced aggression in Syrian hamsters. *Horm. Behav.* **46,** 444–449.

Caldwell, H. K., Lee, H. J., Macbeth, A. H., and Young, W. S., 3rd (2008). Vasopressin: Behavioral roles of an "original" neuropeptide. *Prog. Neurobiol.* **84,** 1–24.

Carr, D. B., and Sesack, S. R. (2000). Projections from the rat prefrontal cortex to the ventral tegmental area: Target specificity in the synaptic associations with mesoaccumbens and mesocortical neurons. *J. Neurosci.* **20,** 3864–3873.

Carter, C. S., DeVries, A. C., and Getz, L. L. (1995). Physiological substrates of mammalian monogamy: The prairie vole model. *Neurosci. Biobehav. Rev.* **19,** 303–314.

Cavanaugh, B. L., and Lonstein, J. S. (2010). Social novelty increases tyrosine hydroxylase immunoreactivity in the extended olfactory amygdala of female prairie voles. *Physiol. Behav.* **100,** 381–386.

Chermack, S. T., Murray, R. L., Walton, M. A., Booth, B. A., Wryobeck, J., and Blow, F. C. (2008). Partner aggression among men and women in substance use disorder treatment: Correlates of psychological and physical aggression and injury. *Drug Alcohol Depend.* **98,** 35–44.

Cheung, S., Ballew, J. R., Moore, K. E., and Lookingland, K. J. (1998). Contribution of dopamine neurons in the medial zona incerta to the innervation of the central nucleus of the amygdala, horizontal diagonal band of Broca and hypothalamic paraventricular nucleus. *Brain Res.* **808,** 174–181.

Choi, G. B., Dong, H. W., Murphy, A. J., Valenzuela, D. M., Yancopoulos, G. D., Swanson, L. W., and Anderson, D. J. (2005). Lhx6 delineates a pathway mediating innate reproductive behaviors from the amygdala to the hypothalamus. *Neuron* **46,** 647–660.

Coccaro, E. F., Kavoussi, R. J., Hauger, R. L., Cooper, T. B., and Ferris, C. F. (1998). Cerebrospinal fluid vasopressin levels: Correlates with aggression and serotonin function in personality-disordered subjects. *Arch. Gen. Psychiatry* **55,** 708–714.

Cooper, M. A., Karom, M., Huhman, K. L., and Albers, H. E. (2005). Repeated agonistic encounters in hamsters modulate AVP V1a receptor binding. *Horm. Behav.* **48,** 545–551.

Curtis, J. T., and Wang, Z. (2003). Forebrain c-fos expression under conditions conducive to pair bonding in female prairie voles (Microtus ochrogaster). *Physiol. Behav.* **80,** 95–101.

Curtis, J. T., and Wang, Z. (2005a). Glucocorticoid receptor involvement in pair bonding in female prairie voles: The effects of acute blockade and interactions with central dopamine reward systems. *Neuroscience* **134,** 369–376.

Curtis, J. T., and Wang, Z. (2005b). Ventral tegmental area involvement in pair bonding in male prairie voles. *Physiol. Behav.* **86,** 338–346.

Curtis, J. T., Hood, A. N., Chen, Y., Cobb, G. P., and Wallace, D. R. (2010). Chronic metals ingestion by prairie voles produces sex-specific deficits in social behavior: An animal model of autism. *Behav. Brain Res.* **213,** 42–49.

Cushing, B. S., Mogekwu, N., Le, W. W., Hoffman, G. E., and Carter, C. S. (2003a). Cohabitation induced Fos immunoreactivity in the monogamous prairie vole. *Brain Res.* **965,** 203–211.

Cushing, B. S., Yamamoto, Y., Hoffman, G. E., and Carter, C. S. (2003b). Central expression of c-Fos in neonatal male and female prairie voles in response to treatment with oxytocin. *Brain Res. Dev. Brain Res.* **143,** 129–136.

DeLeon, K. R., Grimes, J. M., Connor, D. F., and Melloni, R. H., Jr. (2002a). Adolescent cocaine exposure and offensive aggression: Involvement of serotonin neural signaling and innervation in male Syrian hamsters. *Behav. Brain Res.* **133,** 211–220.

DeLeon, K. R., Grimes, J. M., and Melloni, R. H., Jr. (2002b). Repeated anabolic-androgenic steroid treatment during adolescence increases vasopressin V(1A) receptor binding in Syrian hamsters: Correlation with offensive aggression. *Horm. Behav.* **42,** 182–191.

Delville, Y., De Vries, G. J., and Ferris, C. F. (2000). Neural connections of the anterior hypothalamus and agonistic behavior in golden hamsters. *Brain Behav. Evol.* **55,** 53–76.

Demas, G. E., Moffatt, C. A., Drazen, D. L., and Nelson, R. J. (1999). Castration does not inhibit aggressive behavior in adult male prairie voles (Microtus ochrogaster). *Physiol. Behav.* **66,** 59–62.

DeVries, A. C., DeVries, M. B., Taymans, S., and Carter, C. S. (1995). Modulation of pair bonding in female prairie voles (Microtus ochrogaster) by corticosterone. *Proc. Natl. Acad. Sci. USA* **92,** 7744–7748.

Dewsbury, D. A. (1987). The comparative psychology of monogamy. *Nebr. Symp. Motiv.* **35,** 1–50.

Deyo, R. A., Straube, K. T., and Disterhoft, J. F. (1989). Nimodipine facilitates associative learning in aging rabbits. *Science* **243,** 809–811.

Donaldson, Z. R., and Young, L. J. (2008). Oxytocin, vasopressin, and the neurogenetics of sociality. *Science* **322,** 900–904.

El-Bassel, N., Witte, S. S., Wada, T., Gilbert, L., and Wallace, J. (2001). Correlates of partner violence among female street-based sex workers: Substance abuse, history of childhood abuse, and HIV risks. *AIDS Patient Care STDS* **15,** 41–51.

Farooque, R. S., Stout, R. G., and Ernst, F. A. (2005). Heterosexual intimate partner homicide: Review of ten years of clinical experience. *J. Forensic Sci.* **50,** 648–651.

Ferris, C. F., and Potegal, M. (1988). Vasopressin receptor blockade in the anterior hypothalamus suppresses aggression in hamsters. *Physiol. Behav.* **44,** 235–239.

Ferris, C. F., Albers, H. E., Wesolowski, S. M., Goldman, B. D., and Luman, S. E. (1984). Vasopressin injected into the hypothalamus triggers a stereotypic behavior in golden hamsters. *Science* **224,** 521–523.

Ferris, C. F., Axelson, J. F., Martin, A. M., and Roberge, L. F. (1989). Vasopressin immunoreactivity in the anterior hypothalamus is altered during the establishment of dominant/subordinate relationships between hamsters. *Neuroscience* **29,** 675–683.

Ferris, C. F., Melloni, R. H., Jr., Koppel, G., Perry, K. W., Fuller, R. W., and Delville, Y. (1997). Vasopressin/serotonin interactions in the anterior hypothalamus control aggressive behavior in golden hamsters. *J. Neurosci.* **17,** 4331–4340.

Ferris, C. F., Lu, S. F., Messenger, T., Guillon, C. D., Heindel, N., Miller, M., Koppel, G., Robert Bruns, F., and Simon, N. G. (2006). Orally active vasopressin V1a receptor antagonist, SRX251, selectively blocks aggressive behavior. *Pharmacol. Biochem. Behav.* **83,** 169–174.

Gatewood, J. D., Wills, A., Shetty, S., Xu, J., Arnold, A. P., Burgoyne, P. S., and Rissman, E. F. (2006). Sex chromosome complement and gonadal sex influence aggressive and parental behaviors in mice. *J. Neurosci.* **26,** 2335–2342.

Gingrich, B. S., Huot, R. L., Wang, Z., and Insel, T. R. (1997). Differential Fos expression following microinjection of oxytocin or vasopressin in the prairie vole brain. *Ann. N. Y. Acad. Sci.* **807,** 504–505.

Gingrich, B., Liu, Y., Cascio, C., Wang, Z., and Insel, T. R. (2000). Dopamine D2 receptors in the nucleus accumbens are important for social attachment in female prairie voles (Microtus ochrogaster). *Behav. Neurosci.* **114,** 173–183.

Gobrogge, K. L., and Wang, Z. (2009). Neural Circuitry Underlying Pair Bonding-induced Aggression in Monogamous Male Prairie Voles. Society for Neuroscience Abstracts #377.4 Chicago, IL.

Gobrogge, K. L., Liu, Y., Jia, X., and Wang, Z. (2007). Anterior hypothalamic neural activation and neurochemical associations with aggression in pair-bonded male prairie voles. *J. Comp. Neurol.* **502,** 1109–1122.

Gobrogge, K. L., Liu, Y., and Wang, Z. (2008). Dopamine regulation of pair bonding in monogamous prairie voles. *In* "The Neurobiology of the Parental Brain" (R. Bridges, ed.), pp. 345–358. Elsevier, San Diego, CA.

Gobrogge, K. L., Liu, Y., Young, L. J., and Wang, Z. (2009). Anterior hypothalamic vasopressin regulates pair-bonding and drug-induced aggression in a monogamous rodent. *Proc. Natl. Acad. Sci. USA* **106,** 19144–19149.

Goodson, J. L. (2008). Nonapeptides and the evolutionary patterning of sociality. *Prog. Brain Res.* **170**, 3–15.

Grimes, J. M., Ricci, L. A., and Melloni, R. H., Jr. (2007). Alterations in anterior hypothalamic vasopressin, but not serotonin, correlate with the temporal onset of aggressive behavior during adolescent anabolic-androgenic steroid exposure in hamsters (Mesocricetus auratus). *Behav. Neurosci.* **121**, 941–948.

Grippo, A. J., Cushing, B. S., and Carter, C. S. (2007). Depression-like behavior and stressor-induced neuroendocrine activation in female prairie voles exposed to chronic social isolation. *Psychosom. Med.* **69**, 149–157.

Haapasalo, J., and Hamalainen, T. (1996). Childhood family problems and current psychiatric problems among young violent and property offenders. *J. Am. Acad. Child Adolesc. Psychiatry* **35**, 1394–1401.

Hagelstam, C., and Hakkanen, H. (2006). Adolescent homicides in Finland: Offence and offender characteristics. *Forensic Sci. Int.* **164**, 110–115.

Hairston, J. E., Ball, G. F., and Nelson, R. J. (2003). Photoperiodic and temporal influences on chemosensory induction of brain fos expression in female prairie voles. *J. Neuroendocrinol.* **15**, 161–172.

Haller, J., Abraham, I., Zelena, D., Juhasz, G., Makara, G. B., and Kruk, M. R. (1998). Aggressive experience affects the sensitivity of neurons towards pharmacological treatment in the hypothalamic attack area. *Behav. Pharmacol.* **9**, 469–475.

Hammock, E. A., and Young, L. J. (2002). Variation in the vasopressin V1a receptor promoter and expression: Implications for inter- and intraspecific variation in social behaviour. *Eur. J. Neurosci.* **16**, 399–402.

Hammock, E. A., and Young, L. J. (2004). Functional microsatellite polymorphism associated with divergent social structure in vole species. *Mol. Biol. Evol.* **21**, 1057–1063.

Hammock, E. A., and Young, L. J. (2005). Microsatellite instability generates diversity in brain and sociobehavioral traits. *Science* **308**, 1630–1634.

Harrison, R. J., Connor, D. F., Nowak, C., and Melloni, R. H., Jr. (2000a). Chronic low-dose cocaine treatment during adolescence facilitates aggression in hamsters. *Physiol. Behav.* **69**, 555–562.

Harrison, R. J., Connor, D. F., Nowak, C., Nash, K., and Melloni, R. H., Jr. (2000b). Chronic anabolic-androgenic steroid treatment during adolescence increases anterior hypothalamic vasopressin and aggression in intact hamsters. *Psychoneuroendocrinology* **25**, 317–338.

Hostetler, C. M., Harkey, S. L., and Bales, K. L. (2010). D2 antagonist during development decreases anxiety and infanticidal behavior in adult female prairie voles (Microtus ochrogaster). *Behav. Brain Res.* **210**, 127–130.

Insel, T. R. (2010). The challenge of translation in social neuroscience: A review of oxytocin, vasopressin, and affiliative behavior. *Neuron* **65**, 768–779.

Insel, T. R., and Hulihan, T. J. (1995). A gender-specific mechanism for pair bonding: Oxytocin and partner preference formation in monogamous voles. *Behav. Neurosci.* **109**, 782–789.

Insel, T. R., and Shapiro, L. E. (1992). Oxytocin receptor distribution reflects social organization in monogamous and polygamous voles. *Proc. Natl. Acad. Sci. USA* **89**, 5981–5985.

Insel, T. R., Wang, Z., and Ferris, C. F. (1994). Patterns of brain vasopressin receptor distribution associated with social organization in microtine rodents. *J. Neurosci.* **14**, 5381–5392.

Insel, T. R., Preston, S., and Winslow, J. T. (1995). Mating in the monogamous male: Behavioral consequences. *Physiol. Behav.* **57**, 615–627.

Insel, T. R., Winslow, J. T., Wang, Z., and Young, L. J. (1998). Oxytocin, vasopressin, and the neuroendocrine basis of pair bond formation. *Adv. Exp. Med. Biol.* **449**, 215–224.

Israel, S., Lerer, E., Shalev, I., Uzefovsky, F., Reibold, M., Bachner-Melman, R., Granot, R., Bornstein, G., Knafo, A., Yirmiya, N., and Ebstein, R. P. (2008). Molecular genetic studies of the arginine vasopressin 1a receptor (AVPR1a) and the oxytocin receptor (OXTR) in human behaviour: From autism to altruism with some notes in between. *Prog. Brain Res.* **170**, 435–449.

Jackson, D., Burns, R., Trksak, G., Simeone, B., DeLeon, K. R., Connor, D. F., Harrison, R. J., and Melloni, R. H., Jr. (2005). Anterior hypothalamic vasopressin modulates the aggression-stimulating effects of adolescent cocaine exposure in Syrian hamsters. *Neuroscience* **133**, 635–646.

Jannett, F. (1982). Nesting patterns of adult voles, Microtus montanus, in field populations. *J. Mammal.* **63**, 495–498.

Johnson, A. E., Audigier, S., Rossi, F., Jard, S., Tribollet, E., and Barberis, C. (1993). Localization and characterization of vasopressin binding sites in the rat brain using an iodinated linear AVP antagonist. *Brain Res.* **622**, 9–16.

Katz, L. F., Ball, G. F., and Nelson, R. J. (1999). Elevated Fos-like immunoreactivity in the brains of postpartum female prairie voles, Microtus ochrogaster. *Cell Tissue Res.* **298**, 425–435.

Kelley, A. E., and Berridge, K. C. (2002). The neuroscience of natural rewards: Relevance to addictive drugs. *J. Neurosci.* **22**, 3306–3311.

Kim, S. J., Young, L. J., Gonen, D., Veenstra-VanderWeele, J., Courchesne, R., Courchesne, E., Lord, C., Leventhal, B. L., Cook, E. H., Jr., and Insel, T. R. (2002). Transmission disequilibrium testing of arginine vasopressin receptor 1A (AVPR1A) polymorphisms in autism. *Mol. Psychiatry* **7**, 503–507.

Kirkpatrick, B., Kim, J. W., and Insel, T. R. (1994). Limbic system fos expression associated with paternal behavior. *Brain Res.* **658**, 112–118.

Knyshevski, I., Connor, D. F., Harrison, R. J., Ricci, L. A., and Melloni, R. H., Jr. (2005a). Persistent activation of select forebrain regions in aggressive, adolescent cocaine-treated hamsters. *Behav. Brain Res.* **159**, 277–286.

Knyshevski, I., Ricci, L. A., McCann, T. E., and Melloni, R. H., Jr. (2005b). Serotonin type-1A receptors modulate adolescent, cocaine-induced offensive aggression in hamsters. *Physiol. Behav.* **85**, 167–176.

Koolhaas, J. M., Van den Brink, T. H. C., Roozendaal, B., and Boorsma, F. (1990). Medial amygdala and aggressive behavior: Interaction between testosterone and vasopressin. *Aggress. Behav.* **16**, 223–229.

Kozorovitskiy, Y., Hughes, M., Lee, K., and Gould, E. (2006). Fatherhood affects dendritic spines and vasopressin V1a receptors in the primate prefrontal cortex. *Nat. Neurosci.* **9**, 1094–1095.

Kramer, K. M., Choe, C., Carter, C. S., and Cushing, B. S. (2006). Developmental effects of oxytocin on neural activation and neuropeptide release in response to social stimuli. *Horm. Behav.* **49**, 206–214.

Krug, E. G., Dahlberg, L. L., Mercy, J. A., Zwi, A. B., and Lozito, R. (2002). World Report on Violence and Health. World Health Organization, Geneva.

Kruk, M. R. (1991). Ethology and pharmacology of hypothalamic aggression in the rat. *Neurosci. Biobehav. Rev.* **15**, 527–538.

Kruk, M. R., Van der Laan, C. E., Mos, J., Van der Poel, A. M., Meelis, W., and Olivier, B. (1984). Comparison of aggressive behaviour induced by electrical stimulation in the hypothalamus of male and female rats. *Prog. Brain Res.* **61**, 303–314.

Kuptsov, P. A., Pleskacheva, M. G., Voronkova, D. N., Lipp, H. P., and Anokhin, K. V. (2005). Features of the c-Fos gene expression along the hippocampal rostro-caudal axis in common voles after rapid spatial learning. *Zh. Vyssh. Nerv. Deiat. Im. I P Pavlova* **55**, 231–240.

Landgraf, R., Frank, E., Aldag, J. M., Neumann, I. D., Sharer, C. A., Ren, X., Terwilliger, E. F., Niwa, M., Wigger, A., and Young, L. J. (2003). Viral vector-mediated gene transfer of the vole V1a vasopressin receptor in the rat septum: Improved social discrimination and active social behaviour. *Eur. J. Neurosci.* **18**, 403–411.

Lim, M. M., and Young, L. J. (2004). Vasopressin-dependent neural circuits underlying pair bond formation in the monogamous prairie vole. *Neuroscience* **125,** 35–45.

Lim, M. M., Wang, Z., Olazabal, D. E., Ren, X., Terwilliger, E. F., and Young, L. J. (2004). Enhanced partner preference in a promiscuous species by manipulating the expression of a single gene. *Nature* **429,** 754–757.

Lin, D., Boyle, M. P., Dollar, P., Lee, H., Lein, E. S., Perona, P., and Anderson, D. J. (2011). Functional identification of an aggression locus in the mouse hypothalamus. *Nature* **470,** 221–226.

Lindvall, O., and Stenevi, U. (1978). Dopamine and noradrenaline neurons projecting to the septal area in the rat. *Cell Tissue Res.* **190,** 383–407.

Liu, Y., and Wang, Z. (2003). Nucleus accumbens oxytocin and dopamine interact to regulate pair bond formation in female prairie voles. *Neuroscience* **121,** 537–544.

Liu, Y., Curtis, J. T., and Wang, Z. (2001). Vasopressin in the lateral septum regulates pair bond formation in male prairie voles (Microtus ochrogaster). *Behav. Neurosci.* **115,** 910–919.

Liu, Y., Aragona, B. J., Young, K. A., Dietz, D. M., Kabbaj, M., Mazei-Robison, M., Nestler, E. J., and Wang, Z. (2010). Nucleus accumbens dopamine mediates amphetamine-induced impairment of social bonding in a monogamous rodent species. *Proc. Natl. Acad. Sci. USA* **107,** 1217–1222.

Madan, A., Beech, D. J., and Flint, L. (2001). Drugs, guns, and kids: The association between substance use and injury caused by interpersonal violence. *J. Pediatr. Surg.* **36,** 440–442.

Maeda, H., and Mogenson, G. J. (1980). An electrophysiological study of inputs to neurons of the ventral tegmental area from the nucleus accumbens and medial preoptic-anterior hypothalamic areas. *Brain Res.* **197,** 365–377.

McGuire, B., and Novak, M. (1984). A comparison of maternal behavior in the meadow vole (Microtus pennsylvanicus), prairie vole (M. ochrogaster) and pine vole (M. pinetorum). *Anim. Behav.* **32,** 1132–1141.

McGuire, B., and Novak, M. (1986). Parental care and its relation to social organization in the montane vole. *J. Mammal.* **67,** 305–311.

Melloni, R. H., Jr., and Ferris, C. F. (1996). Adolescent anabolic steroid use and aggressive behavior in golden hamsters. *Ann. N. Y. Acad. Sci.* **794,** 372–375.

Melloni, R. H., Jr., Connor, D. F., Hang, P. T., Harrison, R. J., and Ferris, C. F. (1997). Anabolic-androgenic steroid exposure during adolescence and aggressive behavior in golden hamsters. *Physiol. Behav.* **61,** 359–364.

Melloni, R. H., Jr., Connor, D. F., Todtenkopf, M. S., DeLeon, K. R., Sanyal, P., and Harrison, R. J. (2001). Repeated cocaine treatment activates flank marking in adolescent female hamsters. *Physiol. Behav.* **73,** 561–570.

Meyer-Lindenberg, A., Kolachana, B., Gold, B., Olsh, A., Nicodemus, K. K., Mattay, V., Dean, M., and Weinberger, D. R. (2008). Genetic variants in AVPR1A linked to autism predict amygdala activation and personality traits in healthy humans. *Mol. Psychiatry* **10,** 968–975.

Michard-Vanhee, C. (1988). Aggressive behavior induced in female mice by an early single injection of testosterone is genotype dependent. *Behav. Genet.* **18,** 1–12.

Michell, R. H., Kirk, C. J., and Billah, M. M. (1979). Hormonal stimulation of phosphatidylinositol breakdown with particular reference to the hepatic effects of vasopressin. *Biochem. Soc. Trans.* **7,** 861–865.

Miczek, K. A., de Almeida, R. M., Kravitz, E. A., Rissman, E. F., de Boer, S. F., and Raine, A. (2007). Neurobiology of escalated aggression and violence. *J. Neurosci.* **27,** 11803–11806.

Mokuau, N. (2002). Culturally based interventions for substance use and child abuse among native Hawaiians. *Public Health Rep.* **117**(Suppl. 1), S82–S87.

Moore, T. M., Stuart, G. L., Meehan, J. C., Rhatigan, D. L., Hellmuth, J. C., and Keen, S. M. (2008). Drug abuse and aggression between intimate partners: A meta-analytic review. *Clin. Psychol. Rev.* **28,** 247–274.

Motta, S. C., Goto, M., Gouveia, F. V., Baldo, M. V., Canteras, N. S., and Swanson, L. W. (2009). Dissecting the brain's fear system reveals the hypothalamus is critical for responding in subordinate conspecific intruders. *Proc. Natl. Acad. Sci. USA* **106**, 4870–4875.

Murphy, T. H., Worley, P. F., and Baraban, J. M. (1991). L-type voltage-sensitive calcium channels mediate synaptic activation of immediate early genes. *Neuron* **7**, 625–635.

Nesse, R. M., and Berridge, K. C. (1997). Psychoactive drug use in evolutionary perspective. *Science* **278**, 63–66.

Nestler, E. J., and Aghajanian, G. K. (1997). Molecular and cellular basis of addiction. *Science* **278**, 58–63.

Neve, K. A., Seamans, J. K., and Trantham-Davidson, H. (2004). Dopamine receptor signaling. *J. Recept. Signal Transduct. Res.* **24**, 165–205.

Newman, S. W. (1999). The medial extended amygdala in male reproductive behavior. A node in the mammalian social behavior network. *Ann. N. Y. Acad. Sci.* **877**, 242–257.

Northcutt, K. V., and Lonstein, J. S. (2009). Social contact elicits immediate-early gene expression in dopaminergic cells of the male prairie vole extended olfactory amygdala. *Neuroscience* **163**, 9–22.

Northcutt, K. V., Wang, Z., and Lonstein, J. S. (2007). Sex and species differences in tyrosine hydroxylase-synthesizing cells of the rodent olfactory extended amygdala. *J. Comp. Neurol.* **500**, 103–115.

O'Farrell, T. J., and Fals-Stewart, W. (2000). Behavioral couples therapy for alcoholism and drug abuse. *J. Subst. Abuse Treat.* **18**, 51–54.

Oliveras, D., and Novak, M. (1986). A comparison of paternal behavior in the meadow vole, Microtus pennsylvanicus, the pine vole, Microtus pinetorum, and prairie vole, Microtus ochrogaster. *Anim. Behav.* **34**, 519–526.

Ophir, A. G., Wolff, J. O., and Phelps, S. M. (2008). Variation in neural V1aR predicts sexual fidelity and space use among male prairie voles in semi-natural settings. *Proc. Natl. Acad. Sci. USA* **105**, 1249–1254.

Panksepp, J., Knutson, B., and Burgdorf, J. (2002). The role of brain emotional systems in addictions: A neuro-evolutionary perspective and new 'self-report' animal model. *Addiction* **97**, 459–469.

Pitkow, L. J., Sharer, C. A., Ren, X., Insel, T. R., Terwilliger, E. F., and Young, L. J. (2001). Facilitation of affiliation and pair-bond formation by vasopressin receptor gene transfer into the ventral forebrain of a monogamous vole. *J. Neurosci.* **21**, 7392–7396.

Prichard, Z. M., Mackinnon, A. J., Jorm, A. F., and Easteal, S. (2007). AVPR1A and OXTR polymorphisms are associated with sexual and reproductive behavioral phenotypes in humans. *Hum. Mutat.* **28**, 1150. Mutation in brief no. 981. Online.

Raggenbass, M., Goumaz, M., Sermasi, E., Tribollet, E., and Dreifuss, J. J. (1991). Vasopressin generates a persistent voltage-dependent sodium current in a mammalian motoneuron. *J. Neurosci.* **11**, 1609–1616.

Ramamurthi, B. (1988). Stereotactic operation in behaviour disorders. Amygdalotomy and hypothalamotomy. *Acta Neurochir. Suppl.* **44**, 152–157.

Ricci, L. A., Schwartzer, J. J., and Melloni, R. H., Jr. (2009). Alterations in the anterior hypothalamic dopamine system in aggressive adolescent AAS-treated hamsters. *Horm. Behav.* **55**, 348–355.

Riters, L. V., and Panksepp, J. (1997). Effects of vasotocin on aggressive behavior in male Japanese quail. *Ann. N. Y. Acad. Sci.* **807**, 478–480.

Roeling, T. A., Kruk, M. R., Schuurmans, R., and Veening, J. G. (1993). Behavioural responses of bicuculline methiodide injections into the ventral hypothalamus of freely moving, socially interacting rats. *Brain Res.* **615**, 121–127.

Ryding, E., Lindstrom, M., and Traskman-Bendz, L. (2008). The role of dopamine and serotonin in suicidal behaviour and aggression. *Prog. Brain Res.* **172**, 307–315.

Sano, K., Yoshioka, M., Ogashiwa, M., Ishijima, B., and Ohye, C. (1966). Postero-medial hypothalamotomy in the treatment of aggressive behaviors. *Confin. Neurol.* **27**, 164–167.

Schwartzer, J. J., and Melloni, R. H., Jr. (2010a). Anterior hypothalamic dopamine D2 receptors modulate adolescent anabolic/androgenic steroid-induced offensive aggression in the Syrian hamster. *Behav. Pharmacol.* **21**, 314–322.

Schwartzer, J. J., and Melloni, R. H., Jr. (2010b). Dopamine activity in the lateral anterior hypothalamus modulates AAS-induced aggression through D2 but not D5 receptors. *Behav. Neurosci.* **124**, 645–655.

Schwartzer, J. J., Ricci, L. A., and Melloni, R. H., Jr. (2009). Interactions between the dopaminergic and GABAergic neural systems in the lateral anterior hypothalamus of aggressive AAS-treated hamsters. *Behav. Brain Res.* **203**, 15–22.

Siegel, A., Roeling, T. A., Gregg, T. R., and Kruk, M. R. (1999). Neuropharmacology of brain-stimulation-evoked aggression. *Neurosci. Biobehav. Rev.* **23**, 359–389.

Siever, L. J. (2008). Neurobiology of aggression and violence. *Am. J. Psychiatry* **165**, 429–442.

Smeltzer, M. D., Curtis, J. T., Aragona, B. J., and Wang, Z. (2006). Dopamine, oxytocin, and vasopressin receptor binding in the medial prefrontal cortex of monogamous and promiscuous voles. *Neurosci. Lett.* **394**, 146–151.

Son, M. C., and Brinton, R. D. (2001). Regulation and mechanism of L-type calcium channel activation via V1a vasopressin receptor activation in cultured cortical neurons. *Neurobiol. Learn. Mem.* **76**, 388–402.

Spunt, B., Brownstein, H. H., Crimmins, S. M., Langley, S., and Spanjol, K. (1998). Alcohol-related homicides committed by women. *J. Psychoactive Drugs* **30**, 33–43.

Stowe, J. R., Liu, Y., Curtis, J. T., Freeman, M. E., and Wang, Z. (2005). Species differences in anxiety-related responses in male prairie and meadow voles: The effects of social isolation. *Physiol. Behav.* **86**, 369–378.

Stribley, J. M., and Carter, C. S. (1999). Developmental exposure to vasopressin increases aggression in adult prairie voles. *Proc. Natl. Acad. Sci. USA* **96**, 12601–12604.

Swanson, L. W. (1982). The projections of the ventral tegmental area and adjacent regions: A combined fluorescent retrograde tracer and immunofluorescence study in the rat. *Brain Res. Bull.* **9**, 321–353.

Swartz, M. S., Swanson, J. W., Hiday, V. A., Borum, R., Wagner, H. R., and Burns, B. J. (1998). Violence and severe mental illness: The effects of substance abuse and nonadherence to medication. *Am. J. Psychiatry* **155**, 226–231.

Tamarin, R. (1985). Biology of new world Microtus. *Am. Soc. Mamm. Spec. Pub.* 8. Shippensburg, PA.

Thibonnier, M. (1992). Signal transduction of V1-vascular vasopressin receptors. *Regul. Pept.* **38**, 1–11.

Thibonnier, M., Bayer, A. L., Simonson, M. S., and Douglas, J. G. (1992). Effects of amiloride analogues on AVP binding and activation of V1-receptor-expressing cells. *Am. J. Physiol.* **262**, E76–E86.

Thibonnier, M., Goraya, T., and Berti-Mattera, L. (1993). G protein coupling of human platelet V1 vascular vasopressin receptors. *Am. J. Physiol.* **264**, C1336–C1344.

Thibonnier, M., Auzan, C., Madhun, Z., Wilkins, P., Berti-Mattera, L., and Clauser, E. (1994). Molecular cloning, sequencing, and functional expression of a cDNA encoding the human V1a vasopressin receptor. *J. Biol. Chem.* **269**, 3304–3310.

Thompson, R., Gupta, S., Miller, K., Mills, S., and Orr, S. (2004). The effects of vasopressin on human facial responses related to social communication. *Psychoneuroendocrinology* **29**, 35–48.

Tidey, J. W., and Miczek, K. A. (1992). Morphine withdrawal aggression: Modification with D1 and D2 receptor agonists. *Psychopharmacology (Berl.)* **108**, 177–184.

Tubbiola, M. L., and Wysocki, C. J. (1997). FOS immunoreactivity after exposure to conspecific or heterospecific urine: Where are chemosensory cues sorted? *Physiol. Behav.* **62,** 867–870.

Vale, J. R., Ray, D., and Vale, C. A. (1972). The interaction of genotype and exogenous neonatal androgen: Agonistic behavior in female mice. *Behav. Biol.* **7,** 321–334.

Veenema, A. H., Blume, A., Niederle, D., Buwalda, B., and Neumann, I. D. (2006). Effects of early life stress on adult male aggression and hypothalamic vasopressin and serotonin. *Eur. J. Neurosci.* **24,** 1711–1720.

Veenema, A. H., Beiderbeck, D. I., Lukas, M., and Neumann, I. D. (2010). Distinct correlations of vasopressin release within the lateral septum and the bed nucleus of the stria terminalis with the display of intermale aggression. *Horm. Behav.* **58,** 273–281.

Veening, J. G., Coolen, L. M., de Jong, T. R., Joosten, H. W., de Boer, S. F., Koolhaas, J. M., and Olivier, B. (2005). Do similar neural systems subserve aggressive and sexual behaviour in male rats? Insights from c-Fos and pharmacological studies. *Eur. J. Pharmacol.* **526,** 226–239.

Villalba, C., Boyle, P. A., Caliguri, E. J., and De Vries, G. J. (1997). Effects of the selective serotonin reuptake inhibitor fluoxetine on social behaviors in male and female prairie voles (Microtus ochrogaster). *Horm. Behav.* **32,** 184–191.

Walsh, C., MacMillan, H. L., and Jamieson, E. (2003). The relationship between parental substance abuse and child maltreatment: Findings from the Ontario Health Supplement. *Child Abuse Negl.* **27,** 1409–1425.

Walum, H., Westberg, L., Henningsson, S., Neiderhiser, J. M., Reiss, D., Igl, W., Ganiban, J. M., Spotts, E. L., Pedersen, N. L., Eriksson, E., and Lichtenstein, P. (2008). Genetic variation in the vasopressin receptor 1a gene (AVPR1A) associates with pair-bonding behavior in humans. *Proc. Natl. Acad. Sci. USA* **105,** 14153–14156.

Wang, Z. (1995). Species differences in the vasopressin-immunoreactive pathways in the bed nucleus of the stria terminalis and medial amygdaloid nucleus in prairie voles (Microtus ochrogaster) and meadow voles (Microtus pennsylvanicus). *Behav. Neurosci.* **109,** 305–311.

Wang, Z., and Aragona, B. J. (2004). Neurochemical regulation of pair bonding in male prairie voles. *Physiol. Behav.* **83,** 319–328.

Wang, Z., Smith, W., Major, D. E., and De Vries, G. J. (1994). Sex and species differences in the effects of cohabitation on vasopressin messenger RNA expression in the bed nucleus of the stria terminalis in prairie voles (Microtus ochrogaster) and meadow voles (Microtus pennsylvanicus). *Brain Res.* **650,** 212–218.

Wang, Z., Zhou, L., Hulihan, T. J., and Insel, T. R. (1996). Immunoreactivity of central vasopressin and oxytocin pathways in microtine rodents: A quantitative comparative study. *J. Comp. Neurol.* **366,** 726–737.

Wang, Z., Hulihan, T. J., and Insel, T. R. (1997a). Sexual and social experience is associated with different patterns of behavior and neural activation in male prairie voles. *Brain Res.* **767,** 321–332.

Wang, Z., Liu, Y., Young, L. J., and Insel, T. R. (1997b). Developmental changes in forebrain vasopressin receptor binding in prairie voles (Microtus ochrogaster) and montane voles (Microtus montanus). *Ann. N. Y. Acad. Sci.* **807,** 510–513.

Wang, Z., Young, L. J., Liu, Y., and Insel, T. R. (1997c). Species differences in vasopressin receptor binding are evident early in development: Comparative anatomic studies in prairie and montane voles. *J. Comp. Neurol.* **378,** 535–546.

Wang, Z., Young, L. J., De Vries, G. J., and Insel, T. R. (1998). Voles and vasopressin: A review of molecular, cellular, and behavioral studies of pair bonding and paternal behaviors. *Prog. Brain Res.* **119,** 483–499.

Wang, Z., Yu, G., Cascio, C., Liu, Y., Gingrich, B., and Insel, T. R. (1999). Dopamine D2 receptor-mediated regulation of partner preferences in female prairie voles (Microtus ochrogaster): A mechanism for pair bonding? *Behav. Neurosci.* **113,** 602–611.

White, F. J., and Kalivas, P. W. (1998). Neuroadaptations involved in amphetamine and cocaine addiction. *Drug Alcohol Depend.* **51,** 141–153.

Williams, J. R., Carter, C. S., and Insel, T. (1992). Partner preference development in female prairie voles is facilitated by mating or the central infusion of oxytocin. *Ann. N. Y. Acad. Sci.* **652,** 487–489.

Williams, J. R., Insel, T. R., Harbaugh, C. R., and Carter, C. S. (1994). Oxytocin administered centrally facilitates formation of a partner preference in female prairie voles (Microtus ochrogaster). *J. Neuroendocrinol.* **6,** 247–250.

Winslow, J. T., Hastings, N., Carter, C. S., Harbaugh, C. R., and Insel, T. R. (1993). A role for central vasopressin in pair bonding in monogamous prairie voles. *Nature* **365,** 545–548.

Wise, R. A. (2002). Brain reward circuitry: Insights from unsensed incentives. *Neuron* **36,** 229–240.

Yamada, S., Uchida, S., Ohkura, T., Kimura, R., Yamaguchi, M., Suzuki, M., and Yamamoto, M. (1996). Alterations in calcium antagonist receptors and calcium content in senescent brain and attenuation by nimodipine and nicardipine. *J. Pharmacol. Exp. Ther.* **277,** 721–727.

Young, L. J. (1999). Frank A. Beach Award. Oxytocin and vasopressin receptors and species-typical social behaviors. *Horm. Behav.* **36,** 212–221.

Young, L. J., and Wang, Z. (2004). The neurobiology of pair bonding. *Nat. Neurosci.* **7,** 1048–1054.

Young, L. J., Winslow, J. T., Nilsen, R., and Insel, T. R. (1997). Species differences in V1a receptor gene expression in monogamous and nonmonogamous voles: Behavioral consequences. *Behav. Neurosci.* **111,** 599–605.

Young, L. J., Wang, Z., and Insel, T. R. (1998). Neuroendocrine bases of monogamy. *Trends Neurosci.* **21,** 71–75.

Young, L. J., Nilsen, R., Waymire, K. G., MacGregor, G. R., and Insel, T. R. (1999). Increased affiliative response to vasopressin in mice expressing the V1a receptor from a monogamous vole. *Nature* **400,** 766–768.

Young, K. A., Liu, Y., and Wang, Z. (2008). The neurobiology of social attachment: A comparative approach to behavioral, neuroanatomical, and neurochemical studies. *Comp. Biochem. Physiol. C Toxicol. Pharmacol.* **148,** 401–410.

Young, K. A., Gobrogge, K. L., Liu, Y., and Wang, Z. (2011a). The neurobiology of pair bonding: Insights from a socially monogamous rodent. *Front. Neuroendocrinol.* **32,** 53–69.

Young, K. A., Gobrogge, K. L., and Wang, Z. (2011b). The role of mesocorticolimbic dopamine in regulating interactions between drugs of abuse and social behavior. *Neurosci. Biobehav. Rev.* **35,** 498–515.

7

The Neurochemistry of Human Aggression

Rachel Yanowitch and Emil F. Coccaro
Clinical Neuroscience Research Unit, Department of Psychiatry, The University of Chicago Pritzker School of Medicine, Chicago, Illinois, USA

ABSTRACT

Various data from scientific research studies conducted over the past three decades suggest that central neurotransmitters play a key role in the modulation of aggression in all mammalian species, including humans. Specific neurotransmitter systems involved in mammalian aggression include serotonin, dopamine, norepinephrine, GABA, and neuropeptides such as vasopressin and oxytocin. Neurotransmitters not only help to execute basic behavioral components but also serve to modulate these preexisting behavioral states by amplifying or reducing their effects. This chapter reviews the currently available data to present a contemporary view of how central neurotransmitters influence the vulnerability for aggressive behavior and/or initiation of aggressive behavior in social situations. Data reviewed in this chapter include emoiric information from neurochemical, pharmaco-challenge, molecular genetic and neuroimaging studies.   © 2011, Elsevier Inc.

Advances in Genetics, Vol. 75
Copyright 2011, Elsevier Inc. All rights reserved.

0065-2660/11 $35.00
DOI: 10.1016/B978-0-12-380858-5.00005-8

I. INTRODUCTION

Since the late 1970s, data from scientific research studies have suggested that endogenous brain chemicals called neurotransmitters play a key role in the modulation of aggression. Human aggression is a multidimensional behavior that is determined by an amalgamation of biological, genetic, environmental, and psychological factors. Neurotransmitters not only help to execute these basic behavioral components but also serve to modulate these preexisting behavioral states by amplifying or reducing their effects. Genetic abnormalities in a number of neurotransmitter pathways have been implicated in aggression-related disorders. Current and future research aims to understand how these neurotransmitters function both normally and abnormally to mediate aggression and other human behaviors. With the evolution of genetic testing and continued development of neuroimaging technologies such as functional magnetic resonance imaging (fMRI) and positron emission tomography (PET) scanning, the ability of the scientific researcher to investigate the brain's constellation of synapses and neurotransmitters is growing ever more proficient. While it is clear from these studies that neurotransmitters contribute significantly to the predisposition of an individual toward aggressiveness, whether neurotransmitter dysfunction alone is sufficient to cause violent aggression remains unclear.

Aggression may be impulsive or premeditated in nature. In the former case, impulsivity defines, or describes, the aggression. That is, it is the aggression that is impulsive not that the person is aggressive and at other times impulsive, though that may be true as well. Diagnoses associated with impulsive aggression include Intermittent Explosive Disorder (IED) characterized by frequent and problematic impulsive aggressive outbursts, and Borderline Personality Disorder (BPD) characterized by instability in self-image, in interpersonal relationships as well as impulsivity and affect (including anger and aggression). In the latter case, the aggression is planned and carried out in order to achieve some tangible goal. Diagnoses associated with this type of aggression include Antisocial Personality Disorder (AsPD) which is characterized by a pattern of disregard for, and violation of, the rights of others. These types of aggression are not mutually exclusive, however, and some individuals display both types of aggression at different times.

II. SEROTONIN

Serotonin or 5-hydroxytryptamine (5-HT) is a multipurpose monoamine neurotransmitter derived from the amino acid, L-tryptophan, and has been implicated as an important regulator of mood (Kumar et al., 2010; Kunisato et al., 2010; Ruhé et al., 2007), appetite (Curzon, 1991; Dourish, 1995; Lam et al., 2010), gastrointestinal muscle contractility (Gershon, 2004; Xu et al., 2007), self-injurious behavior (Peddeer, 1992), and sleep (Monti, 2010; Monti and Jantos, 2008; Monti and Monti, 2000). With respect to aggression and other behavioral disorders, serotonin action is highly complex and varies depending upon which receptor it is bound to, how much 5-HT is available in the synapse, how much enzymatic activity is present, and whether other agonists or antagonists are available for competitive binding. Both the clinical and molecular data on central 5-HT function in the mammalian brain overwhelmingly suggests that a reduction in 5-HT activity in emotion-modulating brain regions such as the prefrontal cortex and the anterior cingulate cortex leads to a predisposition for impulsive aggressiveness (New et al., 2002; Parsey et al., 2002; Seo et al., 2008; Siever et al., 1999). Research suggests that 5-HT significantly contributes to the genetically determined differences seen in individuals and that its primary mechanism of action is via the genes encoding major components of 5-HT viability in the brain such as the enzymes tryptophan hydroxylase-1 and mono-amine oxidase A (MAOA) (Popova, 2008). Most of the literature discussing 5-HT regulation of aggression focuses on 5-HT metabolite levels and the functional state of 5-HT receptors.

The first literature written on serotonin and impulsive aggressive behavior in human subjects came from two independent research groups (Asberg et al., 1976; Sheard et al., 1976) in 1976. Sheard et al. reported that administration of the putative 5-HT-enhancing agent, lithium carbonate, significantly reduced impulsive aggressive behavior in a prison inmate population. Asberg et al. found that lower concentrations of lumbar CSF 5-hydroxyindoleactic acid (5-HIAA), the most stable 5-HT metabolite in the brain, were correlated with violent and suicidal behavior. In 1979, Brown et al. studied 26 males with significant personality disorder traits (Brown et al., 1979). CSF amine metabolite levels of serotonin (5-HIAA), norepinephrine (3-methoxy-4-hydroxy-phenyl-glycol, MHPG), and dopamine (homovanillic acid, HVA), respectively, were studied. CSF 5-HIAA was significantly negatively correlated with aggression ($r = -0.78$), and MHPG was significantly positively correlated with aggression ($r = 0.64$). Brown et al. replicated this finding for CSF 5-HIAA and extended these findings to include other measures of aggression such as "psychopathic deviance" (i.e., defiance of authority and impulsivity) (Brown et al., 1982; Coccaro and Siever, 2002). Despite the fact that a number of subsequent studies supported the findings of an inverse relationship between aggression and CSF 5-HIAA levels (Kruesi et al., 1990; Lidberg et al., 1985; Limson et al., 1991; Linnoila et al., 1983), additional reports also suggest a direct (Castellanos et al., 1994; Moller et al., 1996; Prochazka and Agren, 2003) or no relationship (Coccaro et al., 1997c; Gardner et al., 1990) between the two. In a recent paper by Coccaro et al., the authors were able to reconcile the disputed data by reconsidering the CSF 5-HIAA levels in the context of (1) the severity of the aggression of the individual and (2) the CSF HVA levels present concomitantly

(Coccaro and Lee, 2010). Under this new paradigm, the results emerged against Brown's preliminary findings: CSF 5-HIAA concentrations varied directly with aggression and CSF HVA concentrations varied inversely. In this model, a deficiency hypothesis of 5-HT for aggressiveness is only fulfilled if presynaptic release of 5-HT is being reduced and there is compensation of postsynaptic 5-HT receptor function (Coccaro, 1998).

Evidence for a model in which postsynaptic 5-HT receptor function is altered by presynaptic reduction begins with Stanley *et al.*'s (1982) report demonstrating a reduced number of presynaptic 5-HT transporter sites in aggressive suicide victims as compared with accident victims (Stanley *et al.*, 1982). The following year, Stanley and Mann published additional results showing increased postsynaptic 5-HT2A receptor sites in suicide subjects (Stanley and Mann, 1983), suggesting that, in addition to modified responsiveness, there may be a change in receptor number as well. In response to these data, researchers designed psychopharmacologic challenge studies in order to further assess pre- and postsynaptic function in premortem subjects (Coccaro, 1998). These pharmacochallenge studies involve the activation of a specific neurotransmitter system through the administration and consequent ligand–receptor interaction of a pharmacologic agent. Subsequent signaling cascades result in physiological events that trigger homeostatic, behavioral, and hormonal alterations that can be measured as an index of the responsiveness of the neurotransmitter system in question (Coccaro and Kavoussi, 1994).

The first report of a correlation between aggression and pharmacochallenge studies were published by Coccaro *et al.*, 1989. In this study, prolactin responses to 60 mg of oral D,L-fenfluramine of 45 males with major affective ($n = 25$) and/or personality ($n = 20$) disorder were compared to those of 18 healthy male controls. D,L-fenfluramine was chosen as a challenge probe because of its properties as a serotonin-releasing agent. Its mechanism of action is the release of serotonin by disrupting vesicular storage of the neurotransmitter and reversing serotonin transporter function (Welch and Lim, 2007). Since prolactin secretion is directly dependent upon 5-HT transmission, recording prolactin levels can provide an indirect but effective measurement of 5-HT activity (Coccaro *et al.*, 1998a). Both groups of subjects demonstrated reduced prolactin responses to D,L-fenfluramine compared to controls. However, significant correlations appeared between reduced prolactin responses to D,L-fenfluramine and history of suicide attempts in all experimental subjects and impulsive aggression in males with personality disorder (Coccaro *et al.*, 1989). These results suggest that altered 5-HT activity, specifically reduced receptor function, is apparent in subjects with aggression-related disorders. In a later study by Coccaro *et al.*, the relationship between life history of aggression and prolactin response to D-fenfluramine and to CSF 5-HIAA concentration was evaluated (Coccaro *et al.*, 1997a). The results were consistent with the altered postsynaptic 5-HT receptor function hypothesis: aggression was significantly and inversely correlated with prolactin responses to D-fenfluramine but not with CSF 5-HIAA levels. Notably, prolactin response to fenfluramine appears to reflect activation of 5-HT2 receptors, likely of the 5-HT2c subtype (Coccaro et al., 2010a). Additional research has revealed that prolactin responses to fenfluramine are also positively correlated with prolactin responses to m-CPP challenge, which assesses 5-HT postsynaptic receptor activation (Coccaro et al., 1997b).

Thus far, seven subtypes of 5-HT receptors have been identified, ranging from 5-HTR1 to 5-HTR7. These receptors have been found to mediate both excitatory and inhibitory inputs in a number of brain regions associated with aggression (Siever, 2008), emotion regulation, and cognition. Inhibition of offensive aggression via agonists of 5-HT1a attenuates various forms of aggression in animals (Ferris et al., 1999; Joppa et al., 1996; Miczek et al., 2004; Olivier et al., 1995; Ricci et al., 2006; White et al., 1991). According to one study by Popova et al., less aggressive rats had higher 5-HT1a receptor expression in the midbrain (Popova et al., 2005), whereas in the frontal cortex, lower aggression was associated with a decrease in 5-HT1a receptor mRNA (Popova et al., 2007). In support of this hypothesis, a recent study showed that high 5-HT1a receptor density corresponded to increased aggressiveness in male Golden hamsters (Cervantes and Delville, 2009). Additional confirmation came from a study in which PET imaging of healthy subjects revealed that aggression is positively correlated to 5-HT1a receptor distribution in the dorsolateral and ventromedial prefrontal cortex, in the orbitofrontal cortex, and in the anterior cingulate cortex (Witte et al., 2009). Support of 5-HT1a's involvement in aggression also comes from animal studies showing that offensive aggression in hamsters is inhibited by 5-HT1a receptors and facilitated by 5-HT3 receptor activation (Cervantes et al., 2010). Agonists of the 5-HT1a and 5-HT1b (5-HT1d in the human) receptors in the medial prefrontal cortex or septal area can increase aggressive behavior under specific conditions (Takahashi et al., 2011). Activation of these two receptors, as well as the 5-HT2a and 5-HT2c receptors in mesocorticolimbic areas, reduces species-typical and other aggressive behaviors. Pathological aggression is reportedly reduced by activation of 5-HT transporters, whereas dysfunction of genes that affect the 5-HT system directly such as MAOA cause an escalation in pathological aggression (Alia-Klein et al., 2008).

With respect to 5-HT2a distribution, PET imaging demonstrates that individuals with IED and current physical aggression have increased receptor density in the orbitofrontal cortex when compared to individuals with IED but no current physical aggression or when compared to individuals who served as healthy controls (Rosell et al., 2010). This is similar to 5-HT1a distribution in the orbitofrontal cortex with the respect to aggression, as noted above (Witte et al., 2009). In a separate study, 5-HT2a receptor-binding activity was investigated in a nearby brain region, the dorsolateral prefrontal cortex, and a pattern different to that seen in the orbitofrontal cortex emerged. The results found that

5-HT2a receptor-binding potentials were lower in the dorsolateral prefrontal cortex in individuals with more severe impulsivity and aggression than in healthy subjects (Meyer *et al.*, 2008). Lower 5-HT2a binding potentials occur at younger ages, when violent behavior is more frequent and is more prominent when impulsivity and aggression are more severe. However, this has not been causally linked; a low binding potential indicates low ligand-receptor-binding interaction and therefore the cause of these reduced binding potentials require further investigation. In a novel study led by Soloff *et al.*, gender differences were identified in 5-HT2a availability with respect to aggression, negativism, and suspiciousness, highlighting a potential for gender biases and a need to control for them when conducting research (Soloff *et al.*, 2010).

Advances in molecular biology and neuroimaging have allowed for experimental studies in which 5-HT activity can be altered by tryptophan manipulation and subsequent brain activity and behavior monitored. These studies have long noted that 5-HT in the central nervous system (CNS), as well as in the periphery (e.g., through assessment of 5-HT transporter binding sites on the blood platelet) is reduced in aggressive behavior (Coccaro *et al.*, 2010a). Platelet 5-HTT sites are structurally identical to corresponding sites on central 5-HT neurons (Lesch *et al.*, 1993) and are therefore appropriate for further hypothesis testing. Preliminary studies by Stoff *et al.* found that lowered tryptophan levels and ingestion of alcohol were associated with increased aggression and lower 5-HTT binding (B) by H^3-imipramine in normal adult males, suggesting that low 5-HT levels may be involved in the etiology of aggression and particularly, alcohol-induced violence (Pihl *et al.*, 1995). Similarly, Birmaher *et al.*, reported that a reduction in platelet H^3-imipramine (B_{max}) was associated with aggression in children and adolescents (Birmaher *et al.*, 1990). Two studies by Coccaro *et al.* have also demonstrated that the number of 5-HTT binding sites assessed by platelet H^3-paroxetine is inversely related to aggression (Coccaro *et al.*, 1996, 2010b). Individuals with IED also had fewer 5-HT transporter platelet binding sites than comparable personality disordered subjects without IED; measures of impulsivity did not correlate with 5-HTT binding in these studies.

Animal and clinical studies have highlighted that impulsive aggression and its comorbid psychiatric disorders may result from a failure of the 5-HT system to communicate properly with other neurotransmitter systems, particularly that of dopamine (De Simoni *et al.*, 1987). Specifically, failure of the dopamine and serotonin systems to successfully interact in the prefrontal cortex may underlie impulsive aggression (Seo *et al.*, 2008). Van Erp and Miczek recently reported that increased aggressive behavior in male Long-Evans rats was related to both increased dopamine in the nucleus accumbens and reduced 5-HT levels in the frontal cortex (Van Erp and Miczek, 2000). Previous studies have illustrated that serotonergic and dopaminergic systems are tightly linked (Daw *et al.*, 2002; Kapur and Remington, 1996; Wong *et al.*, 1995), and it is thought that subnormal serotonergic function may lead to dopaminergic hyper-activity, which in turn leads to impulsive and aggressive behavior (Seo *et al.*, 2008).

III. DOPAMINE

Dopamine (DA) is a catecholamine neurotransmitter that acts both on the central and the sympathetic branch of the peripheral nervous systems. DA in the CNS has been linked to cognition (Browman *et al.*, 2005; Heijtz *et al.*, 2007), movement (Devos *et al.*, 2003), sleep (Dzirasa *et al.*, 2006; Lima *et al.*, 2008), mood (Brown and Gershon, 1993; Diehl and Gershon, 1992), attention (Nieoullon, 2002), and learning and memory (Arias-Carrión and Pöppel, 2007; Denenberg *et al.*, 2004). Additionally, DA has developed a well-established and essential role as the neurotransmitter responsible for reward pathways involved in drug use (Pettit and Justice, 1991; Ranaldi *et al.*, 1999; Weiss *et al.*, 1992), eating (Hernandez and Hoebel, 1988), and sexual behavior (Hull *et al.*, 1993; Pfaus *et al.*, 1990). In patients with frontotemporal dementia, increased dopaminergic neurotransmission and serotonergic modulation of dopaminergic activity is, respectively, associated with agitated and aggressive behavior (Engelborghs *et al.*, 2008), suggesting DA function contributes to the aggressive behavioral state. While much of what we know about dopamine and its biological effects remains to be determined, it is clear from the literature and data so far that dopamine is a neurotransmitter with a multitude of behavioral, physiological, and psychological capabilities.

From a molecular perspective, research into DA function in aggressive individuals has revealed a spectrum of genetic variability that is linked to a number of polymorphisms in DA-specific genes. Led by Elena L. Grigorenko of Yale University's Child Study Center, a coalition of scientists in 2010 found positive correlates between genetic polymorphisms in four genes involved in DA turnover and behavior pathology (Grigorenko *et al.*, 2010). The four genes investigated included catechol-O-methyl-transferase (COMT), involved in catecholamine metabolism; dopamine beta hydroxylase (DbH), responsible for dopamine conversion; and MAOA and MAOB, both involved in the degradation of DA and/or other neurotransmitters. In this study, blood samples from 179 adolescent offender males sentenced to a juvenile detention center in a large capital city in Northern Russia were compared for genetic analysis to those of two control groups of Russian male adolescents ($n = 182$; $n = 60$). While no single dopaminergic polymorphism revealed a definitive causal link to conduct disorder, criminality, aggression, or delinquency, combination of variants across two (COMT and DbH), three (COMT, DbH, and MAOB), or all four (COMT, DbH, MAOA, and MAOB) of the DA-specific genes investigated showed positive correlations with the behavioral traits in question. Nemoda *et al.*

found similar results among patients with borderline personality disorder in a separate study done in 2010 using young adults from low-to-moderate income households ($n = 99$) and major depressive or bipolar patients ($n = 136$) (Nemoda et al., 2010). The results of this study found that a promoter variant in the dopamine D4 receptor may be involved in the development of BPD traits, including aggression. The DA D4 receptor has been postulated as a candidate nexus for BPD because of its preferential expression in the prefrontal cortex (Oak et al., 2000), and its noted role in novelty-seeking and impulsivity (Munafò et al., 2008). Data from both experimental groups showed polymorphisms in COMT and the DA transporter (DAT1) of the dopamine D2 receptor were directly related to self-injurious and impulsive behavior, both BPD traits. Other reports have confirmed genetic abnormalities with COMT in the presence of BPD, including a recent study citing an over-representation of the low activity Met/Met genotype of the gene in BPD patients ($n = 161$) (Tadić et al., 2009). Interestingly, COMT and DAT1 are similarly implicated in bipolar and major depressive disorder (Joyce et al., 2006), suggesting DA dysfunction may encompass a much larger behavioral and physiological state.

In 2008, Couppis and Kennedy published novel findings that found dopamine to be a reward for aggressive behavior in mice (Couppis and Kennedy, 2008). The authors had hypothesized that aggression could be linked to the dopaminergic receptors of the nucleus accumbens (NAc), citing their reputation as the most strongly implicated neurotransmitter in positive reinforcement (Wise, 2004) and reward behavior. The results of their studies showed that administration of a D1-like (D1 and D5) receptor antagonist (SCH-23390), or a D2-like (D2, D3, and D4) receptor antagonist (sulperide) into the NAc significantly reduced aggression responses when compared to administration outside of the NAc. In addition to the reductions in aggression, concomitant reductions in mobility were seen in these first studies. These results showing a simultaneous reduction in aggression and DA levels were consistent with previous reports that had suggested increased DA in the NAc led to increased aggression (Van Erp and Miczek, 2000). After Couppis and Kennedy's initial publication, Schwartzer and Melloni reported similar findings that dopamine activity primarily mediated by D2 receptors was involved in modulating anabolic/androgenic steroid-induced offensive aggression in Syrian hamsters (Schwartzer and Melloni, 2010b). Interestingly, administration of the D2-like DA antagonist into the anterior hypothalamus (AH) rather than the NAc produced no side effects of reduced mobility: in a follow-up experiment, the authors reported that treatment of male Syrian hamsters with the D2-like receptor antagonist eticlopride in the AH results in dose-dependent suppression of aggression behaviors without causing mobility changes (Schwartzer and Melloni, 2010a). Conversely, injection of SCH-23390 into the AH reduced aggressiveness but showed simultaneous changes in sociability and mobility.

Postmortem studies revealed sparse population of GAD_{67} (a GABA production marker) neurons distributed within the D5 receptors of the lateral AH. Based on these findings, the authors conclude that D5 receptors in the lateral AH modulate non-GABAergic pathways that may indirectly influence aggression behavior. Future aggression studies should aim to better understand the role of dopaminergic activity in the hypothalamus and other limbic structures that are in part physiologically responsible for emotion and behavior regulation.

IV. NOREPINEPHRINE (NORADRENALINE)

Synthesized from tyrosine-derived dopamine via dopamine decarboxylase and β-hydroxylase (Sofuoglu and Sewell, 2009), norepinephrine (NE) is both a catecholamine neurotransmitter and a stimulant stress hormone. As a stress hormone, NE primarily targets brain regions responsible for attention such as the amygdala and works in conjunction with epinephrine (adrenaline) to produce the "fight-or-flight" response (Tanaka et al., 2000). During times of high stress, this response increases heart rate, releases glucose from energy stores, and increases blood flow to skeletal muscle in an attempt to increase the oxygen supply to the brain. When released from the locus ceruleus, NE also works to actively suppress neuroinflammation that may potentially cause damage to the brain (Heneka et al., 2010).

One of the earliest reports relating aggression to norepinephrine emerged in a 1972 publication by Thoa and colleagues in *Science* magazine (Thoa et al., 1972). In this study, rats that received an intraventricular injection of 90 mg of 6-hydroxydopamine (a neurotoxic agent used to selectively target dopaminergic or noradrenergic neurons) showed increased shock-induced aggression and reduced brain norepinephrine while dopamine levels remained unaltered. This inverse relationship between norepinephrine availability and shock-induced aggression suggests that the behavioral trait is partially modulated by noradrenergic function. In the mid-1980s, Pucilowski and colleagues launched a series of studies that confirmed NE was intimately related to aggression. The first paper, dating from 1985, demonstrated that chemically induced muricide could be in part suppressed by norepinephrine (Pucilowski and Valzelli, 1985). A second study, published shortly thereafter, showed that bilateral microinjections of hydroxydopamine into the nuclei loci coerulei of male Wistar rats resulted in decreased mesencephalic and striatal norepinephrine levels as well as marked increased aggression (Pucilowski et al., 1986). In 1987, a similar study by the same group *gave* microinjections of the NE-depleting toxin N-(2-chloroethyl)-N-ethyl-2-bromobenzylamine (DSP-4) with or without apomorphine into the amygdala. Here, the results clearly showed that NE was able to markedly reduce apomorphine-induced aggression (Pucilowski et al., 1987). The data also showed that norepinephrine significantly reduced locomotor activity and to a lesser degree, sensitivity to pain.

In 1998, a novel research experiment authored by Spivak et al. reported that neuroleptic-resistant chronic schizophrenic patients maintained on clozapine for 1 year had significantly less aggression ($p < 0.01$), higher plasma NE levels ($p < 0.0001$), and higher serum triglycerides ($p < 0.01$) than patients treated with classical antipsychotic agents for the same period of time (Spivak et al., 1998). Clozapine is a dibenzodiazepine antipsychotic used for the treatment of schizophrenia, and its mechanism of action is primarily anchored by binding to dopamine and serotonin receptors (Naheed and Green, 2001). Based on these findings, the authors conclude that the antiaggressive/antisuicidal activity of clozapine may be a direct result of elevated plasma NE levels (Spivak et al., 1998). Further, this inverse relationship between increased NE levels and decreased aggression confirms what was first seen by Thoa and colleagues in the 1972 Science reports and substantiates a role for NE in modulating behavioral aggression.

Although it is clear that alterations in NE levels affect aggressiveness directly, there is also evidence to suggest that norepinephrine plays an indirect role in modulating serotonergic-mediated impulsive aggression. In 1991, Siever and Davis published findings that suggested decreased NE activity leads to depression, suicide, and self-directed aggression whereas increases in NE activity can lead to irritability and violent aggression (Siever and Davis, 1991). In the same year, Coccaro et al., published additional findings that demonstrated that greater growth hormone responses to the NE agonist clonidine were positively correlated with behavioral irritability (Coccaro et al., 1991). This finding was not replicated in a subsequent study by the same author (Coccaro and Kavoussi, 2010), but another study from this group reported an inverse relationship between plasma MHPG and life history of aggression in personality disordered subjects (Coccaro et al., 2003). In this case, presynaptic NE (reflected by plasma MHPG) could be associated with higher postsynaptic NE receptor sensitivity, and thus, increased fight or flight responses when presented with an aversive stimulus.

V. GABA

While gamma-aminobutyric acid (GABA) has a critical and well-defined function in vertebrate neurological systems, its role in behavioral aggression is not as prominent as 5-HT, DA, and NE. GABA is a primary neurotransmitter in the CNS, and is known as chief inhibitory neurotransmitter (de Almeida et al., 2005). GABA-related activity and dysfunction has been associated with schizophrenia (Kantrowitz et al., 2009; Wassef et al., 1999), epilepsy (Snodgrass, 1992; Treiman, 2001), and pain and nociception (Enna and McCarson, 2006; Sawynok, 1984).

In 2007, researchers at the Model Organism Research Center of the Shanghai Institutes for Biological Sciences found that GABA transporter subtype 1-deficient (or GAT1$^{-/-}$) mice exhibit lower behavioral aggressiveness compared to wild-type mice (Coccaro et al., 1998b). Deficiency in GABA transporter function, analogous to inhibition of GABA transporters, leads to an increase in synaptic GABA. Later, Takahashi et al. found that pharmacological activation of GABA (B), but not GABA(A), receptors in the dorsal raphé nucleus significantly increased aggression (Takahashi et al., 2010). The authors theorized that since the majority of forebrain 5-HT originates from the raphé nucleus, GABAergic control of this region could provide an indirect mechanism for escalations in behavioral aggression. A similar study by the same group showed that male CFW mice, treated with the GABA(A) receptor agonist muscimol, had increased aggressive tendencies following alcohol consumption compared to mice given water. These results demonstrate that GABA(A), but not GABA(B), receptors in the dorsal raphé nucleus are one of the neurobiological targets of alcohol-induced aggression (Coid et al., 1983), and illustrate a functional role for GABA in modulating aggressive behavior.

In addition to its affiliations with serotonin, it has also been demonstrated that GABA is associated with the dopaminergic systems as well and that this relationship may influence displays of aggression. GABAergic interneurons in various brain regions including the AH are commonly found to express dopamine D2 receptors (Gerfen et al., 1990; Santana et al., 2009). Based on this observation and the known presence of DA D2 receptors in the AH (Ricci et al., 2009), Schwartzer et al. postulated that DA D2 receptor activity may be modulating behavioral aggression through direct inhibition of GABA in the AH (Schwartzer et al., 2009). The authors found that adolescent male Syrian hamsters exposed to anabolic-androgenic steroids had DA-stimulated increased aggression marked by the removal of GABAergic inhibition in the lateral AH.

Human studies of GABA and aggression are limited but include two studies in personality disordered subjects from the laboratory of Coccaro et al. In the first study, Lee et al. demonstrated a direct relationship between CSF GABA and measures of impulsivity and history of suicide attempt (but not aggression) in personality disordered subjects (Lee et al., 2008). In the second study, the growth hormone (GH) response to the GABA(B) receptor agonist, baclofen, was found to be inversely correlated with measures of impulsivity (but not aggression) (Lee et al., in press). Taken together, these studies suggest that elevated central GABA may lead to, or be associated with, a reduction of GABA(B) receptors and that this reduction in downstream GABA(B) mediated activity is associated with increased liability to impulsive behavior. As such, these data are consistent with the work of Takahashi et al. (2010).

VI. PEPTIDES

Limited published data suggest relationships between human aggression and central vasopressin, oxytocin, and opiates. (Coccaro *et al.*, 1998b) first reported a positive correlation between CSF vasopressin concentration and life history of aggression in male and female subjects with personality disorders. This relationship was confined to males and remained even after the inverse correlation between CSF vasopressin and a collateral assessment of serotonin function (i.e., PRL response to FEN) was accounted for. Later, Lee *et al.* (2009) reported an inverse relationship between CSF oxytocin and life history of aggression in an overlapping group of subjects. CSF vasopressin and CSF oxytocin were inversely correlated, but CSF oxytocin continued to be related to aggression even after the influence of CSF vasopressin on aggression was controlled for. This lab has also noted a positive correlation between CSF Neuropeptide Y and CSF Substance P in these same subjects. In addition, circulating levels of metenkephalins have been associated with self-injurious behaviors in one study (Coid *et al.*, 1983). Postmortem studies of violent suicide victims have found greater number of mu receptors in the brain. In healthy volunteers, administration of codeine (Spiga *et al.*, 1990) or morphine (Berman *et al.*, 1993) heightened aggression on laboratory measures. These studies suggest that increased opioid activity may increase the likelihood of aggressive behavior. In fact, naltrexone, an opioid antagonist, attenuates self-injurious behavior in autistic and retarded patients (Sandman *et al.*, 1990, 2000).

VII. CONCLUSION

The neurobiology of aggression is clearly complex. However, we now know more about the biological underpinnings of this behavior than ever before and this knowledge points the way to possible strategies for treatment. Many agents appear to have therapeutic efficacy but many only work on the brain 5-HT system. In the upcoming years, we look to the development of agents that work on non-5-HT systems (e.g., vasopressin, oxytocin, etc.) so that we may have a more varied toolbox with which to treat individuals with problematic aggressive behavior.

References

Alia-Klein, N., Goldstein, R. Z., Kriplani, A., Logan, J., Tomasi, D., Williams, B., Telang, F., Shumay, E., Biegon, A., Craig, I. W., Henn, F., Wang, G. J., Volkow, N. D., and Fowler, J. S. (2008). Brain monoamine oxidase. A activity predicts trait aggression. *J. Neurosci.* **28**(19), 5099–5104.

Arias-Carrión, O., and Pöppel, E. (2007). Dopamine, learning and reward-seeking behavior. *Act. Neurobiol. Exp.* **67**(4), 481–488.

Asberg, M., Traksman, L., and Thoren, P. (1976). 5-HIAA in the cerebrospinal fluid: A biochemical suicide predictor. *Arch. Gen. Psychiatry* **33**, 1193–1197.

Berman, M., Taylor, S., and Marged, B. (1993). Morphine and human aggression. *Addict. Behav.* **18**, 263–268.

Birmaher, B., Stanley, M., Greenhill, L., Twomey, J., Gavrilescu, A., and Rabinovich, H. (1990). Platelet imipramine binding in children and adolescents with impulsive behavior. *J. Am. Acad. Child Adolesc. Psychiatry* **29**, 914–918.

Browman, K. E., Curzon, P., Pan, J. B., Molesky, A. L., Komater, V. A., Decker, M. W., Brioni, J. D., Moreland, R. B., *et al.* (2005). A-412997, a selective dopamine D4 agonist, improves cognitive performance in rats. *Pharmacol. Biochem. Behav.* **82**(1), 148–155.

Brown, A. S., and Gershon, S. (1993). Dopamine and depression. *J. Neural Transm. Gen. Sect.* **91**(2–3), 75–109. Review.

Brown, G. L., Goodwin, F. K., Ballenger, J. C., Goyer, P. F., and Major, L. F. (1979). Aggression in human correlates with cerebrospinal fluid amine metabolites. *Psychiatry Res.* **1**, 131–139.

Brown, G. L., Ebert, M. H., Goyer, P. F., *et al.* (1982). Aggression, suicide, and serotonin: Relationships to CSF amine metabolites. *Am. J. Psychiatry* **139**, 741–746.

Castellanos, F. X., Elia, J., Kreusi, M. J. P., Gulotta, C. S., Mefford, I. N., Potter, W. Z., Ritchie, G. F., and Rapoport, J. L. (1994). Cerebrospinal fluid monoamine metabolites in boys with attention deficit hyperactivity disorder. *Psychiatry Res.* **52**, 305–316.

Cervantes, M. C., and Delville, Y. (2009). Serotonin 5-HT1A and 5-HT3 receptors in an impulsive-aggressive phenotype. *Behav. Neurosci.* **123**(3), 589–598.

Cervantes, M. C., Biggs, E. A., and Delville, Y. (2010). Differential responses to serotonin receptor ligands in an impulsive-aggressive phenotype. *Behav. Neurosci.* **124**(4), 455–469.

Coccaro, E. F. (1998). Central neurotransmitter function in human aggression and impulsivity. *In* "Neurobiology and Clinical Views on Aggression and Impulsivity" (M. Maes and E. F. Coccaro, eds.), pp. 143–168. Chichester, Wiley, UK.

Coccaro, E. F., and Kavoussi, R. J. (1994). The neuropsychopharmacologic challenge in biological psychiatry. *Clin. Chem.* **40**, 319–327.

Coccaro, E. F., and Kavoussi, R. J. (2010). GH response to intravenous clonidine challenge: Absence of relationship with behavioral irritability, aggression, or impulsivity in human subjects. *Psychiatry Res.* **178**, 443–445.

Coccaro, E. F., and Lee, R. (2010). Cerebrospinal fluid 5-hydroxyindolacetic acid and homovanillic acid: Reciprocal relationships with impulsive aggression in human subjects. *J. Neural Transm.* **117**(2), 241–248.

Coccaro, E. F., and Siever, L. J. (2002). Pathophysiology and treatment of aggression. *In* "Psychopharmacology: The Fifth Generation of Progress" (K. L. Davis, D. Charney, J. T. Coyle, and C. Nemeroff, eds.), pp. 1709–1723. Lippincott Williams & Wilkins, Philadelphia.

Coccaro, E. F., Siever, L. J., Klar, H. M., Maurer, G., Cochrane, K., Mohs, R. C., and Davis, K. L. (1989). Serotonergic studies in affective and personality disorder: Correlates with suicidal and impulsive aggressive behavior. *Arch. Gen. Psychiatry* **46**, 587–599.

Coccaro, E. F., Lawrence, T., Trestman, R., Gabriel, S., Klar, H. M., and Siever, L. J. (1991). Growth hormone responses to intravenous clonidine challenge correlate with behavioral irritability in psychiatric patients and healthy volunteers. *Psychiatry Res.* **39**(2), 129–139.

Coccaro, E. F., Kavoussi, R. J., Sheline, Y. I., Lish, J. D., and Csernansky, J. G. (1996). Impulsive aggression in personality disorder: Correlates with 3-H-Paroxetine binding in the platelet. *Arch. Gen. Psychiatry* **53**, 531–536.

Coccaro, E. F., Kavoussi, R. J., Cooper, T. B., and Hauger, R. L. (1997a). Central serotonin activity and aggression: Inverse relationship with prolactin response to d-fenfluramine, but Not CSF 5-HIAA concentration, in human subjects. *Am. J. Psychiatry* **154**, 1430–1435.

Coccaro, E. F., Kavoussi, R. J., and Hauger, R. L. (1997b). Serotonin function and antiaggressive responses to fluoxetine: A pilot study. *Biol. Psychiatry* **42**, 546–552.

Coccaro, E. F., Kavoussi, R. J., Cooper, T. B., and Hauger, R. L. (1997c). Central serotonin and aggression: Inverse relationship with prolactin response to d-fenfluramine, but not with CSF 5-HIAA concentration in human subjects. *Am. J. Psychiatry* **154**, 1430–1435.

Coccaro, E. F., Kavoussi, R. J., Cooper, T. B., and Hauger, R. (1998a). Acute tryptophan depletion attenuates the prolactin response to d-fenfluramine challenge in healthy human subjects. *Psychopharmacology* **138**, 9–15.

Coccaro, E. F., Kavoussi, R. J., Hauger, R. L., Cooper, T. B., and Ferris, C. F. (1998b). Cerebrospinal fluid vasopressin: Correlates with aggression and serotonin function in personality disordered subjects. *Arch. Gen. Psychiatry* **55**, 708–714.

Coccaro, E. F., Lee, R., and McCloskey, M. (2003). Norepinephrine function in personality disorder: Plasma free MHPG correlates inversely with life history of aggression. *CNS Spectr.* **8**, 731–736.

Coccaro, E. F., Lee, R., and Kavoussi, R. J. (2010a). Aggression, Suicidality, and Intermittent Explosive Disorder: Serotonergic correlates in personality disorder and healthy control subjects. *Neuropsychopharmacology* **35**, 435–444.

Coccaro, E. F., Lee, R., and Kavoussi, R. J. (2010b). Inverse relationship between numbers of 5-HT transporter binding sites and life history of aggression and intermittent explosive disorder. *J. Psychiatr. Res.* **44**(3), 137–142.

Coid, J., Allolio, B., and Rees, L. H. (1983). Raised plasma metenkephalin in patients who habitually mutilate themselves. *Lancet* **2**, 545–546.

Couppis, M. H., and Kennedy, C. H. (2008). The rewarding effect of aggression is reduced by nucleus accumbens dopamine receptor antagonism in mice. *Psychopharmacology (Berl.)* **197**(3), 449–456.

Curzon, G. (1991). Effects of tryptophan and of 5-hydroxytryptamine receptor subtype agonists on feeding. *Adv. Exp. Med. Biol.* **294**, 377–388.

Daw, N. D., Kakade, S., and Dayan, P. (2002). Opponent interactions between serotonin and dopamine. *Neural Netw.* **15**, 603–616.

de Almeida, R. M., Ferrari, P. F., Parmigiani, S., and Miczek, K. A. (2005). Escalated aggressive behavior: Dopamine, serotonin and GABA. *Eur. J. Pharmacol.* **526**, 51–64.

De Simoni, M. G., Dal Toso, G., Fodritto, F., Sokola, A., and Algeri, S. (1987). Modulation of striatal dopamine metabolism by the activity of dorsal raphe serotonergic afferences. *Brain Res.* **411**, 81–88.

Denenberg, V. H., Kim, D. S., and Palmiter, R. D. (2004). The role of dopamine in learning, memory, and performance of a water escape task. *Behav. Brain Res.* **148**(1–2), 73–78.

Devos, D., Labyt, E., Derambure, P., Bourriez, J. L., Cassim, F., Guieu, J. D., Destée, A., and Defebvre, L. (2003). Effect of L-dopa on the pattern of movement-related (de)synchronisation in advanced Parkinson's disease. *Neurophysiol. Clin.* **33**(5), 203–212.

Diehl, D. J., and Gershon, S. (1992). The role of dopamine in mood disorders. *Compr. Psychiatry* **33**(2), 115–120. Review.

Dourish, C. T. (1995). Multiple serotonin receptors: Opportunities for new treatments for obesity? *Obes. Res.* **3**(Suppl. 4), 449S–462S.

Dzirasa, K., Ribeiro, S., Costa, R., Santos, L. M., Lin, S. C., Grosmark, A., Sotnikova, T. D., Gainetdinov, R. R., Caron, M. G., and Nicolelis, M. A. (2006). Dopaminergic control of sleep-wake states. *J. Neurosci.* **26**(41), 10577–10589.

Engelborghs, S., Vloeberghs, E., Le Bastard, N., Van Buggenhout, M., Mariën, P., Somers, N., Nagels, G., Pickut, B. A., and De Deyn, P. P. (2008). The dopaminergic neurotransmitter system is associated with aggression and agitation in frontotemporal dementia. *Neurochem. Int.* **52**(6), 1052–1060.

Enna, S. J., and McCarson, K. E. (2006). The role of GABA in the mediation and perception of pain. *Adv. Pharmacol.* **54**, 1–27. Review.

Ferris, C. F., Stolberg, T., and Delville, Y. (1999). Serotonin regulation of aggressive behavior in male golden hamsters Mesocricetus auratus. *Behav. Neurosci.* **113**, 804–815.

Gardner, D. L., Lucas, P. B., and Cowdry, R. W. (1990). CSF metabolites in borderline personality disorder compared with normal controls. *Biol. Psychiatry* **28**, 247–254.

Gerfen, C. R., Engber, T. M., Mahan, L. C., Susel, Z., Chase, T. N., Monsma, F. J., Jr., *et al.* (1990). D1 and D2 dopamine receptor-regulated gene expression of striatonigral and striatopallidal neurons. *Science* **250**, 1429–1432.

Gershon, M. D. (2004). Review article: Serotonin receptors and transporters—Roles in normal and abnormal gastrointestinal motility. *Aliment. Pharmacol. Ther.* **20**(Suppl. 7), 3–14.

Grigorenko, E. L., De Young, C. G., Eastman, M., Getchell, M., Haeffel, G. J., Klinteberg, B., Koposov, R. A., Oreland, L., Pakstis, A. J., Ponomarev, O. A., Ruchkin, V. V., Singh, J. P., and Yrigollen, C. M. (2010). Aggressive behavior, related conduct problems, and variation in genes affecting dopamine turnover. *Aggress. Behav.* **36**(3), 158–176.

Heijtz, R. D., Kolb, B., and Forssberg, H. (2007). Motor inhibitory role of dopamine D1 receptors: Implications for ADHD. *Physiol. Behav.* **92**(1–2), 155–160.

Heneka, M. T., Nadrigny, F., Regen, T., Martinez-Hernandez, A., Dumitrescu-Ozimek, L., Terwel, D., Jardanhazi-Kurutz, D., Walter, J., Kirchhoff, F., Hanisch, U. K., and Kummer, M. P. (2010). Locus ceruleus controls Alzheimer's disease pathology by modulating microglial functions through norepinephrine. *Proc. Natl. Acad. Sci. USA* **17**, 6058–6063.

Hernandez, L., and Hoebel, B. G. (1988). Food reward and cocaine increase extracellular dopamine in the nucleus accumbens as measured by microdialysis. *Life Sci.* **42**, 1705–1712.

Hull, E. M., Eaton, R. C., Moses, J., and Lorrain, D. (1993). Copulation increases dopamine activity in the medial preoptic area of male rats. *Life Sci.* **52**, 935–940.

Joppa, M. A., Rowe, R. K., and Meisel, R. L. (1996). Effects of serotonin 1A or 1B receptor agonists on social aggression in male and female Syrian hamsters. *Pharmacol. Biochem. Behav.* **58**, 349–353.

Joyce, P. R., McHugh, P. C., McKenzie, J. M., Sullivan, P. F., Mulder, R. T., Luty, S. E., Carter, J. D., Frampton, C. M., Robert Cloninger, C., Miller, A. M., and Kennedy, M. A. (2006). A dopamine transporter polymorphism is a risk factor for borderline personality disorder in depressed patients. *Psychol. Med.* **36**(6), 807–813.

Kantrowitz, J., Citrome, L., and Javitt, D. (2009). GABA(B) receptors, schizophrenia and sleep dysfunction: A review of the relationship and its potential clinical and therapeutic implications. *CNS Drugs* **23**(8), 681–691.

Kapur, S., and Remington, G. (1996). Serotonin-dopamine interaction and its relevance to schizophrenia. *Am. J. Psychiatry* **153**, 436–476.

Kruesi, M. J. P., Rapoport, J. L., Hamberger, S., Hibbs, E., Potter, W. Z., Lenane, M., and Brown, G. L. (1990). Cerebrospinal fluid metabolites, aggression, and impulsivity in disruptive behavior disorders of children and adolescents. *Arch. Gen. Psychiatry* **47**, 419–462.

Kumar, K. K., Tung, S., and Iqbal, J. (2010). Bone loss in anorexia nervosa: Leptin, serotonin, and the sympathetic nervous system. *Ann. N. Y. Acad. Sci.* **1211**, 51–65.

Kunisato, Y., Okamoto, Y., Okada, G., Aoyama, S., Demoto, Y., Munakata, A., Nomura, M., Onoda, K., and Yamawaki, S. (2010). Modulation of default-mode network activity by acute tryptophan depletion is associated with mood change: A resting state functional magnetic resonance imaging study. *Neurosci. Res.* **69**, 129–134.

Lam, D. D., Garfield, A. S., Marston, O. J., Shaw, J., and Heisler, L. K. (2010). Brain serotonin system in the coordination of food intake and body weight. *Pharmacol. Biochem. Behav.* **97**(1), 84–91.

Lee, R., Petty, F., and Coccaro, E. F. (2008). Cerebrospinal fluid GABA concentration: Relationship with impulsivity and history of suicidal behavior, but not aggression, in human subjects. *J. Psychiatric. Res.* **43**, 353–359.

Lee, R., Ferris, C., Van de Kar, L. D., and Coccaro, E. F. (2009). Cerebrospinal fluid oxytocin, life history of aggression, and personality disorder. *Psychoneuroendocrinology* **34**, 1567–1573.

Lee, R., Chong, B., and Coccaro, E. F. (2011). Growth hormone responses to GABA-B receptor challenge with baclofen and impulsivity in healthy control and personality disorder subjects. *Psychopharmacology (Berl)* **215**(1), 41–48.

Lesch, K. P., Wolozin, B. L., Murphy, D. L., and Reiderer, P. (1993). Primary structure of the human platelet serotonin uptake site: Identity with the brain serotonin transporter. *J. Neurochem.* **60,** 2319–2322.

Lidberg, L., Tuck, J. R., Asberg, M., Scalia-Tomba, G. P., and Bertilsson, L. (1985). Homicide, suicide and CSF 5-HIAA. *Acta Psych. Scand.* **71,** 230–236.

Lima, M. M., Andersen, M. L., Reksidler, A. B., Silva, A., Zager, A., Zanata, S. M., Vital, M. A., and Tufik, S. (2008). Blockage of dopaminergic D(2) receptors produces decrease of REM but not of slow wave sleep in rats after REM sleep deprivation. *Behav. Brain Res.* **188**(2), 406–411.

Limson, R., Goldman, D., Roy, A., Lamparski, D., Ravitz, B., Adinoff, B., and Linnoila, M. (1991). Personality and cerebrospinal fluid monoamine metabolites in alcoholics and controls. *Arch. Gen. Psychiatry* **48,** 437–441.

Linnoila, M., Virkkunen, M., Scheinin, M., Nuutila, A., Rimon, R., and Goodwin, F. K. (1983). Low cerebrospinal fluid 5-hydroxyIndolacetic acid concentration differentiates impulsive from nonimpulsive violent behavior. *Life Sci.* **33,** 2609–2614.

Meyer, J. H., Wilson, A. A., Rusjan, P., Clark, M., Houle, S., Woodside, S., Arrowood, J., Martin, K., and Colleton, M. (2008). Serotonin2A receptor-binding potential in people with aggressive and violent behaviour. *J. Psychiatry Neurosci.* **33**(6), 499–508.

Miczek, K. A., Faccidomo, S., de Almeida, R. M. M., Bannai, M., Fish, E. W., and DeBold, J. F. (2004). Escalated aggressive behavior: New pharmacotherapeutic approaches and opportunities. *Ann. N. Y. Acad. Sci.* **1036,** 336–355.

Moller, S. E., Mortensen, E. L., Breum, L., Alling, C., Larsen, O. G., Boge-Rasmussen, T., Jensen, C., and Bennicke, K. (1996). Aggression and personality: Association with amino acids and monoamine metabolites. *Psychol. Med.* **26,** 323–331.

Monti, J. M. (2010). The role of dorsal raphe nucleus serotonergic and non-serotonergic neurons, and of their receptors, in regulating waking and rapid eye movement (REM) sleep. *Sleep Med. Rev.* **14**(5), 319–327.

Monti, J. M., and Jantos, H. (2008). The roles of dopamine and serotonin, and of their receptors, in regulating sleep and waking. *Prog. Brain Res.* **172,** 625–646.

Monti, J. M., and Monti, D. (2000). Role of dorsal raphe nucleus serotonin 5-HT1A receptor in the regulation of REM sleep. *Life Sci.* **66**(21), 1999–2012.

Munafò, M. R., Yalcin, B., Willis-Owen, S. A., and Flint, J. (2008). Association of the dopamine D4 receptor (DRD4) gene and approach-related personality traits: Meta-analysis and new data. *Biol. Psychiatry* **63**(2), 197–206.

Naheed, M., and Green, B. (2001). Focus on clozapine. *Curr. Med. Res. Opin.* **17**(3), 223–229.

Nemoda, Z., Lyons-Ruth, K., Szekely, A., Bertha, E., Faludi, G., and Sasvari-Szekely, M. (2010). Association between dopaminergic polymorphisms and borderline personality traits among at-risk young adults and psychiatric inpatients. *Behav. Brain Funct.* **6,** 4.

New, A. S., Hazlett, E. A., Buchsbaum, M. S., Goodman, M., Reynolds, D., Mitropoulou, V., Sprung, L., Shaw, R. B., Jr., Koenigsberg, H., Platholi, J., Silverman, J., and Siever, L. J. (2002). Blunted prefrontal cortical 18fluorodeoxyglucose positron emission tomography response to meta-chlorophenylpiperazine in impulsive aggression. *Arch. Gen. Psychiatry* **59**(7), 621–629.

Nieoullon, A. (2002). Dopamine and the regulation of cognition and attention. *Prog. Neurobiol.* **67**(1), 53–83. Review.

Oak, J. N., Oldenhof, J., and Van Tol, H. H. (2000). The dopamine D(4) receptor: One decade of research. *Eur. J. Pharmacol.* **405**(1–3), 303–327. Review.

Olivier, B., Mos, J., van Oorschot, R., and Hen, R. (1995). Serotonin receptors and animal models of aggressive behavior. *Pharmacopsychiatry* **28,** 80–90.

Parsey, R. V., Oquendo, M. A., Simpson, N. R., Ogden, R. T., Van Heertum, R., Arango, V., and Mann, J. J. (2002). Effects of sex, age, and aggressive traits in man on brain serotonin 5-HT1A receptor-binding potential measured by PET using [C-11]WAY-100635. *Brain Res.* **954**, 173–182.

Peddeer, J. (1992). Psychoanalytic views of aggression: Some theoretical problems. *Br. J. Med. Psychol.* **65**, 95–106.

Pettit, H. O., and Justice, J. B. (1991). Effect of dose on cocaine self-administration behavior and dopamine levels in the nucleus accumbens. *Brain Res.* **539**, 94–102.

Pfaus, J. G., Damsma, G., Nomikos, G. G., Wenkstern, D. G., Blaha, C. D., Phillips, A. G., and Fibiger, H. C. (1990). Sexual behavior enhances central dopamine transmission in the male rat. *Brain Res.* **530**, 345–348.

Pihl, R.-O., Young, S., Harden, P., Plotnick, S., Chamberlain, B., and Ervin, F. R. (1995). Acute effect of altered tryptophan levels and alcohol on aggression in normal human males. *Psychopharmacology* **119**, 353–360.

Popova, N. K. (2008). From gene to aggressive behavior: The role of brain serotonin. *Neurosci. Behav. Physiol.* **38**(5), 471–475.

Popova, N. K., Naumenko, V. S., Plyusnina, I. Z., and Kulikov, A. V. (2005). Reduction in 5-HT1A receptor density, 5-HT1A mRNA expression, and functional correlates for 5-HT1A receptors in genetically defined aggressive rats. *J. Neurosci. Res.* **80**, 286–292.

Popova, N. K., Naumenko, V. S., and Plyusnina, I. Z. (2007). Involvement of brain serotonin 5-HT1A receptors in genetic predisposition to aggressive behavior. *Neurosci. Behav. Physiol.* **37**, 631–635.

Prochazka, H., and Agren, H. (2003). Self-rated aggression and cerebral monoaminergic turnover. Sex differences in patients with persistent depressive disorder. *Eur. Arch Psychiatry Clin. Neurosci.* **253**(4), 185–192.

Pucilowski, O., and Valzelli, L. (1985). Evidence of norepinephrine-mediated suppression of para-chlorophenylalanine-induced muricidal behavior. *Pharmacol. Res. Commun.* **17**, 983–989.

Pucilowski, O., Kozak, W., and Valzelli, L. (1986). Effect of 6-OHDA injected into the locus coeruleus on apomorphine-induced aggression. *Pharmacol. Biochem. Behav.* **24**, 773–775.

Pucilowski, O., Trzaskowska, E., Kostowski, W., and Valzelli, L. (1987). Norepinephrine-mediated suppression of apomorphine-induced aggression and locomotor activity in the rat amygdala. *Pharmacol. Biochem. Behav.* **26**(2), 217–222.

Ranaldi, R., Pocock, D., Zereik, R., and Wise, R. A. (1999). Dopamine fluctuations in the nucleus accumbens during maintenance, extinction, and reinstatement of intravenous d-amphetamine self-administration. *J. Neurosci.* **19**, 4102–4109.

Ricci, L. A., Rasakham, K., Grimes, J. M., and Melloni, R. H., Jr. (2006). Serotonin-1A receptor activity and expression modulate adolescent anabolic/androgenic steroid-induced aggression in hamsters. *Pharmacol. Biochem. Behav.* **85**, 1–11.

Ricci, L. A., Schwartzer, J. J., and Melloni, R. H., Jr. (2009). Alterations in the anterior hypothalamic dopamine system in aggressive adolescent AAS-treated hamsters. *Horm. Behav.* **55**, 348–355.

Rosell, D. R., Thompson, J. L., Slifstein, M., Xu, X., Frankle, W. G., New, A. S., Goodman, M., Weinstein, S. R., Laruelle, M., Abi-Dargham, A., and Siever, L. J. (2010). Increased serotonin 2A receptor availability in the orbitofrontal cortex of physically aggressive personality disordered patients. *Biol. Psychiatry* **67**(12), 1154–1162.

Ruhé, H. G., Mason, N. S., and Schene, A. H. (2007). Mood is indirectly related to serotonin, norepinephrine and dopamine levels in humans: A meta-analysis of monoamine depletion studies. *Mol. Psychiatry* **12**(4), 331–359.

Sandman, C. A., Barron, J. L., and Colman, H. (1990). An orally administered opiate blocker, naltrexone, attenuates self-injurious behavior. *Am. J. Ment. Retard.* **95**, 93–102.

Sandman, C. A., Hedrick, W., Taylor, D. V., Marion, S. D., Touchette, P., Barron, J. L., Martinezzi, V., Steinberg, R. M., and Crinella, F. M. (2000). Long-term effects of naltrexone on self-injurious behavior. *Am. J. Ment. Retard.* **105,** 103–117.

Santana, N., Mengod, G., and Artigas, F. (2009). Quantitative analysis of the expression of dopamine D1 and D2 receptors in pyramidal and GABAergic neurons of the rat prefrontal cortex. *Cereb. Cortex* **19,** 849–860.

Sawynok, J. (1984). GABAergic mechanisms in antinociception. *Prog. Neuropsychopharmacol. Biol. Psychiatry* **8**(4–6), 581–586.

Schwartzer, J. J., and Melloni, R. H., Jr. (2010a). Dopamine activity in the lateral anterior hypothalamus modulates AAS-induced aggression through D2 but not D5 receptors. *Behav. Neurosci.* **124**(5), 645–655.

Schwartzer, J. J., and Melloni, R. H., Jr. (2010b). Anterior hypothalamic dopamine D2 receptors modulate adolescent anabolic/androgenic steroid-induced offensive aggression in the Syrian hamster. *Behav. Pharmacol.* **21**(4), 314–322.

Schwartzer, J. J., Ricci, L. A., and Melloni, R. H., Jr. (2009). Interactions between the dopaminergic and GABAergic neural systems in the lateral anterior hypothalamus of aggressive AAS-treated hamsters. *Behav. Brain Res.* **203**(1), 15–22.

Seo, D., Patrick, C. J., and Kennealy, P. J. (2008). Role of serotonin and dopamine system interactions in the neurobiology of impulsive aggression and its comorbidity with other clinical disorders. *Aggress Violent Behav.* **13**(5), 383–395.

Sheard, M., Manini, J., Bridges, C., and Wapner, A. (1976). The effect of lithium on impulsive aggressive behavior in man. *Am. J. Psychiatry* **133,** 1409–1413.

Siever, L. J. (2008). Neurobiology of aggression and violence. *Am. J. Psychiatry* **165**(4), 429–442.

Siever, L. J., and Davis, K. L. (1991). A psychobiological perspective on the personality disorders. *Am. J. Psychiatry* **148**(12), 1647–1658. Review.

Siever, L. J., Buchsbaum, M. S., New, A. S., Spiegel-Cohen, J., Wei, C. T., Hazlett, E., Sevin, E., Nunn, M., and Mitropoulou, V. (1999). D1-Fenfluramine response in impulsive personality disorder assessed with 18F-deoxyglucose positron emission tomography. *Neuropsychopharmacology* **20,** 413–423.

Snodgrass, S. R. (1992). GABA and epilepsy: Their complex relationship and the evolution of our understanding. *J. Child Neurol.* **7**(1), 77–86. Review.

Sofuoglu, M., and Sewell, R. A. (2009). Norepinephrine and stimulant addiction. *Addict. Biol.* **14**(2), 119–129. Review.

Soloff, P. H., Price, J. C., Mason, N. S., Becker, C., and Meltzer, C. C. (2010). Gender, personality, and serotonin-2A receptor-binding in healthy subjects. *Psychiatry Res.* **181**(1), 77–84.

Spiga, R., Cherek, D. R., Roache, J. D., and Cowan, K. A. (1990). The effects of codeine on human aggressive responding. *Int. Clin. Psychopharmacol.* **5,** 195–204.

Spivak, B., Roitman, S., Vered, Y., Mester, R., Graff, E., Talmon, Y., Guy, N., Gonen, N., and Weizman, A. (1998). Diminished suicidal and aggressive behavior, high plasma norepinephrine levels, and serum triglyceride levels in chronic neuroleptic-resistant schizophrenic patients maintained on clozapine. *Clin. Neuropharmacol.* **21**(4), 245–250.

Stanley, M., and Mann, J. J. (1983). Increased serotonin—2 binding sites in frontal cortex of suicide victims. *Lancet* **1,** 214–216.

Stanley, M. S., Viggilio, J., and Gershon, S. (1982). Tritiated imipramine binding sites are decreased in the frontal cortex of suicides. *Science* **216,** 1337–1339.

Tadić, A., Victor, A., Başkaya, O., von Cube, R., Hoch, J., Kouti, I., Anicker, N. J., Höppner, W., Lieb, K., and Dahmen, N. (2009). Interaction between gene variants of the serotonin transporter promoter region (5-HTTLPR) and catechol O-methyltransferase (COMT) in borderline personality disorder. *Am. J. Med. Genet. B.* **150,** 487–495.

Takahashi, A., Shimamoto, A., Boyson, C. O., DeBold, J. F., and Miczek, K. A. (2010). GABA(B) receptor modulation of serotonin neurons in the dorsal raphé nucleus and escalation of aggression in mice. *J. Neurosci.* **30**(35), 11771–11780.

Takahashi, A., Quadros, I. M., de Almeida, R. M., and Miczek, K. A. (2011). Brain serotonin receptors and transporters: Initiation vs. termination of escalated aggression. *Psychopharmacology (Berl.)* **213**(2–3), 183–212.

Tanaka, M., Yoshida, M., Emoto, H., and Ishii, H. (2000). Noradrenaline systems in the hypothalamus, amygdala and locus coeruleus are involved in the provocation of anxiety: Basic studies. *Eur. J. Pharmacol.* **405**(1–3), 397–406. Review.

Thoa, N. B., Eichelman, B., Richardson, J. S., and Jacobowitz, D. (1972). 6-Hydroxydopa depletion of brain norepinephrine and the function of aggressive behavior. *Science* **178**(56), 75–77.

Treiman, D. M. (2001). GABA ergic mechanisms in epilepsy. *Epilepsia* **42**(Suppl. 3), 8–12. Review.

Van Erp, A. M. M., and Miczek, K. A. (2000). Aggressive behavior, increased accumbal dopamine, and decreased cortical serotonin in rats. *J. Neurosci.* **20**(24), 9320–9325.

Wassef, A. A., Dott, S. G., Harris, A., Brown, A., O'Boyle, M., Meyer, W. J., 3rd, and Rose, R. M. (1999). Critical review of GABA-ergic drugs in the treatment of schizophrenia. *J. Clin. Psychopharmacol.* **19**(3), 222–232. Review.

Weiss, F., Paulus, M., Lorang, M. T., and Koob, G. F. (1992). Increases in extracellular dopamine in the nucleus accumbens by cocaine are inversely related to basal levels: Effects of acute and repeated administration. *J. Neurosci.* **12**, 4372–4380.

Welch, J. T., and Lim, D. S. (2007). The synthesis and biological activity of pentafluorosulfanyl analogs of fluoxetine, fenfluramine, and norfenfluramine. *Bioorg. Med. Chem.* **15**(21), 6659–6666.

White, S. M., Kucharik, R. F., and Moyer, J. A. (1991). Effects of serotonergic agents on isolation-induced aggression. *Pharmacol. Biochem. Behav.* **39**, 729–736.

Wise, R. A. (2004). Dopamine, learning, and motivation. *Nat. Rev. Neurosci.* **5**, 483–495.

Witte, A. V., Flöel, A., Stein, P., Savli, M., Mien, L. K., Wadsak, W., Spindelegger, C., Moser, U., Fink, M., Hahn, A., Mitterhauser, M., Kletter, K., Kasper, S., and Lanzenberger, R. (2009). Aggression is related to frontal serotonin-1A receptor distribution as revealed by PET in healthy subjects. *Hum. Brain Mapp.* **30**(8), 2558–2570. Erratum in: *Hum. Brain Mapp.* 2010 Feb; 31(2): 339.

Wong, P. T., Feng, H., and Teo, W. L. (1995). Interaction of the dopaminergic and serotonergic systems in the rat striatum: Effects of selective antagonists and uptake inhibitors. *Neurosci. Res.* **23**, 115–119.

Xu, L., Yu, B. P., Chen, J. G., and Luo, H. S. (2007). Mechanisms mediating serotonin-induced contraction of colonic myocytes. *Clin. Exp. Pharmacol. Physiol.* **34**(1–2), 120–128.

8

Human Aggression Across the Lifespan: Genetic Propensities and Environmental Moderators

Catherine Tuvblad and Laura A. Baker

University of Southern California, Los Angeles, California, USA

ABSTRACT

This chapter reviews the recent evidence of genetic and environmental influences on human aggression. Findings from a large selection of the twin and adoption studies that have investigated the genetic and environmental architecture of

Advances in Genetics, Vol. 75
Copyright 2011, Elsevier Inc. All rights reserved.

0065-2660/11 $35.00
DOI: 10.1016/B978-0-12-380858-5.00007-1

aggressive behavior are summarized. These studies together show that about half (50%) of the variance in aggressive behavior is explained by genetic influences in both males and females, with the remaining 50% of the variance being explained by environmental factors not shared by family members. Form of aggression (reactive, proactive, direct/physical, indirect/relational), method of assessment (laboratory observation, self-report, ratings by parents and teachers), and age of the subjects—all seem to be significant moderators of the magnitude of genetic and environmental influences on aggressive behavior. Neither study design (twin vs. sibling adoption design) nor sex (male vs. female) seems to impact the magnitude of the genetic and environmental influences on aggression. There is also some evidence of gene-environment interaction (G × E) from both twin/adoption studies and molecular genetic studies. Various measures of family adversity and social disadvantage have been found to moderate genetic influences on aggressive behavior. Findings from these G × E studies suggest that not all individuals will be affected to the same degree by experiences and exposures, and that genetic predispositions may have different effects depending on the environment.   © 2011, Elsevier Inc.

Are all humans innately and equally capable of inflicting harm on others? Do we learn by our various experiences to manipulate and even harm others for our own personal gain; or conversely, to be kind and benevolent, offering help even at costs to ourselves? Although these fundamental questions pertaining to the nature of human aggression have plagued scientists and laypersons for centuries, some answers can be found in research spanning the last few decades.

The early experiments of Milgram (1963) made it clear that, under certain circumstances, individuals can be coaxed into aggression and violence. The presence of a strict authority and removal of personal responsibility for one's actions can result in aggressive behaviors that inflict harm on others. The infamous Stanford prison experiment (Haney et al., 1973) also demonstrated that the propensity toward violence and aggression can be elicited—extremely and unexpectedly—in situations, where a legitimized ideology and a powerful authority can lead to impressionability and obedience.

Yet, while these powerful studies revealed the importance of social factors in inducing aggressive behaviors, not all individuals responded in an equally aggressive manner. In Milgram's (1963) first set of experiments, while 65% (26 of 40) of participants complied with the instruction to administer what they believed to be a final, massive 450-volt shock, the remaining 35% did not comply. Many of those who engaged in the aggressive behavior stated they were very uncomfortable doing so, and every participant reportedly questioned the experiment at some point or refused money promised for their study participation (Milgram, 1963). Although the studies by Milgram and Zimbardo provide clear evidence for the role of environment and social situations in affecting aggressive behavior, there are, nonetheless, large individual differences in the propensity for violence and aggression, even under these extreme circumstances.

What factors contribute to individual differences in aggression? Behavioral genetic studies of family members' resemblance for aggressive behavior help shed light on the matter. Twin and adoption studies agree with the experimental literature on aggression, which shows that a large effect of environmental factors is evident, particularly of the nonshared variety. Yet, there is also plenty of evidence, based on a variety of definitions of aggressive behavior from children to adults, for genetic propensity toward aggression (see reviews by Burt, 2009; Miles and Carey, 1997; Rhee and Waldman, 2002). Although few behavioral genetic studies have explicitly examined the question of gene by environment (G × E) interactions, we contend that such interactions are likely to exist and that the genetic propensity for aggression should exert its effects more strongly in some situations than others. Consistent with the early findings of Milgram and Zimbardo, individual genetic predispositions should moderate the extent to which aggression can be elicited, even in extreme situations such as these infamous studies. Our view is that while many, if not most, humans may have the potential for aggression and violence under the right circumstances, not all individuals will succumb to these behaviors under the same circumstances.

This chapter will review recent evidence of genetic and environmental influences on human aggression, with particular attention to several key questions and issues. We first consider how estimates of the relative importance of genetic effects (i.e., heritability) may vary across forms of aggression and the way in which it is measured. As detailed in other chapters of this volume, there are numerous definitions of aggression. Some definitions distinguish between reactive and proactive forms (Dodge *et al.*, 1997; Raine *et al.*, 2006), and others consider direct and indirect forms of aggression (e.g., physical vs. relational; Lahey *et al.*, 2004; Tackett *et al.*, 2009). Some definitions may include extreme criminal violence, such as assault, rape, and murder, although these extreme behaviors are relatively rare and have not been studied extensively in genetically informative designs. Measures of aggression can include self-reporting, teacher and parent reports (particularly for young children), and official records from schools or the justice system. This review focuses on twin and sibling adoption studies of aggressive behavior measured as a trait within the wider population. We compare effect sizes (heritability) across these various definitions and ways of measuring aggression. We also consider how heritability estimates may vary across both age and gender. Given higher levels of aggression in males across the lifespan, one obvious question concerns whether genetic propensities are of greater importance in one sex and how these differences might vary across age. We consider a variety of measurable environmental factors that might moderate these genetic influences and which may thus lead to G × E interactions for aggressive behavior. Although direct tests of G × E interactions have been relatively rare in the behavioral genetic literature on human aggression, it is likely that such interactions exist, given their robust effects in other forms of antisocial behavior (e.g., property criminal offending; Cloninger *et al.*, 1982).

Finally, we briefly review evidence for specific genetic influences in aggression by summarizing some of the more recent findings from molecular genetic studies. These effects are reviewed in detail elsewhere in this volume, so our focus here is on how a few specific genes may be involved in $G \times E$ interactions.

I. HERITABILITY OF AGGRESSION: TWIN AND ADOPTION STUDIES

Behavioral genetic research relies on the different levels of genetic relatedness between family members in order to estimate the relative contribution of heritable and environmental factors to individual differences in a phenotype of interest. Major research designs include: (a) studies of twins raised together and (b) studies of adopted individuals and their biological and adoptive family members. Although designs combining both approaches are the most powerful for separating genetic and environmental effects in human behavior, such studies of twins separated at birth and raised apart are rare and have not studied aggressive behavior extensively. Nonetheless, there are a handful of adoption studies and over two dozen studies of twins raised together which have specifically examined the genetic and environmental influence in aggression in nonselected samples from Northern America and Europe that are reasonably representative of the general population.

In the classical twin design, monozygotic (identical) twins share their common environment and they are assumed to share 100% of their genes. Dizygotic (fraternal) twins also share their common environment and they are assumed to share on average 50% of their genes. By comparing the resemblance for aggressive behavior between monozygotic and dizygotic twins, the total phenotypic variance of aggression can be divided into additive genetic factors (or heritability, h^2), shared environmental factors (c^2), and nonshared environmental factors (e^2). Shared environmental factors refer to nongenetic influences that contribute to similarity within pairs of twins. Nonshared environmental factors are those individual experiences that cause siblings to differ in their levels of aggressive behavior. Heritability is the proportion of total phenotypic variance due to genetic variation (Neale and Cardon, 1992). Genetic influences may also be divided into those that are additive (i.e., allelic effects add up across loci) and those that are nonadditive (i.e., due to dominance or epistasis). In twin studies, however, it is not possible to estimate both additive and nonadditive genetic effects (d^2) simultaneously with shared twin environment effects. The twin correlations summarized in Table 8.2 can be used to estimate the genetic and environmental influences to aggressive behavior. Twice the difference between the MZ and DZ correlations provides an estimate of the relative contribution of additive genetic influences to aggressive behavior [$h^2 = 2(r_{MZ} - r_{DZ})$]. The contribution of the nonadditive genetic effects due to dominance or epistasis (d^2) is obtained by subtracting four times the DZ correlation from twice the MZ correlation ($d^2 = 2r_{MZ} - 4r_{DZ}$). The proportion of the variance that is due to shared environmental influence is given by subtracting the MZ correlation from twice the DZ correlation ($c^2 = 2r_{DZ} - r_{MZ}$). Finally, the contribution of the non-shared environmental influences can be obtained by subtracting the MZ correlation from unit correlation ($e^2 = 1 - r_{MZ}$) (Posthuma *et al.*, 2003). Many twin studies do not specifically examine or test for nonadditive genetic effects and instead report heritability estimates based on additive effects only. However, some twin studies compare models with additive effects and nonadditive effects versus models with additive genetic effects and shared environment.

In sibling adoption studies, the correlation between adoptive siblings is compared with the correlation between biological siblings to estimate the influence of genetic and environmental factors on aggressive behavior (Plomin *et al.*, 2001). Resemblance between adoptive siblings for measures of aggression is indicative of shared (or common) family environment, while the extent to which biological sibling resemblance exceeds that of adoptive siblings is taken as evidence of heritable genetic influences for aggressive behavior.

There have been a few meta-analyses of twin and adoption studies of aggressive behavior and the wider construct of antisocial behavior. In one early meta-analysis of 24 twin and adoption studies, heritable influences explained about half of the total variance in aggressive behavior and the nonshared environment explained the remaining 50% (Miles and Carey, 1997). Rhee and Waldman (2002) also summarized the results from 51 twin and adoption studies on criminal behavior, delinquency, psychopathy, conduct disorder, and antisocial personality disorder, as well as aggressive behavior, in children, adolescents, and adults. Genetic factors explained 41% of the variance in antisocial behavior, 16% was explained by shared environmental influences, and the remaining 43% of variance was explained by nonshared environmental factors. A more recent review focused on 19 twin and adoption studies using child and adolescent samples; studies including adult subjects were excluded. Heritability was found to explain 65%, shared environment explained 5%, and the nonshared environment explained the remaining 30% of the variance in aggressive behavior (Burt, 2009). Both Burt (2009) and Rhee and Waldman (2002) examined nonadditive genetic effects, but only Rhee and Waldman (2002) found significant nonadditive genetic effects for antisocial behavior. It is noteworthy that genetic influences are consistently found across these reviews, while shared environmental influences are comparatively small or nonexistent. Family similarity in aggressive and antisocial behavior, therefore, is primarily the result of shared genes, not environment.

Tables 8.1 and 8.2, respectively, summarize a large selection of twin and sibling adoption studies which have specifically examined the genetic and environmental influences on aggressive behavior in child, adolescent, and adult samples. Several studies use prospective, longitudinal designs, and large samples, and three of the twin studies were designed, in particular, to study

Table 8.1. Effect Sizes for Aggressive Behavior from Adoption Studies

Study (author, year)	Aggression measure	Informant	Age in years	Sex	Biological siblings r(N)	Adoptive siblings r(N)	h^2	c^2
Dutch adoptees (van den Oord et al., 1994)	Aggression (CBCL)	Parent ratings	10–15 Mean=12.4	M	0.40 (30)	0.02 (44)	0.52	0.00
				F	0.45 (35)	0.21 (48)	0.32	0.25
				MF	0.38 (46)	0.05 (129)		
Dutch adoptees (van der Valk et al., 1998)	Aggression (CBCL)	Parent ratings	10–15	M+F	0.42 (111)	0.13 (221)	0.61	0.13
Colorado adoptees (Deater-Deckard and Plomin, 1999)	Aggression (CBCL)	Parent ratings	13–18	M+F	0.36 (152)	0.26 (156)	0.52	0.12
			7, 9, 10, 11, 12 Mean=9.5	M+F	0.39 (94)	0.26 (78)	0.24	0.27
	Aggression (TRF)	Teacher ratings		M+F	0.25 (188)	−0.06 (156)	0.49	0.00
Unknown (Parker, 1989)[a]	Aggression (as reported in Rhee and Waldman, 2002; Burt, 2009)	Parent ratings	4–7	M+F	0.44 (66)	0.47 (45)		

M, Male; F, Female; h^2, heritability; c^2, shared environment; CBCL, Child Behavior Checklist (Achenbach, 1991b); TRF, Teacher Report Form (Achenbach, 1991a).

[a]Genetic and shared environmental estimates were not reported by the authors.

Table 8.2. Effect Sizes (Correlations) for Aggressive Behavior from Twin Studies

Study sample (author, year)	Aggression measure	Assessment method	Age in years	Sex	MZ $r(N)$	DZ $r(N)$	h^2[a]	c^2	Sex limitation effects
Boston twins (Scarr, 1966)	Aggression (ACL)	Parent ratings	6–10	F	0.35 (24)	−0.08 (28)	0.40[a]	—	N/A
Missouri twins (Owen and Sines, 1970)	Aggressive reaction (MCPS)	Lab observation	6–14	M / F	0.09 (10) / 0.58 (8)	−0.24 (11) / 0.22 (13)	0.44[a]	—	Not tested
California twins (Rahe et al., 1978)	Aggression (ACL)	Self-report	42–56 Mean=48	M	0.31 (93)	0.21 (97)	0.56[a]	—	N/A
Colorado twins (O'Connor et al., 1980)	Aggression/bullying (PSR)	Parent ratings	Mean=7.6	M+F	0.72 (52)	0.42 (32)	—	—	Not tested
London twins, UK (Rushton et al., 1986)	Aggression (IBS)	Self-report	19–60 Mean=30	M+F	0.40 (296)	0.04 (179)	0.72[a]	—	Not tested
California preschool twins	Aggression (CBCL)	Parent ratings	Mean=5.2	M+F	0.78 (21)	0.31 (17)	0.94[a]	—	Not tested
Ghodesian-Carpey and Baker, 1987	Aggression (MOCL)	Mothers' observations	Mean=5.2	M+F	0.65 (21)	0.35 (17)	0.60[a]	—	
Philadelphia twins (Meininger et al., 1988)	Impatience/aggression Competitive achievement striving	Teacher rating	6–11	M+F	0.67 (71) / 0.63 (71)	0.11 (34) / 0.13 (34)	1.12[a] / 1.00[a]	— / —	Not tested
Minnesota twins (McGue et al., 1993)[c]	Aggression (MPQ)	Self-report	Mean=19.8	M+F	0.61 (79)	−0.09 (48)			Not tested
	Aggression (MPQ)	Self-report	Mean=29.6	M+F	0.58 (79)	−0.14 (48)			Not tested
Midwest twins (Cates et al., 1993)	BDHI—assault BDHI—indirect hostility	Self-report Self-report	Mean=42.5 Mean=42.5	F F	0.07 (77) 0.40 (77)	0.41 (21) 0.01 (21)	0.00 0.78	— —	N/A N/A
	BDHI—verbal hostility	Self-report	Mean=42.5	F	0.41 (77)	0.06 (21)	0.70	—	N/A
Colorado twins (Schmitz et al., 1995)	Aggression (CBCL)	Parent rating	2–3 4–11	M+F	0.68 (77) / 0.79 (66)	0.40 (183) / 0.41 (137)	0.52 / 0.55	0.16 / 0.19	Not tested

(Continues)

Table 8.2. (Continued)

Study sample (author, year)	Aggression measure	Assessment method	Age in years	Sex	MZ r(N)	DZ r(N)	h^2	c^2	Sex limitation effects
Ohio twins, Western Reserve Twin Project (Edelbrock et al., 1995)	Aggression (CBCL)	Parent rating	7–15 Mean = 11.0	M+F	0.75 (99)	0.45 (82)	0.60	0.15	Not tested
Dutch twins (van den Oord et al., 1996)	Aggression (CBCL)	Parent rating	3	M	0.81 (210)	0.49 (236)	0.69	0.12	Not tested
				F	0.83 (265)	0.49 (238)			
				MF		0.45 (409)			
Minnesota twins (Finkel and McGue, 1997)	Aggression (MPQ)	Self-report	27–64 Mean = 37.8	M	0.37 (220)	0.12 (165)	0.35	0.00	NS quantitative sex differences
				F	0.39 (406)	0.14 (352)	0.39	0.00	
				MF		0.12 (114)			
VET twins (Coccaro et al., 1997)	BDHI—assault	Self-report	Mean = 44.1	M	0.50 (182)	0.19 (118)	0.47	0.00	N/A
	BDHI—indirect hostility	Self-report		M	0.42 (182)	0.02 (118)	0.40	0.00	N/A
	BDHI—verbal hostility	Self-report		M	0.28 (182)	0.07 (118)	0.28	0.00	N/A
Swedish twins (TCHAD; Eley et al., 1999)	Aggression (CBCL)	Parent rating	8–9	M	0.72 (176)	0.41 (182)	0.70	0.07	NS quantitative sex differences, NS qualitative sex differences
				F	0.82 (160)	0.45 (194)			
				MF		0.41 (310)			
UK twins (sample obtained from Register of Child Twins; Eley et al., 1999)	Aggression (CBCL)	Parent rating	12	M	0.68 (99)	0.45 (93)	0.69	0.04	NS quantitative sex differences, NS qualitative sex differences
				F	0.77 (124)	0.44 (80)			
				MF		0.27 (95)			
Virginia twins (Simonoff et al., 1998)	Aggression (physical)	Parent rating	8–16	M+F	0.76 (268)	0.46 (166)	0.58	0.18	Not tested
		Self-report	8–16	M+F	0.31 (268)	0.22 (166)	0.21	0.11	Not tested

Sample	Measure	Method	Age	Sex					
Missouri twins (Hudziak et al., 2000)	Aggression (CBCL)	Parent rating	8–12	M	0.77 (129)	0.50 (156)	0.77	0.00	Not tested
				F	0.73 (91)	0.40 (115)	0.70		
Dutch twins (Hudziak et al., 2003)	Aggression (TRF)	Teacher rating	7	M	0.72 (181)	0.33 (160)	0.69	0.00	NS quantitative sex differences
				F	0.71 (214)	0.33 (151)			
				MF		0.26 (330)			
			10	M	0.73 (153)	0.41 (140)	0.72	0.00	Significant quantitative sex differences
				F	0.73 (202)	0.25 (125)	0.21	0.49	
				MF		0.17 (283)			
Dutch twins (van Beijsterveldt et al., 2003)[c]	Aggression (CBCL)	Parent rating	3	M	0.81 (1055)	0.55 (1066)			Significant quantitative sex differences
				F	0.82 (1226)	0.53 (997)			
				MF		0.48 (2144)			
			7	M	0.83 (927)	0.48 (1069)			Significant quantitative sex differences
				F	0.84 (898)	0.53 (858)			
				MF		0.51 (1723)			
			10	M	0.84 (526)	0.50 (621)			Significant quantitative sex differences
				F	0.79 (471)	0.55 (458)			
				MF		0.47 (907)			
			12	M	0.86 (289)	0.45 (317)			Significant quantitative sex differences
				F	0.83 (237)	0.55 (233)			
				MF		0.57 (433)			
Canadian twins (Dionne et al., 2003)	Aggression (physical)	Parent rating	1.5	M + F	0.59 (107)	0.28 (174)	0.58	0.00	Not tested
UK (E-risk) twins (Taylor, 2004)	Aggression (CBCL)	Parent rating	5	M + F	0.73 (602)	0.24 (514)	0.72	0.00	Not tested
South Wales twins (Button et al., 2004)	Aggression (IAB)	Self-report	11–18 Mean = 13.8	M + F	0.64 (115)	0.40 (143)	0.68	0.00	Not tested
Finnish twins (Vierikko et al., 2004)	Aggression (MPNI)	Parent rating	12	M	0.72 (260)	0.59 (292)	0.14	0.75	Significant quantitative sex differences
				F	0.78 (300)	0.53 (278)	0.54	0.25	
				MF		0.58 (517)			

(Continues)

Table 8.2. (*Continued*)

Study sample (author, year)	Aggression measure	Assessment method	Age in years	Sex	MZ r(N)	DZ r(N)	h^2	c^2	Sex limitation effects
Dutch twins (Polderman et al., 2006)	Aggression (TRF)	Teacher rating (same teacher)	5	M+F	0.84 (67)	0.43 (59)	0.49	0.00	Not tested
		Teacher rating (different teachers)	5	M+F	0.40 (45)	0.21 (44)			
Colorado twins (Haberstick et al., 2006a)	Aggression (CBCL)	Parent rating	7[b]	M	0.74 (69)	0.56 (76)	0.79	0.00	NS quantitative sex differences
				F	0.79 (91)	0.44 (62)			
			9	M	0.57 (73)	0.50 (63)	0.76	0.00	NS quantitative sex differences
				F	0.76 (75)	0.55 (60)			
			10	M	0.77 (58)	0.47 (52)	0.76	0.00	NS quantitative sex differences
				F	0.70 (67)	0.56 (57)			
			11	M	0.64 (58)	0.41 (55)	0.84	0.00	NS quantitative sex differences
				F	0.86 (56)	0.59 (49)			
			12	M	0.68 (69)	0.45 (61)	0.79	0.00	NS quantitative sex differences
				F	0.83 (78)	0.45 (65)			
	Aggression (TRF)	Teacher rating	7	M	0.63 (71)	0.39 (70)	0.58	0.00	NS quantitative sex differences
				F	0.56 (79)	0.34 (62)			
			8	M	0.58 (66)	0.46 (62)	0.61	0.00	NS quantitative sex differences
				F	0.70 (70)	0.41 (60)			
			9	M	0.54 (63)	0.29 (59)	0.59	0.00	NS quantitative sex differences
				F	0.67 (74)	0.37 (53)			
			10	M	0.39 (63)	0.44 (56)	0.43	0.00	NS quantitative sex differences
				F	0.49 (64)	0.18 (54)			
			11	M	0.56 (68)	0.12 (54)	0.52	0.00	NS quantitative sex differences
				F	0.50 (70)	0.45 (54)			
			12	M	0.35 (55)	0.32 (39)	0.42	0.00	NS quantitative sex differences
				F	0.48 (60)	0.24 (49)			

Study	Measure	Method	Age	Sex	MZ (n)	DZ (n)			Sex differences
Ad-Health (Cho et al., 2006)	Aggression	Self-report	12–19	M	0.47 (141)	0.29 (131)	0.50	0.00	Not tested
				F	0.47 (141)	0.27 (114)	0.30	0.00	
				MF		0.21 (197)			
Colorado twins (Gelhorn et al., 2006)	Aggression (DISC items)	Self-report	11–18 Mean = 14.5	M+F	0.47 (531)	0.27 (569)	0.49	0.00	Not tested
				MF		0.28 (212)			
Finnish twins (von der Pahlen et al., 2008)	Aggression (BPAQ)	Self-report	18–33	M	0.45 (190)	0.22 (167 +321 sibs)	0.70	0.00	Not tested
				F	0.52 (608)	0.18 (387 +1838 sibs)	0.69	0.00	
				MF		0.20 (508 +1559 sibs)			
Norwegian twins (Czajkowski et al., 2008)	Passive aggression (DSM-IV)	Self-report	19–36 Mean = 28.2	M	0.35 (221)	0.45 (116)	0.14	0.18	NS quantitative sex differences, NS qualitative sex differences
				F	0.30 (448)	0.19 (261)			
				MF		0.21 (340)			
Tennessee twins (Tackett et al., 2009)	Relational aggression (CAPS)	Self-report	9–18	M	0.54 (356)	0.39 (328)	0.49	0.00	NS quantitative sex differences
				F	0.41 (376)	0.36 (332)			
				MF		0.16 (589)			
		Parent rating	9–18	M	0.66 (356)	0.61 (328)	0.21	0.46	Significant quantitative sex differences
				F	0.65 (376)	0.35 (332)	0.42	0.22	
				MF		0.48 (589)			
California twins—RFAB cohort (Baker et al., 2008)	Reactive aggression (RPQ)	Parent rating	9–10	M	0.48 (141)	0.35 (87)	0.26	0.27	NS quantitative sex differences
				F	0.60 (142)	0.46 (98)			
				MF		0.50 (151)			
		Self-report	9–10	M	0.38 (138)	0.28 (83)	0.38	0.00	Significant quantitative sex differences
				F	0.37 (139)	0.38 (96)	0.00	0.36	
				MF		0.08 (146)			
		Teacher rating	9–10	M	0.59 (67)	0.49 (45)	0.20	0.43	NS quantitative sex differences
				F	0.70 (68)	0.43 (45)			
				MF		0.60 (62)			

(Continues)

Table 8.2. (*Continued*)

Study sample (author, year)	Aggression measure	Assessment method	Age in years	Sex	MZ r(N)	DZ r(N)	h^2	c^2	Sex limitation effects
	Proactive aggression (RPQ)	Parent rating	9–10	M	0.61 (141)	0.34 (87)	0.32	0.21	NS quantitative sex differences
				F	0.57 (142)	0.48 (98)			
				MF		0.55 (151)			
		Self-report	9–10	M	0.60 (138)	0.34 (83)	0.50	0.00	Significant quantitative sex differences
				F	0.12 (139)	0.28 (96)	0.00	0.14	
				MF		0.14 (146)			
		Teacher rating	9–10	M	0.56 (67)	0.42 (45)	0.45	0.14	NS quantitative sex differences
				F	0.74 (68)	0.38 (45)			
				MF		0.35 (62)			
California twins—RFAB cohort (follow-up) (Tuvblad et al., 2009)	Reactive aggression (RPQ)	Parent rating	11–14	M	0.49 (102)	0.33 (55)	0.43	0.15	NS quantitative sex differences
				F	0.58 (98)	0.38 (77)			
				MF		0.42 (103)			
	Proactive aggression (RPQ)	Parent rating	11–14	M	0.55 (102)	0.35 (55)	0.48	0.08	NS quantitative sex differences
				F	0.46 (98)	0.27 (77)			
				MF		0.40 (103)			
Weighted average				M	0.66	0.42			
				F	0.63	0.35			
				MF	0.59	0.38			
				M+F		0.28			

MZ, monozygotic; DZ, dizygotic; M, male twin pairs; F, female twin pairs; MF, male–female twin pairs; M + F, male and female pairs combined; h^2, heritability; c^2, shared environment; CBCL, Child Behavior Checklist (Achenbach, 1991b) [20 items scored as 0 (not true), 1 (somewhat true), and 2 (very true), e.g., bragging and boasting, argues a lot, cruelty or meanness to other, disobedience (home and school)]; MOCL, Mothers' Observational Checklist, including the following behaviors: rejection, destructiveness, negativism, noncompliance, teasing, physical negative, insult, verbal threat, yelling; ACL, Adjective Checklist (Gough, 1960) [consists of 300 adjectives that yields 26 scales]; MCPS, Missouri Children's Picture Series (Sines et al., 1966) [consists of 238 line drawings, each portrays the figure of a child engaged in some activity or situation, the subject is required to sort the cards into two groups, those that look like fun and those that do not look like fun]; PSR, Parent Symptom Ratings (Conners, 1970) [includes six aggression items: bullying, hits or kicks other children, mean, sassy to grown-ups, fights constantly, picks on other children]; IBS, Interpersonal Behavior Survey (Mauger, 1980) [includes items such as: "some people think I have a violent temper" or "I try not to give people a hard time"]; The Mathews Youth Test for Health (MYTH; Mathew and Angulo, 1980) [developed to measure Type A behavior in school-aged children. The instrument consists of 17 items characterized by overt type A behavior and yields two subscales: impatience/aggression and competitive achievement striving]; BDHI, Buss-Durkee Hostility Inventory (Buss and Durkee, 1957) [contains of three subscales: the assault scale (10 items on physical aggression), the verbal hostility scale (13 items on verbal aggression), and the indirect hostility scale (nine items on indirect or undirected or displaced aggression)]; MPQ, Minnesota Personality Questionnaire (Tellegen, unpublished) [physically aggressive, vindictive, likes violent scenes, higher order factor, negative emotionality]; Physical Aggression scale, (Simonoff et al., 1998) [items on physical aggression, extortion, public fight, use of weapon in a fight, cruelty to animal, thrown objects at people, carried a weapon, sworn at teacher, based on Olweus, 1989]; Physical Aggression scale (Dionne et al., 2003) is a 37 item check list on which parents reported whether the child engaged, sometimes engaged, often engaged in a behavior. Based on factor analysis, 10 of the 37 behaviors were determined as direct physical aggression, for example, is cruel toward others, bullies other children, bites others, kicks, fights, takes things away from others, pushes, threatens to hit; IAB, instrument of aggressive behavior (Olweus, 1989) [contains two subscales: aggressive and nonaggressive antisocial behavior. The aggression scale contains 11 items of direct verbal and physical aggression, e.g., swearing at a teacher, bullying]; MPNI, Multidimensional Peer Nomination Inventory (Pulikkinen et al., 1999) [contains of 38 items and the aggression subscale contains of six items, e.g., calls people names, may hurt other kids, bullying, goes around telling people's secrets to others]; Ad-Health [aggression is based on four items, got into a serious physical fight, hurt someone badly enough they needed medical care, used to threaten to use a weapon to get something from someone, took part in a gang fight]; DISC, Diagnostic Interview Schedule for Children (Shaffer et al., 2000) [nine items on physical aggression and five items on verbal aggression, e.g., I cannot help getting into arguments when people disagree with me, I have threatened people I know, I get into fights a little bit more often than average people]; BPAQ, Buss and Perry Aggression Questionnaire (Buss and Perry, 1992) [the Norwegian version of the Structured Interview for DSM-IV personality (Pfohl et al., 1997) [the instrument is a semi-structured diagnostic interview for the assessment of all DSM axis II disorders, including passive-aggressive personality disorder]; CAPS, Child and Adolescent Psychopathology Scale (Lahey et al., 2004) [relational aggression was assessed via the CAPS, a structured interview assessing DSM-IV symptoms of common childhood disorders. Seven items measured relational aggression, for example, tried to keep kids he/she does not like outside his/her friend group, spread rumors to make others stop liking someone, stopped talking to people because he/she was mad at them, teased other people in a mean way]; RPQ, Reactive and Proactive Questionnaire (Raine et al., 2006). [The RPQ is a validated 23-item questionnaire designed to measure reactive and proactive aggression in children and adolescents from the age of 8. The RPQ includes 11 reactive items (e.g., "he/she damages things when he/she is mad"; "he/she gets mad or hits others when they tease him/her") and 12 proactive items (e.g., "he/she threatens and bullies other kids"; "he/she damages or breaks things for fun"). The items in the RPQ have a three-point response format: 0 = never, 1 = sometimes, 2 = often.]; RFAB, USC twin study of Risk Factors for Antisocial Behavior; TCHAD: Twin Study of Child and Adolescent Development.

[a]Heritability estimate is based on either Holzinger's H or Falconer equation and did not report shared environmental influences.

[b]Parent reported CBCL ratings were not collected at age 8.

[c]Genetic and shared environmental estimates were not reported by the authors.

aggressive and antisocial outcomes. All three of these studies are ongoing. One of these is the University of Southern California Twin Study of Risk Factors for Antisocial Behavior (RFAB), which is a prospective study of the interplay of genetic, environmental, social, and biological (psychophysiological) factors on the development of antisocial and aggressive behavior from childhood to emerging adulthood. The project includes more than 750 twin pairs studied on several occasions, at ages 9–10, 11–13, 14–16, and 17–18 years (Baker et al., 2006). A second major twin study is the Environmental Risk Longitudinal Twin Study (E-risk study) in the United Kingdom. The E-risk study involves data on more than 1000 twin pairs at ages 5, 7, and 12 with the special focus on what factors in the home, family, school, and neighborhood (i.e., environmental risks) promote children's aggression (Moffitt, 2002). The Minnesota Study of Twins and Families (MFTS) is a third major longitudinal twin study that specifically investigates antisocial behavior and substance use across development. MTFS was established in 1989 using same-sexed twin pairs aged 11 or 17. Five hundred additional 11-year-old twin pairs were added in 2000. All twins of those ages, who were born in Minnesota, as identified by birth registry data, were invited to participate. Participants are asked about academic ability, personality, and interests; family and social relationships; mental and physical health; and physiological measurements. Of particular interest are prevalence of psychopathology, substance abuse, divorce, leadership, and other traits and behaviors related to mental and physical health, relationships, and religiosity (Iacono et al., 2006; Keyes et al., 2009).

Before reviewing the twin and sibling adoption studies on aggressive behavior presented in Tables 8.1 and 8.2, it is important to consider the ways in which the phenotype of aggressive behavior is defined and measured. The various instruments utilized in the studies reviewed in this chapter are summarized in Tables 8.1 and 8.2, to provide a clear idea of the nature of the aggressive behavior phenotype being investigated. By and large, the Child Behavior Checklist (CBCL; Achenbach, 1991b) has been used more often than any other single instrument in behavioral genetic studies of aggression. Although self-report version of the CBCL is available for older adolescents and young adults (Youth Self Report (YSR); Achenbach, 1991c), studies more commonly rely on parent or teacher (Teacher's Report Form (TRF); Achenbach, 1991a) rating versions. The aggressive behavior subscale of the CBCL includes 20 items on which the child is rated. These include defiance, argumentativeness, physical aggression, and cruelty toward others. Although there are a handful of other instruments that also yield single aggressive behavior scores, two instruments provide multiple scales: the Reactive and Proactive Aggression Questionnaire (RPQ; Raine et al., 2006), which provides separate scales for aggressive reactions to provocation and more planned or proactive forms of aggression; and the Buss-Durkee Hostility Inventory (BDHI; Buss and Durkee, 1957), which yields several subscales of aggression, including assault, verbal, and indirect hostility.

The studies summarized in Tables 8.1 and 8.2 vary on how aggressive behavior was defined (i.e., physical, verbal, relational, reactive, proactive, indirect, bullying) and measured (observation, self-report, parent/caregiver, teacher). A wide range of ages were included, from preschool children to adults; however, the vast majority of studies have used childhood samples (i.e., 12 years of age or younger) which explains why the CBCL is so frequently used to assess aggressive behavior. Correlations for biological and adoptive siblings (Table 8.1), and MZ and DZ twins (Table 8.2) are shown for each study. Most studies reported correlations separately for same-sex pairs of males (M), females (F), and opposite-sex pairs (MF); however, a few studies involve correlations for samples of male and female pairs combined. We review the key questions concerning the genetic influence (heritability) of human aggression based on the effect sizes reported for these studies. We also examine various potential moderators of these effects, including sex, age, method of assessment, form of aggression, study design (twin vs. sibling adoption design), and various social factors and circumstances that may exacerbate or ameliorate the genetic risk for aggression from one person to the next.

A. Does heritability vary depending on sex?

Since it is well documented that males are much more likely than females to engage in most forms of aggressive behavior (Moffitt et al., 2001; Rutter et al., 2003), it is also of interest to examine whether the same genetic and environmental influences are important in both sexes and whether the magnitude of these effects differs between males and females.

In the classical twin design, genetic and environmental variance components for aggressive behavior can be estimated using data from same-sex MZ and DZ twins. Apart from estimating genetic and environmental effects on aggression, it is also possible to investigate whether sex-specific genetic or environmental influences are important. Such effects are referred to as sex-limitation or sex-limited effects. There are two primary questions about sex limitation in genetic research, one being whether there are *qualitative* differences between males and females, such that different genes and/or environmental influences operate in the two sexes, and whether *quantitative* differences exist in the relative magnitude of influences across sexes. To assess whether the magnitude of genetic and environmental effects in aggressive behavior differ between males and females (i.e., quantitative sex differences), only data from same-sex twin pairs are required. However, to determine whether or not it is the same set of genes or shared environmental experiences that influences aggressive behavior in males and females (i.e., qualitative sex differences), data from opposite-sex twin pairs are also needed. If qualitatively different genetic influences are important for aggressive behavior in males and females, then the opposite-sex twins will be less genetically similar for the trait than DZ twins.

Not all twin studies have examined sex-limited effects, either qualitative or quantitative, and several studies combined males and females when computing twin correlations, making it impossible to evaluate these effects based on published results shown in Table 8.2. Nonetheless, quantitative sex differences can be easily evaluated across at least 18 studies in Table 8.2, which present separate twin or sibling correlations by sex. Among these, there are a dozen studies that also include MF, which allow investigation of qualitative sex differences. The average twin correlations across these 18 studies, weighted by their respective sample sizes, shows quite similar twin correlations for both identical ($r_{MZ-Males} = 0.66$; $r_{MZ-Females} = 0.63$) and nonidentical same-sex pairs ($r_{DZ-Males} = 0.42$; $r_{DZ-Females} = 0.35$), indicating that there are no appreciable quantitative sex differences in aggressive behavior. This is consistent with the individual results across studies which formally tested for quantitative sex differences (e.g., Baker *et al.*, 2008; Czajkowski *et al.*, 2008; Eley *et al.*, 1999; Finkel and McGue, 1997; Tackett *et al.*, 2009; Tuvblad *et al.*, 2009). As indicated in Table 8.2, only a small handful of studies have reported significant differences in heritability of aggression for males and females (and these are primarily for younger samples; e.g., Hudziak *et al.*, 2003; van Beijsterveldt *et al.*, 2003; Vierikko *et al.*, 2004). The lack of quantitative sex differences is also well in line with what was reported in a recent meta-analysis summarizing 19 twin and family/adoption studies, whereby genetic influences were found to explain 54% of the variance in aggressive behavior in boys and 57% of the variance in girls (Burt, 2009).

There is no evidence of qualitative sex differences either, given that the weighted twin correlation for MF ($DZ_{Male-Female}$) is 0.38, which is quite similar to the same-sex DZ twin correlations (0.42 in males and 0.35 in females).

In spite of the consistent sex difference in mean levels of aggression, the underlying etiologies of aggressive behavior appear to be remarkably similar for both sexes. There may still be biological and social differences between the sexes that might account for the greater mean levels of aggression observed in males, yet the same genes and the same environmental factors appear to explain individual differences in aggression within each sex to the same degree. One interesting question that has not been addressed, however, is to what extent there may be sex differences in *moderators* of genetic factors. In other words, there may be different circumstances or experiences in males and females that lead to greater expression of genetic predispositions for aggression. For example, sexual jealousy might trigger genetic propensity for aggression to a greater extent in males than females, while threats to resources might be a more important moderator of genetic influences in females compared to males, as discussed in Chapter 9. Other moderators are discussed later in II.A, although more research is clearly warranted to explore the degree to which they may be sex specific.

B. Does heritability change across age?

Although genetic studies of aggression have spanned from childhood to adulthood, most studies included in Tables 8.1 and 8.2 involved children 12 years of age or younger. This suggests that more studies examining the heritability of aggressive behavior in adolescents and adults are needed. Keeping this in mind, it is useful to examine the magnitude of twin correlations across age groups, which span from early childhood to middle-age adults. These correlations are summarized in Fig. 8.1, according to five age groups (early childhood, age 1.5–6 years; middle childhood, age 7–10; adolescence 11–15; late adolescence/young adulthood, age 16–26; and adulthood, age 27–48; Fig. 8.1). These results show that aggressive behavior is clearly influenced by genetic factors across the lifespan, given the fact that the MZ correlations exceed those for DZ pairs at all ages. (The

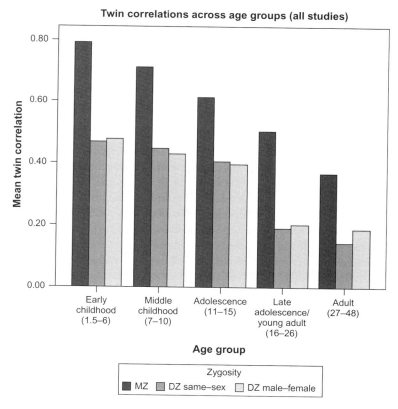

Figure 8.1. Twin correlations across age groups (all studies). (For color version of this figure, the reader is referred to the Web version of this chapter.)

lack of qualitative sex differences is also evident across the life span, in that the DZ correlation is comparable for same-sex and MF pairs across ages.) However, both MZ and DZ correlations decline steadily across development, suggesting the waning importance of shared environmental effects from childhood to adolescence and then adulthood. The DZ correlation exceeds half the value of the MZ correlation (taken as evidence for shared environment) only in early childhood, but not in later age groups. The pattern shown in Fig. 8.1 is evident in individual studies as well. Aggressive behavior in childhood is influenced by genetic factors in all studies, and most of these studies also report shared environmental influences (Table 8.2; e.g., Baker et al., 2008; Eley et al., 1999; Hudziak et al., 2003; Schmitz et al., 1995; Simonoff et al., 1998; Tuvblad et al., 2009; van den Oord et al., 1996; Vierikko et al., 2004; but for an exception see Dionne et al., 2003; Taylor, 2004). Studies including adolescent twins (~younger than 19 years of age) do not in most cases report finding any shared environmental influences (Table 8.2; e.g., Button et al., 2004; Cho et al., 2006; Gelhorn et al., 2006; Tackett et al., 2009). Similarly, studies including adult twins do not report finding any shared environmental influences (Table 8.2; e.g., Coccaro et al., 1997; Finkel and McGue, 1997; von der Pahlen et al., 2008; but for an exception see Czajkowski et al., 2008). The overall pattern across the studies presented in Table 8.2 and Fig. 8.1 indicates that genetic influences for aggressive behavior become increasingly more important, while shared environmental effects become less so as children develop from childhood, through adolescence, and into adulthood. Similarly, findings from a recent meta-analysis reported that genetic influences increased from 55.2% at ages 1–5 years to 62.7% at ages 6–10 years and 62.9% at ages 11–18 years. At the same time, shared environmental influences were decreasing from 18.7% at ages 1–5 years to 13.9% at ages 6–10 years and 2.7% at ages 11–18 years (Burt, 2009). This pattern of decrease in shared environment, and a concomitant increase in heritability during development, is relatively common for personality traits and cognitive abilities (Bartels et al., 2002; Loehlin, 1992; Plomin et al., 2001; Scarr and McCartney, 1983), and has also been found for other phenotypes including prosocial behavior (Knafo and Plomin, 2006).

It should also be kept in mind, however, that methods of assessing aggression vary across age, such that studies of children tend to rely on ratings by teachers and parents, while studies of adults (and some older adolescents) rely more heavily on self-report methods. The confound between method of assessment and age of the subjects has made it difficult in prior studies and meta-analyses to disentangle age effects on heritability from differences that arise from different methods of assessment. Increasing heritability estimates from child to adulthood could therefore also be explained by different methods of assessment as well (e.g., parental bias may lead to overestimation of shared environmental effects and thus attenuate heritability estimates in childhood).

C. Do heritabilities vary across methods of assessment?

It is important to examine the magnitude of twin correlations across methods of assessment, as heritability estimates may vary depending on who is rating the subject. This is especially important given the age trends found for heritable influences in Fig. 8.1, since different methods of assessment tend to be employed for different age groups. As previously discussed, studies of younger subjects rely on parent or teacher ratings, while self-report methods are typically used in studies of adults and often adolescents. The twin correlations are summarized in Fig. 8.2, according to laboratory observation, self-reports, teacher ratings, and parent/caregiver ratings. Indeed, twin correlations—and thus the estimates of genetic and environmental influences on aggressive behavior—do appear to vary across method of assessment. According to the twin correlations summarized in Fig. 8.2, the heritability of aggressive behavior based on laboratory observation is approximately 32% $[h^2: 2(r_{MZ} - r_{DZ}) = 2(0.27 - 0.11)]$, dominant genetic effect accounts for approximately 10% $[d^2: 2(r_{MZ} - r_{DZ}) = 2(0.27) - 4(0.11)]$, and the nonshared

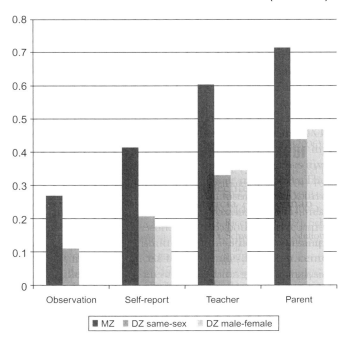

Figure 8.2. Twin correlations across method of assessment (all studies). (For color version of this figure, the reader is referred to the Web version of this chapter.)

environment accounts for the remaining 58% of the variance. The heritability of aggressive behavior based on self-reports is 40% and the nonshared environment accounts for the remaining 60% of the variance. There is no evidence of shared environmental contribution, as the DZ correlation is approximately half the MZ correlation. The heritability of aggressive behavior based on teacher ratings is 54%, shared environmental influences account for 6% [c^2: $2r_{DZ} - r_{MZ} = 2(0.33 - 0.60)$], and the nonshared environment accounts for the remaining 40% of the variance. The heritability of aggressive behavior based on parent/caregiver ratings is 54%, shared environmental influences account for 17%, and the nonshared environment accounts for the remaining 29% of the variance. Thus, parent/caregiver ratings have the largest familial influence, explaining 71% of the variance ($h^2 + c^2 = 0.54\% + 17\%$) in individual differences in aggressive behavior. It should be kept in mind that this is a descriptive approach and that formal modeling is required to determine how well these estimates describe the observed data. Also, this approach does not allow for actual testing of different hypotheses, for example, to test whether it is possible to set any of these effects to zero. It is difficult to discern whether the parent/caregiver rating patterns reflect true shared environment or are instead an artifact of rater bias whereby raters are less able to discriminate between the two twins' aggressive behavior and thus inflate the similarities between them, regardless of zygosity. In fact, when the two co-twins are rated by different teachers (e.g., they are in different classrooms), twin correlations are lower for both MZ and DZ pairs for the wider construct of antisocial behavior (Baker et al., 2007).

A few specific twin studies in Table 8.2, which utilize multiple raters in their design, also illustrate that genetic and environmental influences on aggressive behavior vary depending on method of assessment. For example, twin similarity for relational aggression was influenced only by genetic factors when using self-reported data, explaining 49% of the total variance. When using parent ratings (only biological mothers were used as raters) of relational aggression both genetic and shared environmental influences were important (boys: $h^2 = 21\%$, $c^2 = 46\%$, $e^2 = 33\%$; girls: $h^2 = 42\%$, $c^2 = 22\%$, $e^2 = 36\%$). However, when using youth self-reports, only genetic and nonshared environmental influences were significant ($h^2 = 49\%$, $e^2 = 51\%$; Tackett et al., 2009). A similar pattern was found for reactive and proactive aggression in 9–10-year-old boys, whereby only genetic influences were important for self-reports, but both genes and shared environment were important for teacher and parent ratings (Baker et al., 2008). Aggressive behavior in another sample of twins aged 7–12 years was found to be largely influenced by genetic (or familial) factors (76%–84%), as reported by parents. Data were collected from one or both parents; however, only mother-reported ratings were included in the analyses, as they accounted for the majority of the ratings collected (85.3%–90.1%). In contrast, when teacher ratings were used, aggressive behavior was found to be slightly less influenced by genetic (or familial) factors (42%–61%; Haberstick *et al.*, 2006a). Significant shared environmental effects were not found for either parent or teacher ratings, and were therefore dropped from the models, suggesting that any shared environmental influences are likely to be included in the genetic component. Apart from rater bias, which may result in inflated twin similarity across the board when using single raters for multiple children, the varying patterns of genetic and environmental influence across methods of assessment could be the result of different raters reporting different aspects of the child's aggressive behavior. This could arise in part because individuals behave differently in different situations (e.g., school vs. home) or because some types of aggressive behaviors are more likely to be noticed (e.g., overt forms such as physical aggression) than other types of aggressive behaviors which may be more subtle or covert (e.g., relational aggression). Different raters provide important and unique pieces of information regarding behaviors. Self-reporters are aware of their own motives and behaviors, which may go undetected by their caregivers, teachers, or peers. On the other hand, caregivers or teachers may be able to understand difficult and complex constructs better than children. A teacher is also more likely to compare a child's behavior to his or her peers, whereas a parent is likely to compare a child's behavior to his or her siblings (Bartels *et al.*, 2003). Regardless of the source of these discrepant results across methods of assessment, it is important to keep in mind that when it comes to studies of aggression, it matters who is doing the rating.

D. Do heritabilities vary across forms of aggression?

Different types of aggressive behavior have been investigated across twin and adoption studies, with notable distinctions between reactive and proactive forms of aggression, as well as direct/physical and indirect/relational aggression (Table 8.3). It is likely that there are different etiologies for different forms of aggression; for example, defensive reactions to threatening stimuli may be more environmentally influenced, while more planned, proactive forms may be more genetically influenced (Tuvblad *et al.*, 2009). Comparing heritability estimates collectively across the various measures employed is a reasonable way to address

Table 8.3. Forms of Aggression

Form of aggression	Description
Reactive/hostile/affective	Angry or frustrated responses to a real or perceived threat
Proactive/instrumental	Planning, the motive of the act extends beyond harming the victim
Direct/physical	Intentionally causing pain or harm to the victim
Indirect/relational	Relational social manipulation such as gossip and peer exclusion

this question about whether some kinds of aggression are more heritable than others. When multiple forms of aggressive behavior are measured within the same study, it is also possible to investigate the extent to which the same genes and/or environmental factors are important to different manifestations of aggressive tendencies.

Reactive aggression refers to angry or frustrated responses to a real or perceived threat. This specific type of aggression has been characterized to involve both high emotional arousal, impulsivity, and an inability to regulate or control affect. In contrast, proactive aggression is conceptualized as a more regulated, instrumental form of aggression, with more positive expectancies about the outcomes of aggression (Dodge, 1991; Dodge and Coie, 1987; Schwartz et al., 1998). Although reactive and proactive aggression have each been found to be mainly influenced by genetic and nonshared environmental factors, their genetic correlation is significantly less than 1.0, indicating some genetic specificity for the two forms of aggression. Reactive and proactive aggression each exhibit different developmental patterns in these influences (see Table 8.2; Baker et al., 2008; Tuvblad et al., 2009), that is, the genetic and environmental *stability* in reactive and proactive aggression has been found to differ. In one of the few longitudinal analyses of these constructs, the stability in reactive aggression from childhood to adolescence could be explained by genetic (48%), shared (11%), and nonshared (41%) environmental influences, whereas the continuity in proactive aggression was primarily genetically (85%) mediated (Tuvblad et al., 2009).

Relational forms of aggression, which involve social manipulation such as gossip and peer exclusion, are often more indirect compared to other forms of aggression (Crick and Grotpeter, 1996). Like reactive and proactive forms, relational aggression appears to be influenced by genetic factors, both in self-report (49%) as well as parental reports (boys, 42%; girls, 21%). However, unlike the other more direct forms of aggression, relational aggression is also influenced by shared environmental influences, but only in parental reports (boys, 22%; girls, 46%) and not in self-report measures (Tackett et al., 2009). Together these findings—that is, the less than perfect genetic correlation between reactive and proactive forms, their different developmental etiologies, and the significant shared environmental effects in relational aggression only—provide support for at least some genetic and environmental etiological distinction among different forms of aggression. It should be noted, however, that no study to date has examined the genetic and environmental overlap between relational aggression and other forms such as proactive and reactive aggression.

Other studies based on multifactorial measures of aggression, such as the BDHI, suggest some variability in the heritability estimates across subscales, although the patterns are not entirely consistent across studies. For example, "indirect hostility" showed the lowest heritability (28%) in one study of adult twins (Coccaro et al., 1997), compared to modest heritability for "verbal hostility" (47%) and "assault" (40%). Yet, Cates et al. (1993) found no genetic influences for assault, but strong heritability for both verbal hostility (78%) and indirect hostility (70%). In multivariate genetic analyses, both studies found some support for genetic specificity for the various subscales, similar to what has been found for reactive and proactive aggression, in that genetic correlations (r_G) among the BDHI subscales were less than unity: $r_G = 0.39$ between indirect hostility and assault, $r_G = 0.60$ between indirect hostility and verbal hostility, $r_G = 0.17$ between verbal hostility and assault (Coccaro et al., 1997), $r_G = 0.35$ between indirect hostility and assault, $r_G = 0.39$ between indirect hostility and verbal hostility, and $r_G = 0.49$ between verbal hostility and assault (Cates et al., 1993). Overall, genetic influences are generally found for most, if not all forms of aggression, although somewhat different genetic factors may be operating across these different forms. The mechanisms that underlie more direct, planned, confrontational, and often physical forms of aggression may to some extent be different than those for reactive or indirect aggressive behaviors.

E. Does heritability vary depending on study design (twins vs. adopted siblings)?

There were only a handful of studies identified examining the heritability of aggressive behavior using the sibling adoption design. Visual inspection of the results from these sibling adoption studies (see Table 8.1) compared to the results from studies including twin samples (Table 8.2) indicate that heritability estimates (i.e., h^2) and the shared environmental estimates (i.e., c^2) for aggressive behavior are very similar. This is also well in line with the results of a meta-analysis on antisocial and aggressive behavior that found no differences between twin and sibling adoption studies ($h^2 = 48\%$; $c^2 = 13\%$, $e^2 = 0.39\%$) (Rhee and Waldman, 2002). Thus, the heritability of aggressive behavior does not seem to vary depending on study design.

F. Criticisms of twin and adoption studies: Assumptions and generalizability

There are several assumptions in both twin and adoption studies that are important to consider when reviewing their findings. In adoption studies, the most important factors are (1) random placement of the adoptees into homes and (2) generalizability. Selective placement or matching (i.e., similarities between adoptive and biological parents) for certain characteristics can lead to inflated correlations between adoptive siblings (and thus overestimated effects of shared environment). Although such matching may occur for physical characteristics (including race), direct selective placement is unlikely to be made for aggressive behavior, per se. (Children with more aggressive or antisocial biological parents would not be placed into homes with more aggressive adoptive parents.) Thus, it is unlikely that the genetic and environmental effects summarized in Table 8.1 are biased in any way as a result of selective placement. In terms of generalizability, it is often the case that adoptive parents tend to be in good health and from higher socioeconomic levels; thus, findings from adoption studies may not always be unquestionably generalized to the entire population (Rutter, 2006). Adopted children may also be at greater risk for aggression compared to nonadoptees, since birth parents giving up their children may have increased rates of disordered behaviors, including substance use, criminal offending, and aggression (Cloninger *et al.*, 1985; Lewis *et al.*, 2001). In the Deater-Deckard and Plomin (1999) study, the adopted children did in fact have higher aggression scores compared to the nonadopted children, consistent with the notion that adoptees may be at higher genetic risk for aggression compared to nonadopted individuals. The elevated levels of aggression in adoptees occurred in spite of the fact that background characteristics of the adoptive families were found to be representative of families with children in the larger Denver area, and that the demographic characteristics of the adoptive grandparents and the adopted children's biological grandparents were similar, with regard to educational and occupational level. Similarly, van der Valk *et al.* (1998) reported mean differences between adoptees and nonadoptees, with adoptees showing higher mean levels in aggressive behavior. About 75% of the adoptees were adopted from Korea, India, Columbia, Indonesia, Bangladesh, or Lebanon and the remaining 25% were adopted from European or other non-European countries in both the van der Valk *et al.* (1998) and the van den Oord *et al.* (1994) studies, and the majority of the adoptive parents had a higher level of occupation. Given the higher aggression scores among adoptees compared to nonadoptees, as well as the somewhat greater affluence and ethnic heterogeneity in at least some of the adoption samples, the generalizability of adoption study results to the wider population could be questioned.

To what extent are the twin study results generalizable to the wider population? Twins and singletons have been found to experience similar rates of psychiatric disorders (e.g., attention-deficit hyperactivity disorder (ADHD), oppositional defiant disorder, conduct disorder) and behavioral and emotional problems (Gjone and Novik, 1995; Moilanen *et al.*, 1999; Simonoff *et al.*, 1997; van den Oord *et al.*, 1995). Findings from the RFAB study show no differences in mean levels between MZ and DZ twins in reactive or proactive aggression (Baker *et al.*, 2008; Tuvblad *et al.*, 2009). It can, therefore, be assumed that twins and singletons display equal rates of aggressive behavior.

There are, however, two ways in which twins differ from singletons: (i) lower birth weight, due to shorter length of gestation (Plomin *et al.*, 2001) and (ii) delayed language development (Rutter and Redshaw, 1991). Birth weight has been found to have a minimal effect on academic performance; for twins this effect was judged relative to what is a normal birth weight for twins and not for singletons (Christensen *et al.*, 2006). However, studies have shown that children with birth complications are more likely to later develop antisocial and aggressive behavior (Raine, 2002), but birth complications may not by themselves predispose antisocial and aggressive behaviors, but will require the presence of an environmental risk factor (e.g., poor parenting, maternal rejection). In other words, the relationship between birth complications and antisocial and aggressive behavior is confounded by environmental risk factors (Hodgins et al., 2001; Raine et al., 1997).

In addition to generalizability, there are several key assumptions of the classical twin design that need to be kept in mind when reviewing findings from these studies. These include (1) the equal environments assumption, (2) random mating, and (3) lack of correlation or interaction between genetic and environmental influences. We briefly review each of these assumptions here—both in general and as they pertain to aggressive behavior in particular—and consider their possible effects on the results summarized across studies.

Perhaps the most important and commonly criticized assumption is the "equal environment assumption" (EEA). In the classical twin design, MZ twins, who are assumed to share 100% of their genes, are compared to DZ twins, who are assumed to on average share 50% of their genes. If MZ twins are more similar than DZ twins, it may be inferred that the difference is caused by genetic effects. To make this inference, however, it is necessary to rely on the EEA. It is assumed that environmentally caused similarity is roughly equal for both MZ and DZ twins. If this assumption is violated, higher correlations among MZ twins may be due to environmental factors, rather than genetic factors, and heritability estimate will be overestimated (Plomin et al., 2001).

Several twin studies of various phenotypes have examined the EEA. One way to test the validity of the EEA is to examine whether a trait of interest is influenced by perceived versus assigned zygosity. The effect of perceived zygosity can be added as a "specified" familial environment in a univariate ACE twin model (Kendler et al., 1993) and if this parameter can be omitted without any significant loss in data fit, it can be assumed that the EEA holds for the phenotype under study. These studies generally report that the EEA assumption holds for numerous phenotypes such as physical activity, eating behavior, psychiatric disorders (e.g., major depression, generalized anxiety disorder, phobia, alcohol, and drug abuse; Eriksson et al., 2006; Hettema et al., 1995; Kendler et al., 1993; Klump et al., 2000; Xian et al., 2000), including child and adolescent psychopathology such as anxiety disorder, ADHD, oppositional defiant disorder, conduct disorder, antisocial behavior (Cronk et al., 2002; Jacobson et al., 2002; Tuvblad et al., 2011) as well as aggressive behavior (Derks et al., 2006).

The assumption of random mating for aggression in the parents of the twins is also important to consider, since nonrandom mating can lead to increased resemblance for DZ but not MZ twin pairs. Assortative mating in the parent generation acts to increase the resemblance between dizygotic twins and thereby bias shared environmental estimates upward and additive genetic effects downward. A significant correlation between spouses for a particular trait is often interpreted as assortative mating (Maes *et al.*, 1998). This assumption is probably violated when it comes to antisocial and aggressive behavior, as significant spouse correlations have been found suggesting that assortative mating exists in this behavioral domain (Krueger *et al.*, 1998; Maes *et al.*, 2007; Taylor *et al.*, 2000). Taylor *et al.* (2000) found that parents of twins were correlated for retrospectively reported delinquency ($r = 0.23$ in families of boys and $r = 0.35$ in families of girls) and concluded that assortative mating is modest in degree. Another study using data from the Dunedin sample in New Zealand (Silva and Stanton, 1996) when the participants were 21-years-old found a correlation ($r = 0.54$) between couple members' reports of antisocial behavior in their peers (i.e., participants were asked how many of their friends had aggression, personal, alcohol, or drug problems, or did things against the law), which was identical to the correlation for couple members' reports of their own antisocial behavior as measured by a variety of offenses (e.g., theft, force, fraud, vice). They concluded that assortative mating for antisocial behavior is substantial and that antisocial individuals tend to cluster in peer groups with similar antisocial peers. As such, assortative mating should to be taken into account when modeling antisocial behavior (Krueger *et al.*, 1998). It is interesting, however, that the shared environmental effects are fairly negligible in twin studies of aggressive behavior, both in the prior meta-analyses as well as in our summary in Table 8.2. Thus, any assortative mating for aggression does not appear to have resulted in severe overestimates of shared environment when considering these studies *en masse*. It is possible, on the other hand, that genetic influences themselves have been underestimated and could be larger than the 50% or so than these meta-analyses and our summary suggest.

It is also assumed in the classical twin design that genetic and environmental influences combine additively (i.e., do not interact) and are uncorrelated. It is possible, however, that some genetic predispositions may be associated with certain kinds of social environments or experiences, leading to a correlation between genes and environments. ($G \times E$ interactions are also discussed at length in a later section of this chapter.) Such $G \times E$ correlations (r_{GE}) can arise in three different ways (Scarr and McCartney, 1983): (i) Passive r_{GE} occurs when genes overlap between parents and their offspring. For example, a child with aggressive parents inherits genetic susceptibility for aggression as well as experiences an adverse rearing environment. An example of passive r_{GE} was reported in a study comparing genetic and environmental influences on mothering. Passive r_{GE} correlations were suggested for mother's positivity and monitoring. For mother's negativity and control, primarily nonpassive r_{GE} correlations were suggested (Neiderhiser *et al.*, 2004). (ii) Evocative/reactive r_{GE} can arise when a specific child characteristic elicits a particular response from the environment. For example, aggressive children tend to elicit more negative affect and harsh discipline from their parents (Ge et al., 1996; O'Connor et al., 1998). In a more recent study, using the classical twin design the association between parental criticism and adolescent antisocial behavior was found to be entirely genetically influenced. Approximately half of the genetic contribution to this association was explained by early adolescent aggression. Thus, child aggression seemed to elicit negative parenting followed by adolescent antisocial behavior, indicating an evocative r_{GE} (Narusyte et al., 2006). (iii) Active r_{GE} is defined as the process whereby an individual actively seeks out environmental situations that are more closely matched to the person's genotype. Active r_{GE} has been suggested in adolescent drinking behavior, specifically among girls (Loehlin, 2010). If the assumption of no $G \times E$ correlation is violated, heritability estimates for aggressive behavior in twin studies could include both additive genetic effects and the effects of $G \times E$ correlation (i.e., heritability estimates are inflated). Apart from these specific examples cited here, few studies have examined the effects of r_{GE} in aggressive behavior, making it difficult to know the extent of their effect on heritability estimates in twin studies.

In conclusion, findings from adoption studies should probably be generalized cautiously to other populations as adoptees tend to show higher scores on aggressive behavior compared to controls. On the other hand, most of the assumptions of the classical twin design seem to hold for aggressive behavior. The EEA has been tested and found to hold for various phenotypes including aggressive behavior, and twins and singletons have been found to display similar scores on aggressive behavior, suggesting that findings from twin studies can be generalized to other populations. Most twin studies report finding little or no shared environmental influences on aggressive behavior, suggesting that random mating is of little importance for aggressive behavior. Only a few studies have examined the influence of $G \times E$ correlation on aggressive behavior, suggesting that more research is needed on this topic before we can draw any firm conclusions. Last, in the classical twin design, it is assumed that genetic and environmental influences combine additively and do not interact. This assumption is probably violated to some extent when it comes to aggressive behavior, as several studies have reported finding significant interaction effects. $G \times E$ interaction is discussed in detail in the next section of this chapter.

II. G × E INTERACTION IN AGGRESSIVE BEHAVIOR

There is a general recognition that genes and environment work together—often in complex ways—to produce wide variations in behavior and psychological function. $G \times E$ interaction, by definition, is a statistical term indicating that genetic effects on a given phenotype depend upon environmental factors or vice versa. Gene expression, for example, can be moderated by an individual's experiences or exposure to certain environments. Likewise, various individuals may respond differently to the same environmental exposure because they have different genotypes. Such genetic sensitivity to the environment has been demonstrated extensively in plant and animal species for a variety of traits. But even though the importance of $G \times E$ interactions in human behavior has long been considered (Eaves, 1984; Mather and Jinks, 1982), $G \times E$ interactions have been rarely reported in human traits until relatively recently. The failure to find $G \times E$ interactions in studies of human characteristics may be due to a number of factors. One likely explanation is related to statistical power. In general, it is difficult to detect $G \times E$ effects due to their low statistical power (Rowe, 2003). For example, behavioral genetic studies rely on genetic relatedness for groups of individuals, rather than on sharing of specific alleles between pairs of relatives. Studies relying on variance partitioning often do not find significant $G \times E$ effects, or find that they explain a very small portion of the total variance, and are thus dropped from further analysis. When $G \times E$ is not taken into account in behavioral genetic studies, heritability estimates will tend to be biased, although the direction of the bias depends on whether the moderating environmental influences are of the shared or nonshared variety.

$G \times E$ interactions can be tested or modeled in behavioral genetic studies using several different study designs (e.g., twin or adoption). The two most frequently used methods testing for $G \times E$ interactions in twin and adoption studies include: (1) a *mean levels approach*, testing whether mean values of a phenotype differ across different combinations of genetic risk and environmental settings and (2) a *moderated variance components approach*, examining whether genetic and environmental variance for a trait varies across different measured environmental settings. These two different methods stem from the same conceptual idea, namely, that genetic effects vary across environments or vice versa. Their interpretations and meanings can be rather different, since one is based on means and the other is based on variances. The mean levels $G \times E$ is perhaps a more traditional approach, is typically presented as a statistical interaction in an ANOVA, and indicates whether particular experiences, exposures, or other conditions may (on average) ameliorate or exacerbate specific genetic effects in groups of individuals with similar genetic and environmental risks. In comparison, the moderated effects approach does not address mean levels of risk, but instead evaluates whether variance in genes or environment differs across various measured conditions. Moderation can occur in either raw variances (V_A, V_C, and V_E) or in relative effects (*i.e.*, h^2, c^2, *and* e^2), and may not necessarily coincide with $G \times E$ interactions found in mean levels. The latter point is important when evaluating $G \times E$ interactions across studies using these different approaches, since different patterns can emerge from them. For example, it is possible that certain adverse environments (e.g., low socioeconomic status (SES)) may lead to some genes exerting stronger effects (mean levels), while overall, the relative variance explained by genes (heritability) may be greater in other environments (e.g., high SES). The approach used for testing G × E interactions can vary across study design, such that adoption designs or studies with measured genes are generally required for the mean levels approach, while the moderated variance components approach may be used in both twin and adoption studies in the absence of measured genes.

In molecular genetic studies, both genes and environment are measured, rather than inferred from correlations among family members. G × E interactions can therefore be tested in the general population, that is, without necessary reliance on a twin or adoption design. There are still advantages, nonetheless, to include measured genes and measured environments in the context of a family-based design, including twins and other siblings as well as parents and offspring.

Evidence of G × E interaction in aggressive behavior has been reported in twin and adoption studies, and more recently in molecular genetic studies. Below is a summary of some of the G × E interaction findings in aggressive behavior from adoption, twin and molecular genetic studies. We also discuss two potential moderators of genetic and environmental influences on aggressive behavior, exposure to media violence, and alcohol use.

A. Potential moderators of genetic influence found in adoption and twin studies

1. Family adversity and social disadvantage

G × E interaction for aggressive behavior has been found in several of the early adoption studies, using a mean levels approach. What these early adoption study findings generally showed was that early adverse environments had a greater negative impact on genetically "higher risk" children. Adopted children with criminal biological parents reared by a family where there was adversity showed higher rates of antisocial and aggressive behavior than adopted children with antisocial biological parents not raised in a home with adversity, and than adopted children raised in adversity who are not at higher genetic risk. For example, the interaction of inherited and postnatal factors was examined in about 800 Swedish men adopted at an early age. When both inherited factors and environmental risk factors were present, 40% were found to be criminal; if only genetic factors were present, 12.1% were criminal; if only environmental factors were present, 6.7% were criminal; and with neither inherited nor environmental factors being present, 2.7% were criminal (Cloninger et al., 1982). The fact that 12.1% plus 6.7% is less than 40% would thus be an indication of G × E interactions. This finding was later replicated in females (Cloninger and Gottesman, 1987). It should be pointed out, however, that in the adoption design, the genetic risk factors themselves are considered in a general way, such that the exact nature of the genes is left unspecified, both in terms of which loci or alleles may be involved and what underlying mechanisms may be involved in the path from genes to phenotype. Similarly, the environmental risk factors as indexed by certain traits in the adoptive parents or characteristics of their home do not necessarily specify the exact nature of the child's experiences or how these lead to various outcomes.

Further, maltreatment places children at risk for psychiatric morbidity, especially conduct problems. However, not all maltreated children will develop conduct problems. A recent twin study tested whether the effect of physical maltreatment on risk for conduct problems was strongest among those who were at high genetic risk for these problems using data from the E-risk study, a representative cohort of 1116 5-year-old British twin pairs and their families. Maltreatment was found to be associated with a greater increase in the probability of developing conduct problems among children who had a high genetic liability for conduct disorder compared to children who had a low genetic liability (Jaffee et al., 2005). This finding is consistent with the $G \times E$ interaction found in adoption studies of antisocial and aggressive behavior, in which genetic effects were more pronounced in adverse environments. This clearly suggests that children in risky environments would benefit from interventions. However, another view of this interaction is that favorable genotypes can play a protective role on children's risk for conduct problems, especially under circumstances of maltreatment.

There are also a few studies based on twin samples that have used the moderated variance components approach to examine whether measured environmental (risk) factors moderate the genetic and environmental variances for aggressive behavior. For example, the heritability of conduct problems was found to be lower in children growing up in dysfunctional families and higher in children growing up in families where dysfunction was absent (Button et al., 2005). Another twin study used DeFries-Fulker regression analysis to examine whether genetic and environmental influences on aggressive behavior varied depending on levels of family warmth (DeFries and Fulker, 1985). Genetic influence on aggressive behavior was found to be higher in schools with higher average levels of family warmth. In contrast, environmental influences (both shared and nonshared) were more important in schools with lower average levels of family warmth (Rowe et al., 1999). These findings suggest that genetic effects are more likely to explain individual differences in aggression in more benign environments, whereas in more disadvantaged environments negative family-related factors and context-dependent risks may play a greater role than genetic predispositions in aggressive and antisocial outcomes.

Many early theories about the causes of delinquency and crime assumed that delinquents come from socially disadvantaged backgrounds. For example, Merton postulated that antisocial behavior resulted from the strain caused by the gap between cultural goals and the means available for their achievement (Merton, 1957). Social disadvantage and poverty constitute a reasonable robust, although not always a strong, indication of an increased risk for antisocial and aggressive behavior, assessed by self-reports and official convictions (Leventhal and Brooks-Gunn, 2000; Rutter et al., 1998). SES has also been found to moderate the relative influence of genetic factors on antisocial and aggressive behavior. In a sample of Swedish 16–17-year-old twins, heritability for antisocial and aggressive behavior was higher in the more affluent neighborhoods (boys, 37%; girls, 69%) compared to the less advantaged neighborhoods (boys, 1%, girls, 61%). Conversely, the shared environment was higher in the less advantaged neighborhoods (boys, 69%; girls, 16%) compared to better-off neighborhoods (boys, 13%; girls, 6%). Following the "social push hypothesis," Raine (2002) would suggest that the genetic factors on antisocial and aggressive behavior are more expressed in a socioeconomically advantaged environment where the environmental risk factors are absent. On the contrary, genetic factors for antisocial behavior will be weaker and the shared environment more important in a socioeconomically disadvantaged environment because the environmental risk factors will "camouflage" the genetic contribution (Tuvblad et al., 2006).

These studies using the moderated variance components approach (e.g., Button et al., 2005; Rowe et al., 1999; Tuvblad et al., 2006), all examine whether an environmental (risk) factor moderates genetic and environmental *variance* on antisocial and aggressive behavior. Findings from these studies show that heritable influences on aggressive behavior vary depending on environmental context, indicating the importance of the environmental risk factors in the development of aggressive behavior as well as for gene expression.

2. Violent media exposure

There is an ongoing debate about whether exposure to violent video games increases aggressive behavior, and it is very possible that exposure to media violence could moderate the influences of genetic and environmental influences on aggressive behavior. One line of research argues that mass media exposures contribute to a child's socialization. A primary process in such socialization is observational learning (Bandura, 1973). Children and adolescents mimic what they see and acquire complicated scripts for behaviors, beliefs about the world, and moral precepts about how to behave in the long run from what they observe (Huesmann, 2010). In contrast, another line of research argues that there is little empirical evidence for a link between media exposure and violence. This line of research argues that media violence cannot have any important psychological effect on the risk for aggressive behavior (Ferguson and Kilburn, 2010).

A recent meta-analysis that included 136 studies examined the effects of violent video games on aggressive behaviors. The evidence suggested that exposure to violent video games is a risk factor for increased aggressive behavior, aggressive cognition, and aggressive affect, and for decreased empathy and prosocial behavior. Moderator analyses showed significant research design effects, weak evidence of cultural differences in susceptibility and type of measurement effects, and no evidence of sex differences in susceptibility. Sensitivity analyses were also carried out and they revealed these effects to be robust, with little evidence of selection (publication) bias (Anderson et al., 2010).

Others studies examining the relationship between violent video games and aggressive acts have found little evidence for a relationship. A recent review included a total of 25 studies comprising 27 independent observations. The corrected overall effect size for all included studies was only $r = 0.08$ (Ferguson and Kilburn, 2009). The mixed findings in the literature clearly suggest that more research is needed to resolve whether there is a link between exposure to violent video games and aggressive behavior. Also, some studies have found that exposure to violent video games only explains a small fraction of the variance. An explanation for this paradox could be that exposure to violent video games moderates the influence of genetic and environmental effects on aggressive behavior, rather than exerting direct effects. No genetically informative studies have examined violent video game exposure as a possible moderator of genetic influence on aggression, however, leaving this as an important area in need of study.

3. Alcohol use

It has long been known that some individuals become aggressive after consuming alcohol, and the relationship of violence and aggression with alcohol is well established (Bushman and Cooper, 1990; White et al., 2001). For example, a review including 130 independent studies found that alcohol was correlated with both criminal and domestic violence (Lipsey et al., 1997). Despite this, there is so far no behavioral genetic study that had examined whether alcohol use moderates the influence of genetic and environmental factors on aggressive behavior.

However, the genetic and environmental relationship among alcohol use and aggressive behavior as well as other disruptive and problem behaviors within the disinhibitory spectrum such as antisocial behavior, ADHD, conduct disorder, impulsive and sensation seeking personality traits has been examined in several large population-based twin studies. On a phenotypic level, disruptive and problem behaviors within the disinhibitory spectrum can be united by a common higher order externalizing factor (Krueger et al., 2002, 2005, 2007). This higher order externalizing factor has been found to be largely influenced by genetic factors. For example, the genetic influences on a common externalizing factor describing conduct disorder, substance use, ADHD, and novelty seeking was found to account for more than 80% of the variation in an adolescent sample (Young et al., 2000). Strong heritable influences on an externalizing factor of antisocial behavior, substance abuse, and conduct disorder has also been found among adults (Kendler *et al.*, 2003). Together these studies provide important insight into our understanding of externalizing behaviors. It seems that behaviors and disorders within the externalizing spectrum, including aggressive behavior, share a common genetic liability.

III. SPECIFIC GENES FOR AGGRESSIVE BEHAVIOR: FINDINGS FROM MOLECULAR GENETIC STUDIES

Increasing evidence suggests the importance of heritable factors in the development of aggressive behavior (Burt, 2009; Miles and Carey, 1997; Rhee and Waldman, 2002). The first study that showed a link between a specific genotype and aggressive behavior examined the genetic material of members of a large Dutch family. This specific family had for decades been found to be prone to violent, aggressive, and impulsive behavior, including fighting, arson, attempted rape, and exhibitionism. Some of the male family members were also intellectually disabled. The aggressive males in this large family were shown to share a mutation in the gene that codes for the enzyme MAO (monoamine oxidase A). MAO breaks down brain chemicals (neurotransmitters) such as serotonin, noradrenaline, and dopamine, which transmit messages from one nerve cell to the next. In the afflicted males, however, a mistake in the coding sequence governing proper production of MAO was detected. As a result, abnormally large quantities of these neurotransmitters were found in the blood of the affected males (Brunner *et al.*, 1993). Although this genetic defect remains the first such link to aggressive behavior in humans, exactly how the genetic defect causes aggressive, impulsive behavior, or mental retardation is not known.

Apart from MAO, only a few candidate genes have been linked to aggressive behavior to date. The candidate genes that have been found to be associated with aggressive behavior in humans have, in many cases, been replicated in animal studies. The majority of these candidate genes are genes of the dopamine, serotonin, and norepinephrine neurotransmitter systems. The dopamine system is involved in mood, motivation and reward, arousal, as well as other behaviors. The serotonin system is involved in impulse control, affect regulation, sleep, and appetite, whereas the epinephrine and norepinephrine system facilitate fight-or-flight reactions and autonomic nervous system activity (Niv and Baker, 2010). For example, dopaminergic candidates, including dopamine receptor *DRD4*, has been found to be involved in ADHD and externalizing behavior, and *DRD2* has been found to be involved in substance abuse and disinhibition (Niv and Baker, 2010). The *DRD3* polymorphism has been found to be associated with impulsivity. This association was significant in violent, but not in nonviolent individuals, and there were no association between *DRD3* and violence per se

(Retz *et al.*, 2003). Dopamine transporter gene *DAT1* has also been linked to ADHD (Waldman *et al.*, 1998), as well as with violent behavior and delinquency in adolescents and young adults (Guo *et al.*, 2007). Cateocholamine-O-methyl-transferase (COMT) has been examined primarily in children and adults with ADHD, and mixed evidence emerged for its association with conduct disorder and aggression (Caspi *et al.*, 2008). Several studies have provided evidence that the low activity *VNTR* alleles of *5HTTLPR* show associations with aggression, violence, aggressive symptoms of conduct disorder, and other forms of externalizing behavior (Haberstick *et al.*, 2006b; Linnoila *et al.*, 1983). Aggressive behavior has also shown associations with *SNPs* of epinephrine and norepinephrine. A recent study linked two *SNPs* of *PNMT* to cognitive and aggressive impulsivity in children and adolescents (Oades *et al.*, 2008).

A. G × E interaction involving specific genes for aggressive behavior

Advances in the field of molecular genetics have also made it possible for researchers to identify G × E interactions much more specifically. One of the most influential studies examining G × E in antisocial and aggressive behavior is Caspi *et al.* (2002), a famous study from 2002. The relationship between a functional polymorphism in the MAO-A gene encoding the neurotransmitter-metabolizing enzyme and early childhood maltreatment was examined in the development of antisocial behavior in males. A significant G × E interaction was detected, in that maltreated boys with a genotype conferring low levels of MAO-A were found to be more likely to later develop antisocial problems, including conduct disorder, adult violent crime, and antisocial personality disorder, than maltreated boys who had a genotype conferring high levels of MAO-A (Caspi *et al.*, 2002). So far, there have only been a few replications of this important finding (Foley *et al.*, 2004; Kim-Cohen *et al.*, 2006; Nilsson *et al.*, 2006). For example, Kim-Cohen *et al.* (2006) found that the MAO-A polymorphism moderated the development of psychopathology after experiencing physical abuse in a sample of 975 seven-year-old boys. This finding was extended to the maltreatment exposure closer in time as the subjects were 7-years-old compared with previous work by Caspi *et al.* (2002) in which the subjects were 26-years-old, and therefore the possibility of a spurious finding by accounting for passive and evocative G × E correlation could be ruled out. Passive G × E correlation, as discussed earlier, refers to the association between the genotype a child inherits from his/her parents and the environment in which the child is raised, and evocative G × E correlation occurs when an individual's (heritable) behavior evokes an environmental response. Further, the authors also conducted a meta-analysis including the following five studies: Caspi *et al.* (2002), Foley *et al.* (2004), Haberstick *et al.* (2005), Kim-Cohen *et al.* (2006), and Nilsson *et al.* (2006). The association between maltreatment and mental health problems was significantly stronger in the group of males with a genotype conferring low versus high MAO-A activity. This provides strong evidence that the MAO-A gene influences vulnerability to environmental stress and that this biological process can be initiated early in life. However, there is at least one published failure to replicate (Haberstick *et al.*, 2005), and this finding has been replicated neither in females (Sjöberg *et al.*, 2007) nor in African Americans (Widom and Brzustowicz, 2006).

A G × E interaction between the *DRD2* A1 allele and risk-level in family environments has been suggested in a sample of adolescents with criminal offenses, the National Longitudinal Study of Adolescent Health (Ad-Health). Polymorphisms in genes related to the neurotransmitter dopamine were associated with age of first police contact and arrests, but only for youth from low-risk family environments. More specifically, among those adolescents with a history of criminal offending, those at greatest risk for later onset were those with the A1 allelic form of the *DRD2* gene, in combination with favorable home environments as defined by maternal attachment, involvement, and engagement (DeLisi *et al.*, 2008). It is important to emphasize that this finding involves the age of onset of first police contact and not the overall risk for offending versus not offending.

There is also some evidence for a G × E interaction in the *5HTTLPR* genotype with adult violence, whereby home violence, familial economic difficulties, and educational or home-life disruptions during childhood were found to predict violent behavior later in life only in individuals with the short promoter alleles present (Reif *et al.*, 2007). A similar G × E interaction between the short allele of *5HTTLPR* and childhood adversity has also been reported for ADHD (Retz *et al.*, 2008).

The ability to detect G × E interactions in molecular genetic studies is both exciting and controversial. The identification of specific genetic markers and specific experiences provides the opportunity to evaluate genetic and environmental risk factors at the individual level. This significantly increases opportunities for developing effective treatments and preventions for antisocial and aggressive behavior as well as other forms of psychopathology, which is exciting. At the same time, increased understanding of individual risks has often been considered cautiously because of the potential for bias and discrimination of those individuals who are identified as being at highest risk for being afflicted with disorders.

IV. CONCLUSIONS

Studies (and meta-analyses) including both twin and adoption samples show that about half (50%) of the variance in aggressive behavior is explained by genetic influences in both males and females, with the remaining 50% of the variance being explained by nonshared environmental factors. Form of aggression (reactive, proactive, direct/physical, indirect/relational), method of assessment (observation, self, teacher, parent/caregiver), and age of the subjects—all seem to be significant moderators of the magnitude of genetic and environmental influences on aggressive behavior. Neither study design (twin vs. sibling adoption design) nor sex, on the other hand, seems to impact the nature or magnitude of these genetic and environmental influences on aggression.

Although we are unaware of any twin or adoption studies of aggression induced in authoritative situations such as in the Milgram or Stanford Prison studies, the vast evidence for genetic influences in most forms of aggression that have been studied could suggest that individual differences in those early studies might have stemmed in part from different genetic propensities in their subjects. Findings from $G \times E$ studies on aggressive behavior suggest that not all individuals will be affected to the same degree by these environmental exposures, and also that not all individuals will be affected to the same degree by the genetic predispositions. Adoption and twin studies rely on relationships between family members when examining $G \times E$ interaction effects, whereas molecular genetic studies are using both a measured environmental (risk) factor and a measured genetic factor. To date, there have only been a few twin/adoption and molecular studies that report finding $G \times E$ in aggressive behavior, either using the mean levels approach or the moderated effects approach. These studies have shown that various measures of family adversity and social disadvantage interact (or act as moderators) with genetic factors on aggressive behavior.

Today, we have the potential to identify genetic risks at the level of specific genes, and identify aspects of the environment that make some individuals more vulnerable than others. Yet, there will always be groups of individuals with the same combination of genetic risk and environmental vulnerability who will not engage in aggressive behavior. So, it is still only an increased (probabilistic) risk and not a biological determinism. In spite of such strong support for a genetic basis to aggressive behavior, the importance of potential interventions which are environmentally based must not be ignored. Environmental interventions could be developed, for example, through family or school-based programs, to reduce aggressive behavior. In fact, a general view held by behavioral genetics researchers is that the best way to understand environment—and hence develop effect treatment interventions—is through genetically informative designs such as twin and family data. By using twin and family data, it is not only possible to estimate the influence of heritable factors on a trait or a phenotype, but also the influence of environmental factors. Modern methods for identifying and understanding $G \times E$ interactions will provide a means for doing exactly this.

References

Achenbach, T. M. (1991a). Manual for Teacher's Report form and 1991 Profile. University of Vermont, Department of Psychiatry, Burlington, VT.

Achenbach, T. M. (1991b). Manual for the Child Behavior Checklist/4-18 and 1991 Profile. University of Vermont, Department of Psychiatry, Burlington, VT.

Achenbach, T. M. (1991c). Manual for the Youth Self-Report and 1991 Profile. University of Vermont, Department of Psychiatry, Burlington, VT.

Anderson, C. A., Shibuya, A., Ihori, N., Swing, E. L., Bushman, B. J., Sakamoto, A., Rothstein, H. R., and Saleem, M. (2010). Violent video game effects on aggression, empathy, and prosocial behavior in eastern and western countries: A meta-analytic review. *Psychol. Bull.* **136**(2), 151–173.

Baker, L. A., Barton, M., Lozano, D. I., Raine, A., and Fowler, J. H. (2006). The southern California twin register at the university of southern California: II. *Twin Res. Hum. Genet.* **9**, 933–940.

Baker, L. A., Jacobson, K. C., Raine, A., Lozano, D. I., and Bezdjian, S. (2007). Genetic and environmental bases of childhood antisocial behavior: A multi-informant twin study. *J. Abnorm. Psychol.* **116**(2), 219–235.

Baker, L. A., Raine, A., Liu, J., and Jacobson, K. C. (2008). Differential genetic and environmental influences on reactive and proactive aggression in children. *J. Abnorm. Child Psychol.* **36**, 1265–1278.

Bandura, A. (1973). Aggression: A Social Learning Theory Analysis. Prentice-Hall, New York.

Bartels, M., Rietveld, M. J., van Baal, C. M., and Boomsma, D. (2002). Genetic and Environmental Influences on the Development of Intelligence. *Behav. Genet.* **32**(4), 237–249.

Bartels, M., Hudziak, J. J., van den Oord, E. J. C. G., van Beijsterveldt, C. E. M., Rietveld, M. J. H., and Boomsma, D. (2003). Co-occurrence of aggressive behavior and rule-breaking behavior at age 12: Multi-rater analyses. *Behav. Genet.* **33**, 607–621.

Brunner, H. G., Nelen, M., Breakefield, X. O., Ropers, H. H., and van Oost, B. A. (1993). Abnormal behavior associated with a point mutation in the structural gene for monoamine oxidase A. *Science* **262**(5133), 578–580.

Burt, A. S. (2009). Are there meaningful etiological differences within antisocial behavior? Results of a meta-analysis. *Clin. Psychol. Rev.* **29**, 163–178.

Bushman, B. J., and Cooper, H. M. (1990). Effects of alcohol on human aggression: An integrative research interview. *Psychol. Bull.* **107**(3), 341–354.

Buss, A. H., and Durkee, A. (1957). An inventory for assessing different kinds of hostility. *J. Consult. Clin. Psychol.* **21**, 343–349.

Buss, A. H., and Perry, M. (1992). The aggression questionnaire. *J. Pers. Soc. Psychol.* **63**, 452–459.

Button, T. M. M., Scourfield, J., Martin, N., and McGuffin, P. (2004). Do aggressive and non-aggressive antisocial behaviors in adolescents result from the same genetic and environmental effects? *Am. J. Med. Genet. B Neuropsychiatr. Genet.* **129 B**, 59–63.

Button, T. M. M., Scourfield, J., Martin, N., Purcell, S., and McGuffin, P. (2005). Family dysfunction interacts with genes in the causation of antisocial symptoms. *Behav. Genet.* **35**, 115–120.

Caspi, A., McClay, J., Moffitt, T. E., Mill, J., Martin, J., Craig, I. W., Taylor, A., and Poulton, R. (2002). Role of genotype in the cycle of violence in maltreated children. *Science* **297**(5582), 851–854.

Caspi, A., Langley, K., Milne, B., Moffitt, T. E., O'Donovan, M., Owen, M. J., Polo Tomas, M., Poulton, R., Rutter, M., Taylor, A., Williams, B., and Thapar, A. (2008). A replicated molecular genetic basis for subtyping antisocial behavior in children with attention-deficit/hyperactivity disorder. *Arch. Gen. Psychiatry* **65**(2), 203–210.

Cates, D. S., Houston, B. K., Vavak, C. R., Crawford, M. H., and Uttley, M. (1993). Heritability of hostility-related emotions, attitudes, and behaviors. *J. Behav. Med.* **16**(3), 237–256.

Cho, H., Guo, G., Iritani, B. J., and Hallfors, D. D. (2006). Genetic contribution to suicidal behaviors and associated risk factors among adolescents in the US. *Prev. Sci.* **7**(3), 303–311.

Christensen, K., Petersen, I., Skytthe, A., Herskind, A. M., McGue, M., and Bingley, P. (2006). Comparison of academic performance of twins and singletons in adolescence: Follow-up study. *Br. Med. J.* **6**, 1–5.

Cloninger, C. R., and Gottesman, I. I. (1987). Genetic and environmental factors in antisocial behavior disorders. *In* "The Causes of Crime: New Biological Approaches" (S. A. Mednick, T. E. Moffitt, and S. A. Stack, eds.), pp. 92–109. Cambridge University Press, Cambridge.

Cloninger, C. R., Sigvardsson, S., Bohman, M., and von Knorring, A. L. (1982). Predisposition to petty criminality in Swedish adoptees. II. Cross-fostering analysis of gene-environment interaction. *Arch. Gen. Psychiatry* **39**, 1242–1247.

Cloninger, C. R., Bohman, M., Sigvardsson, S., and von Knorring, A. L. (1985). Psychopathology in adopted-out children of alcoholics. The Stockholm Adoption Study. *Recent Dev. Alcohol.* **3**, 37–51.

Coccaro, E. F., Bergeman, C. S., Kavoussi, R. J., and Seroczynski, A. D. (1997). Heritability of aggression and irritability: A twin study of the Buss-Durkee aggression scales in adult male subjects. *Biol. Psychiatry* **41**(3), 273–284.

Conners, C. K. (1970). Symptom patterns in hyperkinetic, neurotic, and normal children. *Child Dev.* **41**, 667–682.

Crick, N. R., and Grotpeter, J. K. (1996). Children's treatment by peers: Victims of relational and overt aggression. *Dev. Psychopathol.* **8**, 267–280.

Cronk, N., Slutske, W. S., Madden, P. A., Bucholz, K. K., Reich, W., and Heath, A. (2002). Emotional and behavioral problems among female twins: An evaluation of the equal environments assumption. *J. Am. Acad. Child Adolesc. Psychiatry* **41**, 829–837.

Czajkowski, N., Kendler, K. S., Jacobson, K., Tambs, K., Røysamb, E., and Reichborn-Kjennerud, T. (2008). Passive-aggressive (negativistic) personality disorder: A population-based twin study. *J. Pers. Disord.* **22**(1), 109–122.

Deater-Deckard, K., and Plomin, R. (1999). An adoption study of the etiology of teacher and parent reports of externalizing behavior problems in middle childhood. *Child Dev.* **70**, 144–154.

DeFries, J. C., and Fulker, D. F. (1985). Multiple regression analysis of twin data. *Behav. Genet.* **15**, 467–473.

DeLisi, M., Beaver, K. M., Wright, J. P., and Vaughn, M. G. (2008). The etiology of criminal onset: The enduring salience of nature and nurture. *J. Crim. Justice* **36**, 217–223.

Derks, E. M., Dolan, C. V., and Boomsma, D. I. (2006). A test of the equal environment assumption (EEA) in multivariate twin studies. *Twin Res.* **9**(3), 403–411.

Dionne, G., Tremblay, R., Boivin, M., Laplante, D., and Perusse, D. (2003). Physical aggression and expressive vocabulary in 19-month-old twins. *Dev. Psychol.* **39**(2), 261–273.

Dodge, K. (1991). The structure and function of reactive and proactive aggression. *In* "The Development and Treatment of Childhood Aggression" (D. J. Pepler and K. H. Rubin, eds.), pp. 201–218. Erlbaum, Hillsdale, NJ.

Dodge, K. A., and Coie, J. D. (1987). Social-information processing factors in reactive and proactive aggression in children's peer groups. *J. Pers. Soc. Psychol.* **53**, 1146–1158.

Dodge, K. A., Lochman, J. E., Harnish, J. D., Bates, J. E., and Pettit, G. S. (1997). Reactive and proactive aggression in school children and psychiatrically impaired chronically assaultive youth. *J. Abnorm. Psychol.* **106**, 37–51.

Eaves, L. J. (1984). The resolution of genotype x environment interaction in segregation analysis of nuclear families. *Genet. Epidemiol.* **1**, 215–228.

Edelbrock, C., Rende, R., Plomin, R., and Thompson, L. A. (1995). A twin study of competence and problem behavior in childhood and early adolescence. *J. Child Psychol. Psychiatry* **36**(5), 775–785.

Eley, T. C., Lichtenstein, P., and Stevenson, J. (1999). Sex differences in the etiology of aggressive and nonaggressive antisocial behavior: Results from two twin studies. *Child Dev.* **70**(1), 155–168.

Eriksson, M., Rasmussen, F., and Tynelius, P. (2006). Genetic factors in physical activity and the equal environment assumption—The Swedish young male twin study. *Behav. Genet.* **36**(2), 238–247.

Ferguson, C. J., and Kilburn, J. (2009). The public health risks of media violence: A meta-analytic review. *J. Pediatr.* **154**(5), 759–763.

Ferguson, C. J., and Kilburn, J. (2010). Much ado about nothing: The misestimation and overinterpretation of violent video game effects in eastern and western nations: Comment on Anderson et al. (2010). *Psychol. Bull.* **136**(2), 174–178; discussion 82–87.

Finkel, D., and McGue, M. (1997). Sex differences and nonadditivity in heritability of the Multidimensional Personality Questionnaire Scales. *J. Pers. Soc. Psychol.* **72**(4), 929–938.

Foley, D. L., Eaves, L., Wormley, B., Silberg, J., Maes, H., Kuhn, J., and Riley, B. (2004). Childhood adversity, monoamine oxidase a genotype, and risk for conduct disorder. *Arch. Gen. Psychiatry* **61**, 738–744.

Ge, X., Conger, R. D., Cadoret, R. J., Neiderhiser, J. M., Yates, W. R., Troughton, E., and Stewart, M. A. (1996). The developmental interface between nature and nurture: A mutual influence model of child antisocial behavior and parent behaviors. *Dev. Psychol.* **32**(4), 574–589.

Gelhorn, H. L., Stallings, M. C., Young, S. E., Corely, R. P., Rhee, S. H., Hopfer, C., and Hewitt, J. K. (2006). Common and specific genetic influences on aggressive and nonaggressive conduct disorder domains. *J. Am. Acad. Child Adolesc. Psychiatry* **45**(5), 570–577.

Ghodesian-Carpey, J., and Baker, L. A. (1987). Genetic and environmental influences on aggression in 4- to 7-year-old twins. *Aggress. Behav.* **13**, 173–186.

Gjone, H., and Novik, T. S. (1995). Parental ratings of behaviour problems: A twin study and general population comparison. *J. Child Psychol. Psychiatry* **36**(7), 1213–1224.

Gough, H. G. (1960). The adjective check list as a personality assessment research technique. *Psychol. Rep.* **6**, 107–122.

Guo, G., Roettger, M. E., and Shih, J. C. (2007). Contributions of the *DAT1* and *DRD2* genes to serious and violent delinquency among adolescents and young adults. *Hum. Genet.* **121**, 125–136.

Haberstick, B. C., Lessem, J. M., Hopfer, C. J., Smolen, A., Ehringer, M. A., Timberlake, D., and Hewitt, J. K. (2005). Monoamine oxidase A (MAOA) and antisocial behaviors in the presence of childhood and adolescent maltreatment. *Am. J. Med. Genet. B Neuropsychiatr. Genet.* **135B**(1), 59–64.

Haberstick, B. C., Schmitz, S., Young, S. E., and Hewitt, J. K. (2006a). Genes and developmental stability of aggressive behavior problems at home and school in a community sample of twins aged 7–12. *Behav. Genet.* **36**, 809–819.

Haberstick, B. C., Smolen, A., and Hewitt, J. K. (2006b). Family-based association test of the 5HTTLPR and aggressive behavior in a general population sample of children. *Biol. Psychiatry* **59**, 836–843.

Haney, C., Banks, W. C., and Zimbardo, P. G. (1973). A study of prisoners and guards in a simulated prison. *Naval Res. Rev.* **30**, 4–17.

Hettema, J. M., Neale, M. C., and Kendler, K. S. (1995). Physical similarity and the equal-environment assumption in twin studies of psychiatric disorders. *Behav. Genet.* **25**(4), 327–335.

Hodgins, S., Kratzer, L., and McNeil, T. F. (2001). Obstetric complications, parenting, and risk of criminal behavior. *Arch. Gen. Psychiatry* **58**, 746–752.

Hudziak, J. J., Rudiger, L. P., Neale, M. C., Heath, A. C., and Todd, R. D. (2000). A twin study of inattentive, aggressive, and anxious/depressed behaviors. *J. Am. Acad. Child Adolesc. Psychiatry* **39**(4), 469–476.

Hudziak, J. J., van Beijsterveldt, C. E. M., Bartels, M., Rietveld, M. J. H., Rettew, D. C., Derks, E. M., and Boomsma, D. (2003). Individual differences in aggression: Genetic analyses by age, gender, and informant in 3-, 7-, and 10-year-old Dutch Twins. *Behav. Genet.* **33**(5), 575–589.

Huesmann, L. R. (2010). Nailing the coffin shut on doubts that violent video games stimulate aggression: Comment on Anderson *et al.* (2010). *Psychol. Bull.* **136**(2), 179–181.

Iacono, W. G., McGue, M., and Krueger, R. F. (2006). Minnesota center for twin and family research. *Twin Res. Hum. Genet.* **9**(6), 978–984.

Jacobson, K. C., Prescott, C. A., and Kendler, K. (2002). Sex differences in the genetic and environmental influences on the development of antisocial behavior. *Dev. Psychol.* **14**, 395–416.

Jaffee, S. R., Caspi, A., Moffitt, T. E., Dodge, K. A., Rutter, M., Taylor, A., and Tully, L. A. (2005). Nature X nurture: Genetic vulnerabilities interact with physical maltreatment to promote conduct problems. *Dev. Psychopathol.* **17**(1), 67–84.

Kendler, K., Neale, M. C., Kessler, R. C., Heath, A. C., and Eaves, L. J. (1993). A test of the equal-environment assumption in twin studies of psychiatric illness. *Behav. Genet.* **23**(1), 21–27.

Kendler, K., Prescott, C. A., Myers, J., and Neale, M. C. (2003). The structure of genetic and environmental risk factors for common psychiatric and substance use disorders in men and women. *Arch. Gen. Psychiatry* **60**, 929–937.

Keyes, M. A., Malone, S. M., Elkins, I. J., Legrand, L. N., McGue, M., and Iacono, W. G. (2009). The enrichment study of the Minnesota twin family study: Increasing the yield of twin families at high risk for externalizing psychopathology. *Twin Res. Hum. Genet.* **12**(5), 489–501.

Kim-Cohen, J., Caspi, A., Taylor, A., Williams, B., Newcombe, R., Craig, I. W., and Moffitt, T. E. (2006). MAOA, maltreatment, and gene-environment interaction predicting children´s mental health: New evidence and a meta-analysis. *Mol. Psychiatry* **11**(10), 903–913.

Klump, K. L., Holly, A., Iacono, W. G., McGue, M., and Willson, L. E. (2000). Physical similarity and twin resemblance for eating attitudes and behaviors: A test of the equal environments assumption. *Behav. Genet.* **30**(1), 51–58.

Knafo, A., and Plomin, R. (2006). Prosocial behavior from early to middle childhood: Genetic and environmental influences on stability and change. *Dev. Psychol.* **42**(5), 771–786.

Krueger, R. F., Moffitt, T. E., Caspi, A., Bleske, A., and Silva, P. A. (1998). Assortative mating for antisocial behavior: Developmental and methodological implications. *Behav. Genet.* **28**(3), 173–186.

Krueger, R. F., Hicks, B. M., Patrick, C. J., Carlson, S. R., Iacono, W. G., and McGue, M. (2002). Etiologic connections among substance dependence, antisocial behavior, and personality: Modeling the externalizing spectrum. *J. Abnorm. Psychol.* **111**, 411–424.

Krueger, R. F., Markon, K. E., Patrick, C. J., and Iacono, W. G. (2005). Externalizing psychopathology in adulthood: A dimensional-spectrum conceptualization and its implications for DSM–V. *J. Abnorm. Psychol.* **114**(4), 537–550.

Krueger, R. F., Markon, K. E., Patrick, C. J., Benning, S. D., and Kramer, M. D. (2007). Linking antisocial behavior, substance use, and personality: An integrative quantitative model of the adult externalizing spectrum. *J. Abnorm. Psychol.* **116**(4), 645–666.

Lahey, B. B., Applegate, B., Waldman, I. D., Loft, J. D., Hankin, B. L., and Rick, J. (2004). The structure of child and adolescent psychopathology: Generating new hypotheses. *J. Abnorm. Psychol.* **113**, 358–385.

Leventhal, T., and Brooks-Gunn, J. (2000). The neighbourhoods they live in: The effects of neighbourhood residence on child and adolescent outcomes. *Psychol. Bull.* **126**, 309–337.

Lewis, D. O., Yeager, C. A., Gidlow, B., and Lewis, M. (2001). Six adoptees who murdered: Neuropsychiatric vulnerabilities and characteristics of biological and adoptive parents. *J. Am. Acad. Psychiatry Law* **29**(4), 390–397.

Linnoila, M., Virkkunen, M., Scheinin, M., Nuutila, A., Rimon, R., and Goodwin, F. K. (1983). Low cerebrospinal fluid 5-hydroxyindoleacetic acid concentration differentiates impulsive from non-impulsive violent behavior. *Life Sci.* **33**, 2609–2614.

Lipsey, M. W., Wilson, D. B., Cohen, M. A., and Derzon, J. H. (1997). Is there a causal relationship between alcohol use and violence? A synthesis of evidence. *Recent Dev. Alcohol.* **13**, 245–282.

Loehlin, J. C. (1992). Genes and environment in personality development. Newbury Park SAGE, CA.

Loehlin, J. C. (2010). Is there an active gene-environment correlation in adolescent drinking behavior? *Behav. Genet.* **40**(4), 447–451.

Maes, H. H. M., Neale, M. C., Kendler, K., Hewitt, J. K., Silberg, J. L., Foley, D. L., Meyer, J. M., Rutter, M., Simonoff, E., Pickles, A., and Eaves, L. J. (1998). Assortative mating for major psychiatric diagnosis in two population-based samples. *Psychol. Med.* **28**, 1389–1401.

Maes, H. H., Silberg, J. L., Neale, M. C., and Eaves, L. J. (2007). Genetic and cultural transmission of antisocial behavior: An extended twin parent model. *Twin Res. Hum. Genet.* **10**(1), 136–150.

Mather, K., and Jinks, J. L. (1982). Biometrical Genetics: The Study of Continuous Variation. Chapman & Hall, London.

Mathew, K. A., and Angulo, J. (1980). Measurement of the type A behavior pattern in children: Assessment of children's competitiveness, impatience-anger, and aggression. *Child Dev.* **51**, 466–475.

Mauger, P. A. (1980). Interpersonal Behavior Survey (IBS) Manual. Western Psychological Services, Los Angeles.

McGue, M., Bacon, S., and Lykken, D. T. (1993). Personality stability and change in early adulthood: A behavioral genetic analyses. *Dev. Psychol.* **29**, 96–109.

Meininger, J. C., Hayman, L. L., Coates, P. M., and Gallagher, P. (1988). Genetics or environment? Type A behavior and cardiovascular risk factors in twin children. *Nurs. Res.* **37**(6), 341–346.

Merton, R. K. (1957). Social Theory and Social Structure. The Free Press, New York.

Miles, D. R., and Carey, G. (1997). Genetic and environmental architecture of human aggression. *J. Pers. Soc. Psychol.* **72**(1), 207–217.

Milgram, S. (1963). Behavioral study of obedience. *J. Abnorm. Soc. Psychol.* **67**, 371–378.

Moffitt, T. (2002). Teen-aged mothers in contemporary Britain & the E-risk Study Team (39 authors). *J. Child Psychol. Psychiatry* **43**, 727–742.

Moffitt, T. E., Caspi, A., Rutter, M., and Silva, P. (2001). Sex Differences in Antisocial Behaviour: Conduct Disorder, Delinquency and Violence in the Dunedin Longitudinal Study. Cambridge University Press, Cambridge.

Moilanen, I., Linna, S.-L., Kumpulainen, K., Tamminen, K., Piha, J., and Almqvist, F. (1999). Are twins' behavioural 7emotional problems different from singletons'? *Eur. Child Adolesc. Psychiatry* **8**, 62–67.

Narusyte, J., Andershed, A. K., Neiderhiser, J. M., and Lichtenstein, P. (2006). Aggression as a mediator of genetic contributions to the association between negative parent–child relationships and adolescent antisocial behavior. *Eur. Child Adolesc. Psychiatry* **16**(2), 128–137.

Neale, M. C., and Cardon, L. R. (1992). Methodology for Genetic Studies of Twins and Families. Kluwer Academic Publications, Dordrecht, The Netherlands.

Neiderhiser, J. M., Reiss, D., Pedersen, N. L., Lichtenstein, P., Spotts, E. L., Hansson, K., Cederblad, M., and Elthammar, O. (2004). Genetic and environmental influences on mothering of adolescents: A comparison of two samples. *Dev. Psychol.* **40**(3), 335–351.

Nilsson, K. W., Sjöberg, R. L., Damberg, M., Leppert, J., Ohrvik, J., Alm, P. O., Lindström, L., and Oreland, L. (2006). Role of monoamine oxidase A genotype and psychosocial factors in male adolescent criminal activity. *Mol. Psychiatry* **10**(11), 903–913.

Niv, S., and Baker, L. A. (2010). Genetic marker for antisocial behavior. *In* "The Origins of Antisocial Behavior: A Developmental Perspective" (C. Thomas and K. Pope, eds.). Oxford University Press, New York.

Oades, R. D., Lasky-Su, J., Christiansen, H., Faraone, S. V., Sonuga-Barke, E. J. S., Banaschewski, T., Chem, W., Anney, R. J. L., Buitelaar, J. K., Ebstein, R. P., Franke, B., Gill, M., *et al.* (2008). The influence of serotonin and other genes and impulsive behavioral aggression and cognitive impulsivity in children with attention-deficit/hyperactivity disorder (ADHD): Finding from a family-based association test (FBAT) analysis. *Behav. Brain Funct.* **48**, 4–48.

O'Connor, M., Foch, T., Sherry, T., and Plomin, R. (1980). A twin study of specific behavioral problems of socialization as viewed by parents. *J. Abnorm. Child Psychol.* **8**(2), 189–199.

O'Connor, T. G., Deater-Deckard, K., Fulker, D., Rutter, M., and Plomin, R. (1998). Genotype-environment correlations in late childhood and early adolescence: Antisocial behavioral problems and coercive parenting. *Dev. Psychol.* **34**, 970–981.

Olweus, D. (1989). Prevalence and incidence in the study of antisocial behavior: Definitions and measurements. *In* "Cross-National Research in Self-Reported Crime and Delinquency" (M. W. Klein, ed.). Kluwer Academic Publishers, Dordrecht.

Owen, D. R., and Sines, J. O. (1970). Heritability of personality in children. *Behav. Genet.* **1**(3), 235–248.

Parker, T. (1989). Television Viewing and Aggression in Four and Seven Year Old Children. Boulder, Boulder, CO.

Pfohl, B., Blum, N., and Zimmerman, M. (1997). Structured Interview for DSM-IV Personality: SIDP-IV. American Psychiatric Press, Washington, DC.

Plomin, R., DeFries, J. C., McClearn, G. E., and McGuffin, P. (2001). Behavioral Genetics. Worth Publisher, United States of America.

Polderman, T. J., Posthuma, D., De Sonneville, L. M., Verhulst, F. C., and Boomsma, D. I. (2006). Genetic analyses of teacher ratings of problem behavior in 5-year-old twins. *Twin Res. Hum. Genet.* **9**(1), 122–130.

Posthuma, D., Beem, A. L., de Geus, E. J. C., van Baal, C. M., von Hjelmborg, J. B., Iachine, I., and Boomsma, D. I. (2003). Theory and practice in quantitative genetics. *Twin Res. Hum. Genet.* **6**(5), 361–376.

Pulikkined, L., Kaprio, J., and Rose, R. J. (1999). Peers, teachers, and parents as assessors of the behavioural and emotional problems of twins and their adjustment: The Multidimensional Peer Nomination Inventory. *Twin Res. Hum. Genet.* **2**, 274–285.

Rahe, R. H., Hervig, L., and Rosenman, R. H. (1978). Heritability of type A behavior. *Psychosom. Med.* **40**(6), 478–486.

Raine, A. (2002). Biosocial studies of antisocial and violent behavior in children and adults: A review. *J. Abnorm. Child Psychol.* **30**(4), 311–326.

Raine, A., Brennan, P. A., and Mednick, S. A. (1997). Interaction between birth complications and early maternal rejection in predisposing individuals to adult violence: Specificity to serious, early-onset violence. *Am. J. Psychiatry* **154**(9), 1265–1271.

Raine, A., Dodge, K., Loeber, R., Gatzke-Kopp, L., Lynam, D. R., Reynolds, C., Stouthamer-Loeber, M., and Liu, J. (2006). The reactive-proactive aggression questionnaire: Differential correlates of reactive and proactive aggression in adolescent boys. *Aggress. Behav.* **32**, 159–171.

Reif, A., Rosler, M., Freitag, C. M., Schneider, M., Eujen, A., Kissling, C., Wensler, D., Jacob, C. P., Retz-Junginger, P., Thome, J., Lesch, K.-P., and Retz, W. (2007). Nature and nurture predispose to violent behavior: Serotonergic genes and adverse childhood environment. *Neuropsychopahrmacology* **32**, 2375–2383.

Retz, W., Rosler, M., Supprian, T., Retz-Junginger, P., and Thome, J. (2003). Dopamine D3 receptor gene polymorphism and violent behavior: Relation to impulsiveness and ADHD-related psychopathology. *J. Neural Transm.* **110**, 561–572.

Retz, W., Freitag, C. M., Retz-Junginger, P., Wenzler, D., Schneider, M., Kissling, C., *et al.* (2008). A functional serotonin transporter promoter gene polymorphism increases ADHD symptoms in delinquents: Interaction with adverse childhood environment. *Psychiatry Res.* **158**, 123–131.

Rhee, S. H., and Waldman, I. D. (2002). Genetic and environmental influences on antisocial behavior: A meta-analysis of twin and adoption studies. *Psychol. Bull.* **128**, 490–529.

Rowe, D. C. (2003). Assessing genotype-environment interactions and correlations in the postgenomic era. *In* "Behavioral Genetics in the Postgenomic Era" (R. Plomin, J. C. DeFries, I. W. Craig, and P. McGuffin, eds.), pp. 71–99. American Psychological Association, Washington, DC.

Rowe, D. C., Almeida, D. M., and Jacobson, K. C. (1999). School context and genetic influences on aggression in adolescence. *Psychol. Sci.* **10**(3), 277–280.

Rushton, J. P., Fulker, D. W., Neale, M. C., Nias, D. K., and Eysenck, H. J. (1986). Altruism and aggression: The heritability of individual differences. *J. Pers. Soc. Psychol.* **50**(6), 1192–1198.

Rutter, M. (2006). Genes and behavior. Nature-nurture interplay explained. Blackwell Publishing Ltd, Oxford, UK.

Rutter, M., and Redshaw, J. (1991). Annotation: Growing up as a twin: Twin-singleton differences in psychological development. *J. Child Psychol. Psychiatry* **32**(6), 885–895.

Rutter, M., Giller, H., and Hagell, A. (1998). Antisocial Behavior by Young People. Cambridge University Press, Cambridge, UK.

Rutter, M., Caspi, A., and Moffitt, T. E. (2003). Using sex differences in psychopathology to study causal mechanisms: Unifying issues and research strategies. *J. Child Psychol. Psychiatry* **44**(8), 1092–1115.

Scarr, S. (1966). Genetic factors in activity motivation. *Child Dev.* **37**(3), 663–673.

Scarr, S., and McCartney, K. (1983). How people make their own environments: A theory of genotype-environment effects. *Child Dev.* **54,** 425–435.

Schmitz, S., Fulker, D. W., and Mrazek, D. A. (1995). Problem behavior in early and middle childhood: An initial behavior genetic analysis. *J. Child Psychol. Psychiatry* **36**(8), 1443–1458.

Schwartz, D., Dodge, K., Coie, J. D., Hubbard, J. A., Cillessen, A. H. N., Lemerise, E. A., and Bateman, H. (1998). Social-cognitive and behavioral correlates of aggression and victimization in boys' play groups. *J. Abnorm. Child Psychol.* **26**(6), 431–440.

Shaffer, D., Fisher, P., Lucas, C., and Comer, J. (2000). Diagnostic Interview Schedule for Children-DISC-IV Scoring Manual. Columbia University, NY.

Silva, P. A., and Stanton, W. R. (1996). From child to adult: The Dunedin Multidisciplinary Health and Development Study. Oxford University Press Inc., New York, NY.

Simonoff, E., Pickles, A., Meyer, J., Silberg, J., Maes, H., Loeber, M., Rutter, M., Hewitt, J. K., and Eaves, L. (1997). The Virginia twin study of adolescent behavioral development. *Arch. Gen. Psychiatry* **54,** 801–808.

Simonoff, E., Pickles, A., Meyer, J., Silberg, J., and Maes, H. (1998). Genetic and environmental influences on subtypes of conduct disorder behavior in boys. *J. Abnorm. Child Psychol.* **26**(6), 495–509.

Sines, J. O., Pauker, J. D., and Sines, L. K. (1966). The Development of an Objective, Non-verbal Personality Test for Children. Midwestern Psychological Association, Chicago.

Sjöberg, R. L., Nilsson, K. W., Wargelius, H.-L., Leppert, J., Lindström, L., and Oreland, L. (2007). Adolescent girls and criminal activity: Role of MAOA-LPR genotype and psychosocial factors. *Am. J. Med. Genet. B Neuropsychiatr. Genet.* **144B,** 159–164.

Tackett, J. L., Waldman, I. D., and Lahey, B. B. (2009). Etiology and measurement of relational aggression: A multi-informant behavior genetic investigation. *J. Abnorm. Psychol.* **118**(4), 722–733.

Taylor, A. (2004). The consequences of selective participation on behavioral-genetic findings: evidence from simulated and real data. *Twin Res.* **7**(5), 485–504.

Taylor, J., McGue, M., and Iacono, W. G. (2000). Sex differences, assortative mating, and cultural transmission effects on adolescent delinquency: A twin family study. *J. Child Psychol. Psychiatry* **41**(4), 433–440.

Tellegen, A. (2011). Manual for the Multidimensional Personality Questionnaire University of Minnesota Press, Minneapolis.

Tuvblad, C., Grann, M., and Lichtenstein, P. (2006). Heritability for adolescent antisocial behavior differs with socioeconomic status: Gene-environment interaction. *J. Child Psychol. Psychiatry* **47**(7), 734–743.

Tuvblad, C., Raine, A., Zheng, M., and Baker, L. A. (2009). Genetic and environmental stability differs in reactive and proactive aggression. *Aggress. Behav.* **35**(6), 437–452.

Tuvblad, C., Narusyte, J., Grann, M., Sarnecki, J., and Lichtenstein, P. (2011). The genetic and environmental etiology of antisocial behavior, from early adolescence to emerging adulthood. *Behav. Genet.* **41**(5), 629–640.

van Beijsterveldt, C. E. M., Bartels, M., Hudziak, J. J., and Boomsma, D. (2003). Causes of stability of aggression from early childhood to adolescence: A longitudinal genetic analysis in Dutch twins. *Behav. Genet.* **33,** 591–605.

van den Oord, E. J. C. G., Boomsma, D., and Verhulst, F. C. (1994). A study of problem behaviors in 10- to 15-year-old biologically related and unrelated international adoptees. *Behav. Genet.* **24,** 193–205.

van den Oord, E. J. C. G., Koot, H. M., Boomsma, D. I., Verhulst, F. C., and Orlebeke, J. F. (1995). A twin-singleton comparison of problem behaviour in 2–3-year-olds. *J. Child Psychol. Psychiatry* **36**(3), 449–458.

van den Oord, E. J., Verhulst, F. C., and Boomsma, D. I. (1996). A genetic study of maternal and paternal ratings of problem behaviors in 3-year-old twins. *J. Abnorm. Psychol.* **105**(3), 349–357.

van der Valk, F. C., Verhulst, F. C., Stroet, T. M., and Boomsma, D. (1998). Quantitative genetic analysis of internalizing and externalizing problems in a large sample of 3-year-old twins. *Twin Res. Hum. Genet.* **1,** 25–33.

Vierikko, E., Pulkkinen, L., Kaprio, J., and Rose, R. J. (2004). Genetic and environmental influences on the relationship between aggression and hyperactivity-impulsivity as rated by teachers and parents. *Twin Res. Hum. Genet.* **7**(3), 261–274.

von der Pahlen, B., Santtila, P., Johansson, A., Varjonen, M., Jern, P., Witting, K., and Kenneth Sandnabba, N. (2008). Do the same genetic and environmental effects underlie the covariation of alcohol dependence, smoking, and aggressive behaviour? *Biol. Psychol.* **78**(3), 269–277.

Waldman, I. D., Rowe, D. C., Abramowitz, A., Kozel, S. T., Mohr, J. H., Sherman, S. L., Cleveland, H. H., Sanders, M. L., Gard, J. M. C., and Stever, C. (1998). Association and linkage of the dopamine transporter gene and attention-deficit hyperactivity disorder in children: Heterogeneity owing to diagnostic subtype and severity. *Am. J. Hum. Genet.* **63,** 1767–1776.

White, H. R., Xie, M., Thompson, W., Loeber, R., and Stouthamer-Loeber, M. (2001). Psychopathology as a predictor of adolescent drug use trajectories. *Psychol. Addict. Behav.* **15**(3), 210–218.

Widom, C. S., and Brzustowicz, L. M. (2006). MAOA and the "cycle of violence": Childhood abuse and neglect, MAOA genotype, and risk for violent and antisocial behavior. *Biol. Psychiatry* **1**(1), 60.

Xian, H., Scherrer, J. F., Eisen, S. A., True, W. R., Heath, A. C., Goldberg, J., Lyons, M. J., and Tsuang, M. T. (2000). Self-Reported zygosity and the equal-environments assumption for psychiatric disorders in the Vietnam Era Twin Registry. *Behav. Genet.* **30**(4), 303–310.

Young, S. E., Stallings, M. C., Corley, R. P., Krauter, K. S., and Hewitt, J. K. (2000). Genetic and environmental influences on behavioral disinhibition. *Am. J. Med. Genet.* **96,** 684–695.

9

Perinatal Risk Factors in the Development of Aggression and Violence

Jamie L. LaPrairie, Julia C. Schechter, Brittany A. Robinson, and Patricia A. Brennan

Department of Psychology, Emory University, Atlanta, Georgia, USA

Advances in Genetics, Vol. 75
0065-2660/11 $35.00
DOI: 10.1016/B978-0-12-380858-5.00004-6

ABSTRACT

Over the past several decades, the relative contribution of both environmental and genetic influences in the development of aggression and violence has been explored extensively. Only fairly recently, however, has it become increasingly evident that early perinatal life events may substantially increase the vulnerability toward the development of violent and aggressive behaviors in offspring across the lifespan. Early life risk factors, such as pregnancy and birth complications and intrauterine exposure to environmental toxins, appear to have a profound and enduring impact on the neuroregulatory systems mediating violence and aggression, yet the emergence of later adverse behavioral outcomes appears to be both complex and multidimensional. The present chapter reviews available experimental and clinical findings to provide a framework on perinatal risk factors that are associated with altered developmental trajectories leading to violence and aggression, and also highlights the genetic contributions in the expression of these behaviors. © 2011, Elsevier Inc.

I. INTRODUCTION

Aggressive behavior and violent offending are significant burdens not only to the individual but also to society at large; incurring costs upward of 2.3 million dollars per individual in the most severe cases in the United States (Cohen *et al.*, 2010). In contrast to popular belief, longitudinal epidemiological research indicates that the sudden onset of physical aggression in adolescence or adulthood is unusual (Tremblay, 2008). Rather, physically aggressive behaviors can already be detected by 12 months of age, and their frequency peaks by 2–4 years of age (Côté *et al.*, 2006; Tremblay, 2008). Indeed 85% of children exhibit aggression and tantrums by age 2 (Potegal and Davidson, 2003). In the majority of children, the frequency of physical aggression gradually decreases toward the end of the preschool period, reflecting an increase in cognitive control of behavior, the acquisition of alternate problem-solving strategies, and the influence of socialization (Nagin, and Tremblay, 1999; Tremblay, 2008). A 60-year longitudinal study of juvenile delinquents concluded that very few show a lifespan high frequency of violent offending (Sampson and Laub, 2003), and among those who do express chronic physical aggression, impaired executive functioning is evident in adolescence and early childhood, even after controlling other cognitive deficits (Barker *et al.*, 2007). Longitudinal follow-up studies of elementary school children with continued high levels of physical aggression demonstrate that they are at greater risk for numerous adjustment problems throughout their lifetime, including substance abuse, academic failure, antisocial behavior, suicide, depression, spouse abuse, and neglectful parenting (Broidy *et al.*, 2003; Tremblay *et al.*, 2004).

The neurobiological substrates underlying the individual variability in the development of aggression and violence are many (Loeber and Pardini, 2008), and current research suggests that some of the most relevant risk factors for a trajectory of consistently high aggression are predicted by perinatal life events including pregnancy and birth complications and intrauterine exposure to tobacco, drugs, and alcohol (Brennan and Mednick, 1997; Cannon et al., 2002). Genetic studies of aggressive behavior also indicate that childhood aggression is highly heritable (Brendgen et al., 2005; Hicks et al., 2004). Together, these congenital factors appear to produce neuropsychological deficits in the nervous system of offspring, manifesting as subtle neurocognitive difficulties, altered temperament, delayed motor development, and hyperactivity (Moffitt and Caspi, 2001). The two most commonly characterized developmental trajectories for aggression and violence are: (1) an early-onset persistent offender and (2) a late-onset adolescent-limited offender (Moffitt, 1993; Patterson et al., 1989). In the early-onset offender, behavioral problems typically manifest early in life, develop into serious juvenile delinquency during adolescence, and ultimately evolve into a long-term adult history of criminal behavior. Alternatively, late-onset adolescent-limited individuals typically do not begin offending until middle to late adolescence and cease violent and criminal behavior by their mid-1920s (Moffitt, 1993). These two distinct trajectories are associated with substantial variation in severity, chronicity, etiology, and prognosis. In fact, adolescent-limited offending can be considered normative and functional, given the developmental demands on adolescents in modern society (Moffitt, 1993). Adolescent-limited offending may also be explained by normative changes in brain development in postpubertal children. Life-course-persistent offending, in contrast, is not socially sanctioned, functional, or reflective of normative changes in brain development. Instead, this type of offending is likely caused by genetic and perinatal factors which lead to deficits in neurocognitive and neurophysiological functioning throughout the life course. Prior to reviewing the current research on perinatal factors and violence, it is important to briefly describe the neurophysiological processes that are involved in the regulation of aggression and violence. Then we will describe how these factors may mediate that relationship between perinatal risk and aggressive outcomes. Finally, we will attempt to highlight what is known about the genetic contributions to this developmental process.

II. THE NEUROBIOLOGICAL AND PSYCHOPHYSIOLOGICAL SYSTEMS INVOLVED IN THE REGULATION OF AGGRESSION AND VIOLENCE

The central nervous system (CNS) and autonomic nervous system (ANS) maintain homeostasis and monitor physiological responses in humans from the perinatal period throughout the lifespan. These systems are also involved in the mediation of aggressive, violent, and antisocial behavior. Although neurophysiological systems have been the focal point of studies on aggression for decades (Scarpa and Raine, 1997), significant technological advances over the past 20 years have allowed researchers to advance their investigation of the biological basis underlying violent and aggressive behaviors. Indeed, researchers are now able to ask questions concerning the neural underpinnings of aggression and how the interactions between genes and environment may result in later violence; ultimately leading to a more complete picture of the development of aggression and violence throughout the lifespan.

A. Types of aggressive behavior

Two forms of aggression are present in both experimental animal and human models. Throughout the animal literature, predatory aggression has been differentiated from defensive aggression (Scarpa and Raine, 1997). Similarly, human studies often discriminate between instrumental/proactive aggression and hostile/reactive or impulsive aggression (Crick and Dodge, 1996; Nelson and Trainor, 2007). Reactive aggression is characterized as aggressive behavior occurring in the context of anger and high emotionality. It is often more impulsive, less controlled, and occurring in reaction to a provocation or frustration (Scarpa and Raine, 1997). Instrumental aggression is qualified as being more goal-oriented and relatively nonemotional. Studies suggest that this latter type of aggression is regulated by higher cortical systems rather than brain regions (i.e., limbic systems) associated with impulsiveness (Nelson and Trainor, 2007).

Of note, instrumental aggression is characteristic of psychopathy, a subtype of aggression often associated with particularly low levels of physiological arousal (Raine, 2002a). While psychopathological disorders are outside the scope of this chapter, understanding that individuals may display different types of aggressive behavior is essential to conceptualizing physiological research findings and the development of violence, in general.

B. Neurobiological bases of aggression and violence

Recent technological advances have allowed scientists to observe parallel neurochemical and anatomical correlates that are activated during aggression in both human and nonhuman animals (Nelson and Trainor, 2007). Neurobiological literature regarding aggression appears to focus on substrates that are implicated in either the expression of aggressive behavior or inhibitory control of aggression (Siegel and Victoroff, 2009). Below is a brief synthesis of the animal and human literature that pertains to the neurobiological systems that have been implicated in the development and modulation of aggression and violence.

1. Amygdala

The amygdala is part of the limbic system and is considered to play a central role in both emotional regulation (Joseph, 1999) and fear conditioning (Susman, 2006). Lesion studies in animals have been critical to the understanding of the relationship between the amygdala and aggression. Lesioning of the medial amygdala reduces aggression in rats (Kruk, 1992). Male rhesus monkeys with amygdalar lesions display significant increases in aggressive behavior in a group setting (Machado and Bachevalier, 2006), while decreases in aggressive behavior are observed when animals are tested within a dyad (Emery et al., 2001). An explanation for these divergent findings may be that reintroduction into a group is a more fearful situation, thus leading to increases in amygdalar responsiveness (Nelson and Trainor, 2007).

Functional abnormalities in the amygdala have also been noted in human studies in childhood, adolescence (Marsh et al., 2008; Sterzer et al., 2005), and adulthood (Veit et al., 2002); however, these studies have provided mixed results. The posteroventral medial amygdala appears to be involved in the regulation of reactive aggression, as in defensive situations, while the posterodorsal medial amygdala appears to be associated with instrumental or offensive situations (Swanson, 2000). Coccaro et al. (2007) found that amygdalar activation is positively associated with scores on the Lifetime History of Aggression scale for adults with intermittent explosive disorder and healthy controls. Notably, only individuals with intermittent explosive disorder display increased activation of the amygdala in response to angry faces (Coccaro et al., 2007). Conversely, Sterzer et al. (2005) found a negative correlation between aggression and the left amygdala in response to negative affective images in a group of boys diagnosed with conduct disorder (CD). However, the inverse relationship (i.e., positive correlation between amygdala activation and aggression) was found in youth with comorbid anxiety and depression, symptoms often found to be associated with CD (Loeber et al., 2000). Reduced responsiveness of the amygdala in response to fearful faces has been suggested to reflect impairment in the processing of distress cues (Marsh et al., 2008), which may result in a lack of empathy and increases in instrumental aggressive behavior (Blair, 2001). One explanation for these discrepant findings across samples is that aggressive individuals may appear hyporesponsive when faced with detecting threat or distress, but hyperresponsive to distressful situations that can lead to aggression (Decety et al., 2009). Further, these mixed findings highlight that the amygdala's role in the mediation of aggression may be based largely on the type of aggression expressed.

2. Anterior cingulate cortex

The anterior cingulate cortex (ACC) is also a part of the brain's limbic system and has been implicated in emotional and cognitive processing (Bush et al., 2000). Lesions of the ACC have led to a range of outcomes, including inattention, dysregulation of autonomic functions, and emotional instability (Kennard, 1954; Tow and Whitty, 1953). ACC lesions as a treatment for affective disorders in humans have produced decreased distress and emotional liability (Corkin, 1979). More recently, deactivation in the dorsal ACC, an area associated with cognitive monitoring and behavioral regulation (Bush *et al.*, 2000), has been noted in aggressive youth compared to controls (Sterzer *et al.*, 2005). Reduced activation in the ACC is also associated with "novelty seeking" (Cloninger, 1987), a dimension of temperament encompassing a quick-tempered personality and high impulsivity (Stadler *et al.*, 2007). Thus, reduced activation of the ACC may be a connection between temperament, behavior, and emotion processing (Sterzer *et al.*, 2005).

3. Prefrontal cortex

The prefrontal cortex (PFC) modulates subcortical behavior (Siever, 2008), specifically by inhibiting connections between the amygdala and hypothalamus, thereby resulting in increased aggression (Davidson, 2000). Prefrontal lesions in humans as a result of tumors, trauma, or metabolic disturbances have been instrumental in illustrating the role of the PFC in aggressive behavior (Siever, 2008). The well-known case of Phineas Gage, a stable and dependable railroad worker who became irritable, angry, and showed poor judgment following an accident in which a rod entered his skull at the frontal cortex, is a prime example of the critical role of the PFC in monitoring aggression.

Imaging studies exploring brain functioning have further elucidated the role of the frontal cortices in modulation of aggression. Individuals with lower-than-average baseline activity in the frontal cortex demonstrate higher levels of reactive or impulsive aggression (Coccaro and Kavoussi, 1997; Soloff *et al.*, 2003; Volkow *et al.*, 1995). Moreover, the ventromedial prefrontal cortex (vmPFC) has been specifically implicated in the calculation of reward expectation (Elliott and Deakin, 2005), and increased activation is observed in the vmPFC when errors are made during a reversal learning task (Finger *et al.*, 2008). Reversal learning tasks are designed to frustrate participants and measure their ability to adjust behavior in response to changing reinforcement (i.e., avoid frustration; Rolls *et al.*, 1994), a deficit observed in individuals with psychopathic traits (Blair *et al.*, 2001; Budhani and Blair, 2005). Violations of reward expectations (i.e., expecting but not receiving reinforcement) have been linked to frustration and reactive aggression (Berkowitz, 1989), which may be a result of not achieving an expected reward (Sterzer *et al.*, 2005). Thus, increased vmPFC responses to violations of expectations may indeed be associated with an increased risk of frustration and subsequent aggressive or violent behaviors (Blair, 2010).

Experimental animal studies indicate a connection between the orbito-fronal cortex (OFC) and aggressive behavior, particularly with regard to the interpretation of social cues and behavioral responses in social contexts (Nelson and Trainor, 2007). OFC lesions in male rhesus monkeys increase aggression in dominant but not in subordinate animals (Machado and Bachevalier, 2006). Butter and Snyder (1972) observed similar results, but these effects diminished over several months. In humans, structural abnormalities in the OFC, such as reduced levels of gray-matter volume in youth with CD (Huebner *et al.*, 2008) and early brain damage in this region have also been associated with conduct problems (Anderson *et al.*, 1999). Overall, dysfunction in the prefrontal regions of the brain appears to underlie impaired regulation of affective responses and reduced inhibition of aggression (Davidson, 2000).

4. Hypothalamus

The hypothalamus is another critical brain structure involved in the development of aggression. Research with nonhuman primates suggests that electrical stimulation of the ventromedial hypothalamus is linked to vocal threats and aggressive behaviors in marmosets (Lipp and Hunsperger, 1978), while lesions of the anterior hypothalamus reduce vocal threats toward a male intruder (Lloyd and Dixson, 1988). Electrical stimulation of the anterior hypothalamus increases vocalizations in rhesus monkeys (Robinson, 1967) and aggression toward insubordinate male rhesus monkeys (Alexander and Perachio, 1973). Similar findings have been found with electrical stimulation of the hypothalamus in male rats (Kruk, 1992) and cats (Siegel and Victoroff, 2009). In humans, the frontal cortex inhibits circuits in the hypothalamus that increase aggression (Davidson, 2000). During a period in the mid-twentieth century, electrolytic lesions of the hypothalamus were used as a treatment for "excessive aggression" (Heimburger *et al.*, 1966). Although conclusions from such studies should be interpreted cautiously for both methodological and ethical reasons (Scarpa and Raine, 1997), these lesions were found to inhibit aggression in humans.

C. Neurochemical signals of aggression and violence

Specific signaling molecules have provided additional information about neural circuits underlying aggression. As experimental research has turned to the brain for answers regarding the development of aggression and violence, the neurochemistry of aggressive behavior has also come to the forefront. In general, neurotransmitters in the brain either increase or inhibit aggressive behavior. Below is a brief review; a more extensive review of this area of research can be found in a separate chapter in this volume.

1. Neurotransmitters-serotonin

Given the connection between emotion, cognition, and aggression, it seems fitting that many studies have focused their investigation on serotonergic neurotransmission, a system predominantly involved in the regulation of emotional states. Serotonin (5-HT) receptors in specific regions of the brain, such as the OFC and ACC, are involved in the modulation and suppression of aggressive behavior (Siever, 2008). Moreover, low levels of 5-HT are associated with increased aggression in humans (Chiavegatto *et al.*, 2001) and nonhuman primates (Higley *et al.*, 1992). Many studies have investigated the role of 5-HT in neuronal functioning and aggression. For example, Frankle *et al.* (2005) reported reduced 5-HT transporter distribution in the ACC of patients with aggressive personality disorder compared to healthy controls. Reduced prefrontal activation was observed in response to a serotonergic releasing agent (d, 1-fenfluramine) in individuals with impulsive aggression, such as those diagnosed with borderline personality disorder (Soloff *et al.*, 2003), and in depressed patients with a history of suicidal behavior (Mann *et al.*, 1992). Using positron emission tomography (PET) technology, Parsey *et al.* (2002) found a negative association between scores on the Lifetime History of Aggression scale and 5-HT receptor-binding in the PFC and amygdala. Further, individuals with high levels of impulsive aggression display reduced activation in the PFC (Coccaro and Kavoussi, 1997). Further, New *et al.* (2004) found that individuals with borderline personality disorder, who underwent 12 weeks of selective serotonin reuptake inhibitor (SSRI) treatment, increased baseline PFC activation, which was negatively correlated with ratings of aggression.

2. Neurotransmitters-dopamine

Dopamine (DA) has been implicated in the initiation and exhibition of aggression (de Almeida *et al.*, 2005), yet its precise role remains unclear. Ferrari *et al.* (2003) found that rats can be conditioned to increase dopamine secretion and decrease levels of 5-HT in anticipation of aggressive interactions, whereby DA and 5-HT levels in the nucleus accumbens were measured during and following a confrontation with another rat using microdialysis. Heart rate (HR) also increased 1 h prior to the regularly scheduled interactions. Antagonists of both D_1 and D_2 receptors appear to reduce aggression in male mice (de Almeida *et al.*, 2005). Further, animal studies suggest that the activation of catecholaminergic brainstem neurons (e.g., ventral tegmental area) project to dopaminergic structures in the forebrain, such as the hypothalamus and the limbic system (e.g., amygdala, hippocampus, PFC, and ACC). In humans, both proactive and reactive aggressions are associated with dopamine (Siegel and Victoroff, 2009). In clinical populations, decreased D_1 receptors have been observed in depressed individuals suffering from anger attacks (Dougherty *et al.*, 2006). Psychopharmacological agents most frequently employ compounds that act on dopaminergic systems in the brain (McDougle *et al.*, 1998). Risperdone, a D_2 antagonist, effectively reduces aggression in humans (Nelson and Trainor, 2007). Moreover, haloperidol, another D_2 antagonist, is used in the treatment of aggression in psychotic patients (Fitzgerald, 1999), of violent outbursts in adults with borderline personality disorder and dementia, and of aggression in children and adolescents (Beauchaine *et al.*, 2000).

3. Neurotransmitters-norepinephrine

Norepinephrine (also known as noradrenaline) is a monoamine found in the ANS. It is associated with arousing situations and has been specifically cited in the development of both proactive and reactive aggression (Siegel and Victoroff, 2009). Interestingly, a meta-analysis of central (cerebrospinal fluid) measures of norepinephrine found a negative association between norepinephrine and antisocial behavior (Raine, 1993). Plasma levels of norepinephrine are also associated with induced hostile behavior during experiments with healthy controls (Gerra *et al.*, 1997). Pharmacological manipulation studies of noradrenaline levels and noradrenergic receptors suggest that this catecholamine facilitates the development of aggression (Miczek *et al.*, 2002). Further evidence from the animal literature demonstrates that DA beta-hydroxylase knockout mice are unable to produce noradrenaline and display reduced levels of aggression, but normal levels of anxiety (Marino *et al.*, 2005).

D. Hormones

Hormonal factors have long been studied in relation to aggressive behavior, particularly as scientists sought an explanation for the notable gender differences in the rates of violence. Androgens, and in particular testosterone, have received the most attention in this regard. Research on hormones and aggression has not demonstrated a one to one relationship between these factors. Instead, empirical findings suggest that the type of aggressive behavior and the structure and quality of the social environment are likely important moderators in the association between hormones and aggression.

1. Testosterone

A positive relationship between testosterone and aggression is well established in the animal literature, but less support for this association has emerged in humans (Archer, 1991). Injections of testosterone increase aggression in a variety of animals (Monoghan and Glickman, 1992), and aggression has been positively associated with territorial behavior in birds (Wingfield and Hahn, 1994). Moreover, testosterone increases the display of dominance behaviors in rhesus monkeys (Rose *et al.*, 1971). Further, the castration of lizards (Greenberg and Crews, 1983) and male mice (Vom Saal, 1983) leads to reduced aggression. While some research with humans indicates that testosterone is associated with anger and aggression (Olweus, 1986), evidence from the broader literature is mixed (Archer, 1991). Some studies have revealed positive relationships between testosterone and aggression, some report negative relationships, and no association is found in others (van Bokhoven *et al.*, 2006). Specifically, testosterone measured in cerebrospinal fluid, serum, and saliva have been linked to chronic aggression (Ehrenkranz *et al.*, 1974), violent crimes (Dabbs *et al.*, 1987), antisocial personality disorder (Dabbs and Morris, 1990), and peer ratings of "toughness" (Dabbs *et al.*, 1987). However, many studies have not replicated these results (Bain *et al.*, 1987) and a metanalytic study of community and selected samples found only modest correlations (i.e., between 0.08 and 0.14) between testosterone and aggression (Archer *et al.*, 2005). Mixed findings have similarly been observed in adolescents (Olweus *et al.*, 1988; Susman *et al.*, 1987). These inconsistent results may be partially attributable to social and developmental factors (Rowe *et al.*, 2004), as well as other hormones, such as cortisol (Popma *et al.*, 2007b).

2. Cortisol

Cortisol is the end product of the hypothalamic-pituitary-adrenal (HPA) axis, the human body's stress response system, and is often used as a marker of stress responsiveness. Literature regarding the relationship between cortisol and aggression is mixed. Although numerous studies suggest that lower basal cortisol levels are associated with increased disruptive behaviors in males (Hawes *et al.*, 2009; Popma *et al.*, 2007a), other studies report the opposite, with CD and aggression being associated with higher levels of cortisol (Alink *et al.*, 2008; Van Bokhoven *et al.*, 2005). Still other studies have not established any link between cortisol and aggressive behavior (Alink *et al.*, 2008; Scerbo and Kolko, 1994; van Goozen *et al.*, 2000). A significant amount of research has focused on an association between an underreactive HPA axis and aggressive behaviors. Theoretically, this underreactivity may be associated with reduced comprehension of distress cues, which has been related to reduced empathy and behavioral inhibition (Marsh *et al.*, 2008). Lower levels of cortisol are also linked with reductions in fear responsiveness (Cima *et al.*, 2008), which may lead to persistent aggressive behavior (McBurnett *et al.*, 2000). In reviewing the divergent findings linking cortisol levels and aggression, Hawes *et al.* (2009) proposed two hormonal pathways to antisocial behavior—one which links stress exposure to high cortisol and aggression and the other which links low cortisol to aggression through callous-unemotional traits. In support of this proposal, a recent empirical study found that changes in cortisol levels in response to a stressor were positively associated with reactive (but not proactive) aggression in boys (Lopez-Duran et al., 2009).

3. Oxytocin

Recently, the hormone oxytocin has been investigated as a possible link in the development of aggression. Oxytocin is a hormone that is involved in trust and affiliative behaviors (Insel and Winslow, 1998; Young et al., 2001), thereby reduced oxytocin is thought to be associated with increased aggression. This hypothesis has been supported in the animal literature as oxytocin knockout mice exhibit increased aggressive behavior (Ragnauth et al., 2005). In humans, adults administered oxytocin intranasally are significantly better at identifying happy facial expressions compared to other expressions (Marsh et al., 2010). Oxytocin has also been found to reduce activation in the amygdala (Kirsch, 2005). Taken together, data suggest that deficits in oxytocin may influence mistrust and hostility which can contribute to aggression and violence (Siever, 2008).

E. Autonomic response measures

Autonomic arousal is most commonly measured via HR and electrodermal activity (EDA). In general, research regarding autonomic arousal and aggressive and violent behavior has yielded findings suggesting a pattern of lower baseline levels of autonomic arousal and higher autonomic reactivity in children and adolescents (Patrick, 2008). Although less consistent, research on aggressive adults indicates an increase in autonomic activity in response to a stressor (Patrick, 2008). A met analysis by Lorber (2004) found a reliable but modest association between the autonomic measures of HR and EDA and aggression and conduct problems. Below is a brief review of the specific associations between HR and EDA and aggression and violence.

1. Heart rate and electrodermal activity

Low resting HR is a common phenomenon among aggressive children (Scarpa and Raine, 1997) and is associated with conduct problems throughout childhood, adolescence, and adulthood (Lorber, 2004). Evidence for HR reactivity is less consistent. It appears that aggressive children have increased HR in response to a stressor (Lorber, 2004), and this finding is particularly robust in children displaying reactive aggression rather than proactive aggression (Hubbard *et al.*, 2002). HR reactivity and its association with aggression in adults are also mixed.

EDA refers to small changes in electrical activity of the skin, usually occurring 1–3 s following the onset of a stimulus (Scarpa and Raine, 1997). Lorber (2004) found that conduct problems in childhood were associated with reduced EDA in the absence of stimulation, and reduced EDA during a task, but only in the presence of nonnegative stimuli. This met analysis also revealed a positive association between adult aggression and EDA reactivity. Scarpa and Raine (1997) suggested that EDA under arousal may be associated with specific forms of crime, such as crimes of evasion. Skin conductance levels have also been shown to interact with markers of the HPA axis to predict later externalizing behaviors in children (El-Sheikh *et al.*, 2008). Results indicate that children with higher levels of EDA and higher levels of cortisol display increased levels of externalizing behaviors.

F. Electro cortical response measures

Electroencephalography (EEG) uses electrodes placed around the scalp at specified points to measure electrical activity of the brain. EEG evidence suggests that slow-wave activity during adolescence may predict later antisocial behavior (Raine *et al.*, 1990). These findings have been interpreted to suggest that cortical immaturity leads to reduced inhibition (Volavka, 1999) and increased impulsive behavior. EEG abnormalities, specifically under arousal and cortical immaturity, have been reported in violent recidivistic offenders (Raine *et al.*, 1990).

Event-related potential (ERP) refers to the averaged changes in electrical brain activity in response to a stimulus. Literature on ERP and aggressive behavior is mixed. One consistent finding has been the association between reduced amplitude of the P300 wave response, an ERP elicited by infrequent stimuli, among aggressive and impulsive individuals (Gerstle *et al.*, 1998). The P300 is thought to reflect online updating of cognitive representations (Donchin and Coles, 1988), thus the reduced amplitude may suggest impairment in higher cognitive functioning (e.g., working memory) in these individuals (Patrick, 2008). Reduced P300 has been reported in antisocial personality disorder (Bauer *et al.*, 1994), as well as other disorders that involve impaired impulsivity such as child CD and ADHD, drug dependence, and nicotine dependence (Iacono *et al.*, 2002). Therefore, reduced P300 amplitude may be a reflection of impulse control, a characteristic of reactive aggression (Patrick, 2008).

The physiological findings presented above illustrate the apparent biological underpinnings involved in the development of aggression and violence. Research from physiological and neurobiological studies has provided significant evidence for the biological basis of violent and aggressive behavior.

Taken together, the neurobiological and psychophysiological findings suggest a model in which aggression arises due to dysfunction in brain areas responsible for emotion processing, inhibition, and reactivity (Davidson, 2000). Repeated aggression may contribute to deficits in recognizing and processing of emotion, as well as the regulation and modulation of aggressive behavior in response to threat. Reduced activation and distribution of neurotransmitters in areas of the brain implicated in the development of violent behavior, in combination with dysregulation of the body's hormonal responsiveness, may add to difficulties in controlling aggressive outbursts. Furthermore, reduced arousal to distressing stimuli, as evidenced by a diminished physiological response and electrical activity in the brain may contribute to further deficiencies in appropriate responses to increased arousal. Further research is necessary to provide additional evidence for this suggested model.

As the field progresses, particular attention should be paid to the subtypes of aggressive behavior that may be characterized by unique physiological patterns. Future studies should focus on parsing the different types of aggression to elucidate the specific pathways underlying violent behaviors. A better understanding of these pathways allows for earlier intervention that may be able to reduce or prevent severe aggressive behavior later in life. With the knowledge of the psychophysiology associated with the development of violence throughout the lifetime, we now turn specifically to the perinatal period to better understand how these biological pathways can be altered to prevent or lead to later violence.

III. PERINATAL FACTORS RELATED TO THE DEVELOPMENT OF AGGRESSION

The notion of a mother's health affecting that of her unborn infant is one that has pervaded common knowledge for centuries. Only recently, however, have scientists begun to empirically study the specific factors influencing fetal development. Preterm birth, delivery complications, maternal mental illness, gender, and exposure to drugs, alcohol, and tobacco are all topics that have been explored in relation to their impact on fetal development. These factors have been shown to have maladaptive effects on fetal brain development following prenatal exposure—a finding that makes pregnancy a critical window for the prevention of unfavorable outcomes throughout the lifespan of the offspring. One such risk that has been identified in relation to these perinatal factors is the development of aggression. It has been suggested that through disruptions in neural development, the fetus incurs neuropsychological deficits—namely neural impairments in executive and verbal functioning that may result in an irritable disposition, poor behavioral regulation, or aggression (Brennan et al., 2003). This tendency toward aggression tends to persist throughout adolescence and into adulthood, manifesting itself through externalizing behaviors, internalizing problems (depression, loneliness, etc.), poor peer relationships, psychological disorders (CD, oppositional defiant disorder, etc.), recurring criminal behaviors, and violence. Because of the extensiveness of the risks associated with aggression, it has many potentially negative outcomes both in the developing infant, as well as in the significant economic and social burdens it places on society. In this section, the potential influences of the aforementioned perinatal factors will be explored in their relation to the development of aggression and violence. It is important to note that these risks do not necessitate the development of aggression in offspring but rather are important in understanding certain conditions in which aggressive traits may be more likely to arise.

A. Birth complications

Many studies have found associations between birth complications and negative behavioral outcomes in offspring. Specifically, irritable temperament in childhood, violent offending in adolescence and adulthood, and aggressive behaviors throughout the lifespan are among the offspring outcomes that have been associated with higher rates of birth complications. Birth complications typically refer to the following three factors: (1) prenatal complications, such as hypertension, mental illness, stress, drug and alcohol exposure, and viral infections experienced by the mother; (2) perinatal complications, which include difficult fetal delivery (e.g., breech birth), premature breaking of the membrane, assisted delivery (forceps and cesarean), fetal distress (i.e., difficulty breathing), preeclampsia, and umbilical cord prolapsed; and (3) postnatal complications as indicated by either cyanosis or treatment with oxygen (Liu et al., 2009).

The mechanisms through which birth complications may influence the development of aggression are unknown, but they are hypothesized to involve damage to the PFC, hippocampus, and dopamine systems (Brennan et al., 1997; Cannon et al., 2002; Mednick and Kandel, 1988; Raine, 2002b). More specifically, preeclampsia, maternal bleeding, and maternal infection may cause an inadequate supply of blood to the placenta, fetal hypoxia or anoxia (lack of oxygen), and disrupted development of the hippocampus, dopamine systems, and other parts of the brain (Cannon et al., 2002; Liu, 2004; Mednick and Kandel, 1988). Animal research has supported these findings by suggesting that perinatal complications surrounding anoxia in rats may reduce central dopamine transmission (Brake et al., 2000). Dopaminergic neurotransmission appears to be involved in impulse control, aggression, and violence (Chen et al., 2005; Retz et al., 2003). Animal research also suggests that perinatal complications may limit neurotransmitter functioning in the left PFC—an effect that has been one of the most frequently replicated indicators of violent offending in the brain imaging literature (Henry and Moffitt, 1997). It is important to note, however, that birth complications alone are unlikely to predispose infants to externalizing behavior and aggression. Instead, these complications interact with various psychosocial risk factors (i.e., poverty, poor parenting, parental rejection, negative peer relationships, bad neighborhoods, etc.), and likely genetic factors, to influence aggressive tendencies (Raine *et al.*, 1994).

Prenatal, perinatal, and postnatal health care interventions aimed at reducing birth complications may help to decrease risks of later development of aggression and violence. If women who may be at risk for birth complications are identified and educated, these mothers may be in a better position to take steps toward keeping their pregnancies and their babies healthy. If, however, birth complications do occur, early enrichment programs, that improve cognitive ability or enhance the parent–child relationship, may be effective in preventing the emergence of aggressive and violent behavior in adolescence and adulthood.

B. Preterm birth and low birth weight

Research supporting the role of neuropsychological deficits in mediating birth complications and adverse outcomes is consistent with preterm and low birth weight literature. White *et al.* (1994) have shown that medical and congenital risk factors, such as low birth weight and preterm birth, may lead to neuropsychological deficits and CNS damage that result in an increased likelihood for criminal offending. Evidence also suggests that preterm birth may be involved in the development of externalizing behaviors and aggression, and that these negative behavioral outcomes worsen with age (Bhutta *et al.*, 2002).

Low birth weight and preterm birth serve as strong and consistent predictors of neuropsychological deficits that may result in subsequent aggression and antisocial behavior (McCormick, 1985). Low birth weight infants were three times more likely to experience neurological deficits than controls (McCormick, 1985). Moreover, such CNS deficits (Moffitt, 1993) may manifest themselves in a variety of ways, including temperamental difficulties, cognitive deficits, inattention, antisocial behavior, subnormal growth, learning difficulties, hyperactivity, behavioral problems, poor academic achievement, CNS damage, and psychiatric disorders (Piquero and Tibbetts, 1999).

Preterm birth and low birth weight have been found to be correlated with maternal tobacco use, lack of prenatal care, drug and alcohol use by the mother during pregnancy, low socioeconomic status, poor diet, psychotropic drug use during pregnancy, and low parental educational level (Piquero and Tibbetts, 1999). Because of the range of factors that might cause preterm birth and low birth weight in an infant, preventative efforts are critical.

It is important to note that the effects of preterm birth and low birth weight on psychological functioning and aggression may vary depending on the sex of the child. For example, low birth weight boys exhibit a significantly more aggressive and delinquent acts in comparison to their female counterparts (Ross et al., 1990). There is also considerable evidence suggesting that disadvantaged home environments and maternal interactive style may moderate the relationship between these risks and aggressive behaviors (Piquero and Tibbetts, 1999). Therefore, a number of factors, such as sex, home environment, and maternal interactive style, are involved in the phenotypic expression of aggression in premature and low birth weight infants.

Despite the decrease in infant mortality over the past 30 years, the prevalence of preterm birth and low birth weight has actually increased to approximately 12.5% of births in the United States. Given the prevalence of preterm birth and low birth weight in the United States, it is understandable why risk factors associated with this population are such a major public health concern (Berman and Butler, 2006). Preventative measures and public awareness campaigns focusing on the risks involved in preterm birth and low birth weight are a necessary next step in addressing these issues.

C. Prenatal drug and alcohol exposure

1. Alcohol

Sixteen percent of children born in the United States are exposed prenatally to alcohol, making alcohol the most common neurobehavioral teratogen affecting fetal development (Sood et al., 2001). Overall, children who are prenatally exposed to alcohol are 3.2 times more likely to develop aggression and delinquent behavior than nonexposed children. Further, children exposed to low levels of alcohol prenatally show higher scores for aggressive and externalizing behaviors on the Child Behavior Checklist (CBCL) and children exposed to moderate levels have higher scores on delinquent and total problem behavior on the CBCL (Sood et al., 2001). This suggests a higher threshold for the development of delinquency in children, as opposed to aggressive and externalizing behaviors. However, it also negates claims that low levels of alcohol consumption during pregnancy are tolerable with evidence that suggests that even a small dose may have adverse effects on fetal development. More generally speaking, the literature supports a dose–response continuum where a more heavily exposed fetus shows a greater magnitude of these adverse effects (Driscoll et al., 1990).

Fetal alcohol spectrum disorder (FASD) and fetal alcohol syndrome (FAS) are conditions that may arise when children are prenatally exposed to alcohol. These disorders are characterized by physical and mental birth defects that may result in impaired interpersonal skills and social deficits. Some of the behaviors that are commonly observed among populations of FAS individuals are tendencies to demand attention, interrupt others, lie, show impaired moral judgment (especially with regard to social relationships), overreact to situations with excessively strong emotional responses, monopolize conversations, and demonstrate unawareness of the consequences of one's actions. This population may also suffer major language and social communication deficits, which further hamper their social competence (Kelly *et al.*, 2009).

Animal studies have helped in linking prenatal alcohol exposure to the development of aggressive behaviors. Specifically, evidence suggests that alcohol exposure during prenatal development causes CNS damage. Rats provide an excellent model for understanding the development of aggression, because their social behavior has been shown to follow similar patterns to that of humans. Their social behavior results from a combination of influences including genetic makeup, teratogenic influences, early maternal–infant interactions, and later social learning. Primates, too, offer a suitable model for studying the effects of prenatal alcohol exposure, as their gestation characteristics and early developmental stages are similar to that of humans. In both animals and humans, prenatal alcohol exposure is not considered a singular cause of social deficits, but rather a probabilistic contributor serving as a risk factor for the developing child (Kelly *et al.*, 2000).

Because there are social–familial influences associated with prenatal exposure to alcohol, one might ask how it can be determined that alcohol exposure has any actual teratological effect. Animal models provide a means through which alcohol- and environment-related factors can be separated in an experimental fashion. For example, removing a newborn pup from its "alcohol-using" mother and transferring it to a foster-parent environment results in rates of aggression that are analogous to those remaining in the care of their "alcohol-using" mothers, suggesting that changes in aggressive behavior are initiated by the pup's prenatal exposure to alcohol, rather than by its environment. This type of experiment, for obvious reasons, would be ethically impossible to conduct in human populations (Kelly *et al.*, 2000).

Animal models have contributed essential components to our understanding of the specific mechanisms through which prenatal ethanol (alcohol) exposure and aggression are linked. Rodent studies suggest that the behavioral deficits that result from ethanol exposure *in utero* are linked to ethanol-induced changes in the CNS. These changes in the CNS, however, are not uniform, and some brain regions (i.e., neocortex, hippocampus, cerebellum) are more affected than others. Notably, the HPA axis and beta-endorphin (b-EP) systems become dysregulated and hyperresponsive to social situations, which is demonstrated by heightened and prolonged concentrations of hormones, such as corticosterone (CORT) and adrenocorticotropin (ACTH), as well as elevated plasma levels and reduced pituitary content of b-EP compared to controls. Further, ethanol-exposed neonates show heightened sensitivity to stressors, significantly increased corticotrophin release factor (CRF) biosynthesis and expression, and more prolonged CORT and ACTH elevations during and after stress. These effects, which persist throughout the neonate's lifespan, indicate deficits in pituitary–adrenal response inhibition and in recovery from stress. It is through these mechanisms that prenatal alcohol exposure may manifest in aggressive tendencies and externalizing behaviors in the lives of effected offspring (Weinberg et al., 1996).

2. Drugs

Drug use by pregnant women has increased steadily. Despite general awareness of detrimental effects, drug use during pregnancy continues its upward trend with prevalence estimates ranging from 0.3% to 46%. Prenatal drug exposure is associated with behavioral abnormalities, such as excessive irritability, poor social-attachment behavior, and aggression (Johns et al., 1994). The neurobiological processes through which these deficits emerge primarily involve the effects of drugs on fetal organogenesis, especially fetal brain development (Mayes, 1994). Evidence suggests that an increase in aggressive or violent behaviors associated with prenatal drug use may arise from alterations in the CNS. More specifically, aggressive behaviors can be linked to changes in fetal neurotransmitter systems, particularly within the limbic system (Miller et al., 1991). Cocaine (including crack cocaine) is one of the most commonly studied CNS stimulants in the literature on prenatal drug exposure. Cocaine affects monoaminergic neurotransmitter (dopamine, norepinephrine, and 5-HT) systems in the CNS by blocking the reuptake of dopamine, norepinephrine, and 5-HT leaving more of these neurotransmitters within the synaptic space (and therefore the peripheral blood). An excess in the amount of these neurotransmitters results in psychological effects, such as pleasure and euphoria, as well as specific behaviors and physiological reactions. Physiologically, chronic cocaine use may lead to tolerance whereby increasing amounts of the drug are necessary to achieve a desired effect.

In the developing fetus, monoaminergic neurotransmitters play a critical role in fetal brain development by influencing cell proliferation, neural outgrowth, and synaptogenesis (Lauder, 1988; Mattson, 1988). Cocaine and other drugs may affect these neural processes throughout gestation through their effects on the release and metabolism of monoamines. The importance of monoamine neurons in fetal brain development has been demonstrated in both human and animal models. For example, in the rats' second week of gestation, norepinephrine neurons appear in the locus coeruleus, 5-HT neurons are found in the raphe nuclei, and dopaminergic neurons in the substantia nigra are functional (Lauder and Bloom, 1974). By the end of the second month of human gestation, 5-HT and norepinephrine neurons can be found. In both animals and humans, these monoamine neurons are rapidly generating axonal connections with the forebrain and actively influencing the production and differentiation of cell structure in these regions (Lidov and Molliver, 1982a,b; Wallace and Lauder, 1983).

Further evidence of these effects can be demonstrated by administering cocaine to rats during the early postnatal period when synaptogenesis begins in the forebrain. This early postnatal period in rats is functionally equivalent to the third trimester in human gestation during which axonal and dendritic growth take place. When brain glucose metabolism is used as an indicator of activity, animal models exhibit the greatest percentage change in brain regions with high dopaminergic activity in comparison to untreated controls (Dow-Edwards *et al.*, 1988, 1989). Several brain structures associated with the mesocortical and mesolimbic systems, including the cingulate cortex, PFC, nucleus accumbens, amygdala, septum, ventral tegmental area, and ventral thalamic nucleus, appear highly affected by dopamine activity (Goeders and Smith, 1983; Shepard, 1988). Each of these areas is thought to be involved in an organism's arousal, attention, and ability to regulate anxiety and emotional responses (Mayes, 1994).

The neural processes mentioned above may lead to the development of aggression when abnormalities in fetal brain development later manifest themselves in social and behavioral ways. Basic processes, like the regulation of attention, response to sensory stimuli, and the modulation of mood states may all be linked to prenatal drug exposure through the drug's alterations of neurotransmitter activity. Several studies have found that infants exposed to drugs prenatally are often easily irritable and difficult to engage. Evidence also suggests that prenatal drug exposure results in crying patterns that indicate a general "excitable" tendency within affected infants. Human infants exposed prenatally to cocaine have also shown elevated HRs and norepinephrine levels at 2 months of age; lending further support to the influence cocaine has on monoaminergic systems (Mayes, 1994; Mirochnick *et al.*, 1991). Rodent models have demonstrated an increased susceptibility to stressors, higher vulnerability to the environment, and increased rates of aggressive behaviors in response to social competition among offspring prenatally exposed to drugs (Spear *et al.*, 1998).

The influence of prenatal exposure to drugs and alcohol on social and aggressive behavior has serious implications for crime prevention. Early identification and intervention among infants and children who may be affected by prenatal drug or alcohol exposure is necessary to prevent delinquency and poor social relationships within these populations (Johns *et al.*, 1994). Such steps are also necessary for gaining a better understanding of the behavioral differences that may exist within educational, occupational, and social settings due to prenatal exposure to drugs and alcohol.

D. Smoking

Smoking during pregnancy remains a critical public health concern. Nearly half of all women who smoke continue to do so even while pregnant, despite some women's intentions to refrain from doing so. Despite common knowledge of the adverse effects of maternal smoking during pregnancy among the American public, more than half a million infants per year in the United States are prenatally exposed to maternal smoking (Wakschlag et al., 2002). This is of even greater concern when one considers the failure of public health smoking cessation campaigns for the 10.2% of women in the United States who continue to smoke through their pregnancies. Adverse outcomes, which include low birth weight, premature delivery, spontaneous abortion, and infant mortality, have been the primary focus of these campaigns (Weaver et al., 2007). In comparison, relatively little attention has been paid, from a public health standpoint, to the relationship between prenatal smoking and the development of aggression and violence in offspring.

Prenatal smoking predicts children's likelihood of displaying high aggression from as early as 1.5 years and throughout adulthood (Huijbregts et al., 2008). Several externalizing behaviors, including impulsivity, truancy, hyperactivity, attentional difficulties, and delinquency, have all been found to be associated with maternal prenatal smoking through the fetus' exposure in uterine.

Potential neurobiological mechanisms through which prenatal nicotine exposure may increase the offspring's risk for aggressive behaviors include the HPA axis and the CNS (Brennan et al., 1997). Substantial evidence suggests that nicotine crosses the placental barrier and causes neurotoxicity in the fetus. Neurotoxicity occurs via hypoxic effects on the fetal-placental unit (e.g., reduction of fetal blood flow) and teratological effects on the developing fetal brain. Two recent human studies support this contention, noting associations between maternal prenatal smoking and decreased frontal lobe volumes in infants (Ekblad et al., 2010), and a thinning of the cerebral cortex in adolescents (Toro et al., 2008). Within the HPA axis, nicotine produces a heightened ACTH response to stress in adult rats (Poland et al., 1994). Other studies have found that elevated levels of ACTH increase aggressive and defensive behaviors in both rats and nonhuman primates, suggesting that this hormone may be related to the development of aggression (Higley et al., 1992; Veenema et al., 2007). However, lower levels of ACTH have also been found within human criminal and antisocial populations in comparison to controls, so these results are mixed and should be interpreted with caution (Coccaro and Siever, 2002; Virkkunen et al., 1994).

Nicotinic acetylcholine receptors (nAChRs) are responsible for the regulation of many vital phases of brain maturation. These receptors are present in the brain early in gestation and develop throughout prenatal, postnatal, and adolescent periods, suggesting that nicotinic signaling plays a crucial role in neural development. During these developmental periods, NAChRs are particularly sensitive to environmental stimuli and, as specific nicotine-sensitive receptors, are especially vulnerable to exogenous nicotine. Nicotine affects fetal development primarily through its effect on nicotinic-binding sites in the cerebral cortex. More specifically, nicotine has been found to alter the neocortex, hippocampus, and cerebellum during the early postnatal period within rats (the equivalent of the third trimester in humans; Dwyer et al., 2009). Evidence suggests that prenatal nicotine-induced defects within these particular brain regions may increase the likelihood of dopamine-mediated disorders like attention deficit hyperactivity disorder and substance abuse. Patterns of continuous maternal smoking (i.e., the tendency to smoke in a way that maintains plasma nicotine levels at a steady state) cause more negative effects than more periodic patterns of use, which allow the CNS to recover between episodes. The stimulation to nicotinic receptors interacts with the genes that influence differentiation of cells, causing permanent changes in cell functioning. It has been suggested that these processes disrupt the maturation of the fetal brain and produce adverse effects in fetal development that can later manifest themselves in aggression or violence (Wakschlag et al., 2002).

Animal models demonstrate many of the biological effects of prenatal smoking on neonatal behavior. Rats exposed to nicotine prenatally show deficits in learning and memory, as well as in social behavior. Benowitz (1998) found that nicotine infusion in rats causes interference with neural cell replication and abnormal synaptic activity. These, in turn, produce neuroendocrine and behavioral abnormalities that could potentially lead to aggression. Rodent models have also shown similar adverse effects linked to second hand smoke, as well as maternal use of nicotine replacement therapy (NRT), a pharmacotherapy of smoking cessation thought to be less detrimental than smoking cigarettes during pregnancy (Dwyer et al., 2009). Findings of adverse effects related to maternal use of NRT are particularly disturbing, since (1) NRT does not seem to increase the likelihood of successful smoking cessation during pregnancy and (2) NRT has actually been recommended by a number of public health authorities, including the Food and Drug Administration (Bruin et al., 2010). NRT (as well as cognitive-behavioral therapy (CBT)) has been shown to be effective among nonpregnant smokers, so prevention rather than smoking cessation during pregnancy should be the aim for reducing adverse outcomes attributed to prenatal nicotine exposure. Further, interventionists should keep in mind that any prenatal nicotine exposure, even through modes of transmission not related to smoking, can be detrimental to fetal development.

As with most toxins, the effects of prenatal smoking exposure are dose-dependent and thus strongest among offspring of heavy smoking mothers (≥ 10 cigarettes/day). Further, the effects of prenatal smoking are exacerbated when accompanied by low socioeconomic status, poor parenting, family dysfunction, paternal absence, and parental history of antisocial behavior. However, the relationship still exists even when these variables are controlled for (Huijbregts et al., 2008). Evidence suggests that gender might moderate the relationship between maternal prenatal smoking and externalizing behaviors in that the relationship is stronger among male offspring in predicting CD and stronger among female offspring when predicting substance abuse (Brennan et al., 2002).

E. Maternal psychological stress

Prenatal stress is so common an occurrence that it seems unlikely that it could have any significant effects or unfavorable life-long outcomes on child development. However, it is, in fact, associated with low birth weight, preterm birth, preeclampsia, spontaneous abortion, growth-retardation (specifically reduced head circumference), developmental delays, heightened emotionality, externalizing behaviors, irritability, psychopathology, and deficits in attention, cognition, and neurodevelopment (Clarke *et al.*, 1994, 1996; Gutteling *et al.*, 2005; Mulder *et al.*, 2002). Effects involving birth outcomes are relevant to the development of aggression in the ways previously described. However, prenatal stress, more broadly speaking, also affects fetal neurodevelopment in a different way. Prenatal stress may stem from a variety of sources including, but not limited to inadequate social support, low socioeconomic status, unwanted pregnancy, and sexual, physical, or verbal abuse. These stressors may take the form of one traumatic event, several recurrent ones, or more chronically on a daily basis (Mulder *et al.*, 2002). When the stressor is experienced, the HPA axis and the sympathetic nervous system are activated, as the body's response to a particular threat the individual perceives in her environment. This physiological response has evolutionary value in that it places us in "fight or flight" mode, increasing our awareness of problems that may exist and preparing us to find ways of solving them. This process becomes maladaptive, however, when our perception of a threat or stressor is inconsistent with its actual magnitude and relevance to our lives, and alternatively, when the physiological response following a stressor is prolonged (Clarke *et al.*, 1994, 1996; Gutteling *et al.*, 2005; Mulder *et al.*, 2002).

The connection between prenatal stress and HPA axis activity has been demonstrated in rodent, animal, and human studies, and in all of these, both prenatal stress and heightened HPA axis response have been found to be predictive of the development of aggression in offspring (Clarke *et al.*, 1994, 1996; Gutteling *et al.*, 2005; Mulder *et al.*, 2002). A number of rodent studies have introduced stress to a pregnant mother prenatally using electrical shock, immobilization, or randomly administered bursts of noise. These studies have demonstrated a dysregulation of the HPA axis in both mother and pup that eventually leads to heightened emotionality, hostility, and aggression in the offspring (Clarke and Schneider, 1993; Mulder *et al.*, 2002; Sobrian *et al.*, 1997; Takahashi *et al.*, 1990; Ward and Weisz, 2011). Similarly, studies of the offspring of rhesus monkey mothers exposed to stress from mid- to late-gestation demonstrated low birth weight, impaired neuromotor development, attention deficits, and disturbed behavior (Clarke *et al.*, 1994, 1996; Schneider, 1992a,b). These effects were long-term and persisted even into the adolescent period of development (Clarke *et al.*, 1996). In humans, similar effects of maternal prenatal stress have been reported (Gutteling *et al.*, 2005; Mulder *et al.*, 2002).

HPA axis regulation involves several hormones, including cortico-releasing hormone (CRH), cortisol, (nor)adrenaline, and ACTH, which are released into the bloodstream when a stressor is experienced. Small increases in these hormones within the pregnant mother may lead to disproportionately large increases in fetal hormonal levels. Excessive levels of these hormones may potentially inhibit the growth and development of the nervous system, cause damage to the brain, and produce programming effects on the fetal neuroendo-crine system that lead to the developmental deficits mentioned above. This may be especially true when HPA axis activity is characterized by an exceptionally strong, sustained response to the stressor. Animal models have supported this explanation by demonstrating experimentally that levels of these hormones are higher in neonates prenatally exposed to stress in comparison to controls (Clarke et al., 1994, 1996; Gutteling et al., 2005; Mulder et al., 2002).

It is important to note that there have been gender differences in the findings within this area, namely that hostility and aggressive behaviors appeared to be more prevalent in male offspring than in females. These findings, which are typically found in rodent studies, suggest that males may be more vulnerable to the effects of prenatal stress than their female counterparts (Clarke et al., 1996). However, in general, it seems that prenatal stress may be an important predis-posing factor for a number of behavioral deficits among both male and female offspring, even if to different degrees.

Because stress can be so pervasive in the life of a pregnant mother, it often manifests itself in a variety of ways. For example, stress might lead a mother to engage in smoking or alcohol and substance abuse, which, in turn, can produce fetal neurobehavioral deficits of the kind that have been described in previous sections. To prevent these types of detriments from occurring, it may be necessary to assess women's stress levels in early pregnancy, identify those who are at risk, and provide stress reduction programs throughout their pregnancies. Educating women about the risks involved with prenatal stress, as well as training them in relaxation methods, may be helpful in alleviating these effects. Ensuring that women have the appropriate buffers needed to prevent stress is also essen-tial. Adequate social support and financial resources are just a few factors that may be necessary to ensure a mother's psychological well-being.

F. Environmental context

It is important to note that the prenatal factors discussed above may not necessarily operate in a unidirectional manner. It may often be the case that these factors result in child aggression in the form of coercion or manipulation of the parent in order to obtain something that is wanted. This coercion is likely to elicit a negative response from the parent in the form of either negative rein-forcement (giving in to the child) or positive reinforcement (giving increased attention to the child by chastising or yelling). This reinforcement indirectly encourages and exacerbates the child's behavior, producing an ongoing cycle of reinforced aggressive behaviors throughout the child's lifetime. Further, this cycle may be maintained by preexisting neuropsychological deficits and unintentionally harmful reinforcements from other figures in the child's life, such as teachers, grandparents, and peers.

Throughout this section, it has been continually noted that neither social nor biological elements operate alone in contributing to the development of aggression in offspring. One or the other may predispose a child to developing aggressive behaviors, but it is the "double hazard" of perinatal risk and social disadvantage that places a child at maximal risk for later aggression and externalizing behaviors (Brennan and Mednick, 1997). Understanding the interaction of biological and social risk factors in generating aggression is critical for preventing both personal costs (few positive social relationships, poor job performance, etc.), as well as social costs (crime rate, prison costs, etc.). Thus, researchers and policy makers should make it a goal to identify populations potentially affected by perinatal risk factors in order to more accurately predict who might benefit from interventions aimed at preventing later aggression, violence, and criminal offending.

IV. GENETIC CONTRIBUTIONS

As described in detail elsewhere in this volume, aggressive and violent behavior can in part be accounted for by genetic factors. Because prenatal stress and teratogenic exposures may be linked to genetic risk, it is important to consider potential genetic contributions to the association between perinatal factors and aggression.

A. Genetic factors as explanatory

One prenatal risk factor, that has received recent attention in terms of the potentially confound role of genetic factors, is maternal smoking during pregnancy. For example, twin studies have been utilized to assess whether the relationship between maternal prenatal smoking and offspring externalizing behavior remains significant when controlling for genetic influences. In one such study, the association between maternal smoking and offspring ADHD was found to persist after controlling for genetic influences (Thapar *et al.*, 2003). In a separate twin study, researchers found that genetic effects explained about half of the association between maternal prenatal smoking and child conduct problems; in this study, controls for both genetic influences and parent psychopathology accounted for the initial association in its entirety (Maughan et al., 2004).

Recent advances in infertility treatment (i.e., in vitro fertilization using donor eggs) have also allowed for the use of a prenatal cross fostering design to assess the moderating impact of genetic influences on the maternal prenatal smoking/child externalizing disorder association (Rice et al., 2009). In this novel study, maternal smoking during pregnancy was only associated with child anti-social outcomes in cases where the mother was implanted with her own egg, as opposed to an unrelated donor's egg. These results suggest that genetic factors are a necessary component in the noted relationship between maternal smoking and child antisocial outcomes.

Another novel design strategy has recently been used to evaluate out-comes for siblings discordant for maternal prenatal smoking (D'Onofrio et al., 2010; Lindblad and Hjern, 2010). Results from studies using this design suggest that familial background factors, rather than environmental exposure effects, explain associations between maternal prenatal smoking and externalizing pro-blems. However, as acknowledged by their authors, these sibling discordant design studies did not test for gene by environment interactions, leaving open the possibility that prenatal exposure to maternal smoking may result in exter-nalizing behavior outcomes for offspring at particular genetic risk.

B. Gene by environment (G × E) interactions

Gene by environment interactions have recently been examined in terms of their relevance to perinatal risks and child behavioral outcomes. These studies have primarily focused on polymorphisms linked to the neurotransmitter systems of dopamine, norepinephrine, and 5-HT, which have been described previously in terms of their relevance to both perinatal factors and aggressive outcomes.

1. Monoamine oxidase genotype

Monoamine oxidase (MAO) is a critical enzyme involved in the degradation of neurotransmitters, including norepinephrine, dopamine, and 5-HT. MAO exists in two forms, MAOa and MAOb. The gene that codes MAOa has functional variations that influence the level of MAOa (Sabol et al., 1998), which in turn affects central levels of dopamine and 5-HT and thus, directly regulates behav-ior. Therefore, this gene has been the focus of molecular genetics studies of aggressive behavior in both humans (Manuck et al., 2000) and rodents (Cases et al., 1995) and is the most well-established susceptibility variant for aggression in several species. MAOa genotype appears to influence the development of violent behaviors by altering vulnerability to the effects of early adverse environments (Caspi *et al.*, 2002; Kim-Cohen *et al.*, 2006). Specifically, there is robust evidence that the interaction between MAOa and childhood maltreatment predicts child CD and adult antisocial behavior such that males with low expression of MAOa (L allele), but not males with high expression of MAOa (H allele) are at increased risk (Kim-Cohen *et al.*, 2006; Nilsson *et al.*, 2007). Recent evidence also links MAOa-L and low brain MAOa with trait aggression and neural hypersensitivity to social cues (Alia-Klein *et al.*, 2008; Eisenberger *et al.*, 2007). Interestingly, *MAOa* is an X-linked gene (with males carrying only one allele and females carrying two), which suggests the possibility of sex differences in genetic and epigenetic regulation and may explain the increased average aggressiveness in males in comparison to females.

One recent study has noted an interaction between the MAOa uVNTR (untranslated region variable number of tandem repeats) genotype, gender, and maternal prenatal smoking in the prediction of CD symptoms (Wakschlag *et al.*, 2010). Specifically, boys with the low activity MAOa genotype whose mothers smoked during pregnancy were at an increased risk of CD symptoms, whereas girls with the high activity MAOa genotype whose mothers smoked during pregnancy were at increased risk for hostile attribution bias (a characteristic common to aggressive children) as well as CD symptoms.

2. Genes related to dopaminergic function

Kahn *et al.* (2003) noted an interaction between a *DAT1* genotype and maternal prenatal smoking in the prediction of oppositional and hyperactive symptoms in young children. This finding was replicated in an adolescent sample; however, the $G \times E$ effect was specific to hyperactive-impulsive symptoms in males (Becker *et al.*, 2008). Other studies have failed to replicate this effect for *DAT1* and other dopamine-related genotypes (e.g., Brookes *et al.*, 2006; Langley *et al.*, 2008); however, relatively few studies have been completed in this area.

3. Catechol O-methyltransferase

Catechol O-methyltransferase (COMT) is a key modulator of extracellular dopamine levels in the PFC. A common G/A polymorphism produces a valine-to-methionine amino acid substitution at codons 108 and 158 (Val108/158Met; rs4680), which results in a three- to fourfold variation in COMT activity, whereby the Val and Met alleles confer high and low activity, respectively (Lachman *et al.*, 1996). This well-characterized, functional polymorphism has been associated with atypical neural processing and connectivity in healthy individuals (Dennis *et al.*, 2010), deficits in executive functioning abilities (Tunbridge *et al.*, 2006), and with aggression and serious antisocial behavior in individuals with ADHD (Caspi *et al.*, 2008).

Prenatal exposure to nicotine also leads to persistent abnormalities in neurotransmitter functioning in the cerebrocortical areas of the rat brain (Slotkin *et al.*, 2007). Further, both maternal prenatal smoking and COMT associations with CD appear to be specific to aggressive behavior, rather than covert antisocial behavior (Monuteaux *et al.*, 2006, 2009). Taken together, these findings suggest that the combination of the Val/Val genotype and prenatal exposure to maternal smoking may lead to neural processing deficits that increase vulnerability for aggression.

One study examining the interaction of prenatal risk and COMT variation in the prediction of antisocial outcomes found that birth weight interacted with Val108/158Met to predict antisocial behavior in an ADHD sample (Thapar *et al.*, 2005); however, this finding has had at least one failure to replicate (Sengupta *et al.*, 2006). Another recent study (Brennan *et al.*, 2011) found that individuals with the COMT Val/Val genotype whose mothers also smoked during pregnancy were at an increased risk for aggressive behavior outcomes in adolescence and young adulthood. These G × E interaction findings are preliminary but do suggest a potentially important role for genetics in noted relationships between perinatal risk factors and aggression.

C. The role of epigenetics

The analysis of a G × E interaction is still focused on the "fixed" nature of the genome—the DNA sequence itself. The DNA sequence is the same in every cell of the body and does not change across the lifespan. But there are other characteristics of genetic makeup that are not as fixed, and that have been found to change in response to environmental influences over time. These are known as epigenetic phenomena. Several epigenetic processes have been discovered, but the most commonly studied today is the phenomenon of methylation, a measurable chemical modification to DNA that can directly alter the expression of genes.

Methylation patterns change not only in response to toxins and stress encountered in the environment but also in response to nutritional supplements and parenting sensitivity. In a series of seminal studies in this area, Meaney and colleagues discovered that the quality of the parenting (licking and grooming) that a mother rat provided to her pups during early postnatal development changed the methylation patterns in the hippocampus of her offspring (Kappeler and Meaney, 2010). Specifically, higher levels of licking and grooming made the genes (and the offspring) less responsive to stress in the environment. In contrast, low levels of licking and grooming resulted in offspring whose genetic profile enhanced the release of cortisol in response to stress. Importantly, either strategy might be considered "ideal" parenting, depending upon the environment in which the offspring will have to survive.

Recent research suggests that prenatal factors may also influence DNA methylation patterns in offspring (Radtke *et al.*, 2011). Specifically, maternal reports of abuse during pregnancy were found to be correlated with methylation of the glucocorticoid receptor (GR) gene in offspring ages 10–19. In contrast, maternal experiences of abuse prior to or after pregnancy were not associated with offspring DNA methylation patterns. Importantly, methylation of the GR gene directly impacts the functioning of the HPA axis, which (as noted previously) may, in turn, impact levels of aggression and antisocial behavior.

In summary, preliminary evidence suggests that high versus low risk genotypes may moderate the effects of perinatal exposures by influencing an individual's susceptibility or resistance to these environmental experiences. In addition, perinatal risk factors are associated with epigenetic changes that are evident in and may influence later development. Complex behaviors, like aggression, are likely based on interactions of numerous genes and numerous environmental factors. Future molecular genetics and epigenetic programming studies should attempt to unravel the interplay between genes and environment. Such knowledge will provide a clearer understanding of the role of early risk factors in the development of aggression and how they can be used as intervention targets to alter developmental trajectories that lead to a lifetime of violence.

V. CONCLUSIONS

It is highly evident based on experimental and clinical studies that deleterious perinatal exposures can have a profound and enduring impact on the neuroregulatory systems that mediate violence and aggression. Early adverse perinatal experiences, in combination with predisposing genetic factors, combine with unstable family environments to substantially increase the vulnerability for a trajectory of delinquent and aggressive behavior throughout the lifespan; however, these outcomes are both complex and multidimensional. Future studies should focus on genetic risk factors, as well as novel interventions that may mitigate or prevent the deleterious effects of an adverse perinatal environment on the development of aggression. Effective interventions should target prenatal maternal mental and physical health-related behaviors, address parenting behaviors during critical stages of child development (i.e., infancy, early childhood, and adolescence), as well as focus on child cognitive and social enrichment during pre- and elementary-school years. As we are just beginning to understand the complexity of the intergenerational transmission of these problems during pregnancy and early childhood, it is important as a field to focus on the origin, early development, and prevention of aggression and violence to prevent vulnerable families and at-risk children from a lifetime of adversity.

References

Alexander, M., and Perachio, A. A. (1973). The influence of target sex and dominance on evoked attack in Rhesus monkeys. *Am. J. Phys. Anthropol.* **38**(2), 543–547.

Alia-Klein, N., Goldstein, R. Z., Kriplani, A., Logan, J., Tomasi, D., Williams, B., Telang, F., Shumay, E., Biegon, A., Craig, I. W., Henn, F., and Wang, G. (2008). Brain monoamine oxidase A activity predicts trait aggression. *J. Neurosci.* **28**(19), 5099–5104.

Alink, L. R., van Ijzendoorn, M. H., Bakermans-Kranenburg, M. J., Mesman, J., Juffer, F., and Koot, H. M. (2008). Cortisol and externalizing behavior in children and adolescents: Mixed meta-analytic evidence for the inverse relation of basal cortisol and cortisol reactivity with externalizing behavior. *Dev. Psychobiol.* **50**(5), 427–450.

Anderson, S. W., Bechara, A., Damasio, H., Tranel, D., and Damasio, A. R. (1999). Impairment of social and moral behavior related to early damage in human prefrontal cortex. *Nat. Neurosci.* **2**(11), 1032–1037.

Archer, J. (1991). The influence of testosterone on human aggression. *Br. J. Psychol.* **82**(1), 1–28.

Archer, J., Graham-Kevan, N., and Davies, M. (2005). Testosterone and aggression: A reanalysis of Book, Starzyk, and Quinsey's (2001) study. *Aggress. Violent Behav.* **10**(2), 241–261.

Bain, J., Langevin, R., Dickey, R., and Ben-Aron, M. (1987). Sex hormones in murderers and assaulters. *Behav. Sci. Law* **5**, 95–101.

Barker, E. D., Séguin, J. R., White, H. R., Bates, M. E., Lacourse, E., Carbonneau, R., and Tremblay, R. E. (2007). Developmental trajectories of male physical violence and theft: Relations to neurocognitive performance. *Arch. Gen. Psychiatry* **64**(5), 592–599.

Bauer, L. O., O'Connor, S., and Hesselbrock, V. M. (1994). Frontal p300 decrements in antisocial personality disorder. *Alcohol. Clin. Exp. Res.* **18**, 1300–1305.

Beauchaine, T. P., Gartner, J., and Hagen, B. (2000). Comorbid depression and heart rate variability as predictors of aggressive and hyperactive symptom responsiveness during inpatient treatment of conduct-disordered, ADHD boys. *Aggress. Behav.* **26**(6), 425–441.

Becker, K., El-Faddagh, M., Schmidt, M. H., Esser, G., and Laucht, M. (2008). Interaction of dopamine transporter genotype with prenatal smoke exposure on ADHD symptoms. *J. Pediatr.* **152**, 263–269.

Benowitz, N. L. (1998). Pharmacology of nicotine. *In* "Handbook of Substance Abuse: Neurobehavioral Pharmacology" (R. E. Tarter, R. T. Ammerman, and P. J. Ott, eds.), pp. 283–297. Plenum Press, New York, NY.

Berkowitz, L. (1989). Frustration-aggression hypothesis: Examination and reformulation. *Psychol. Bull.* **106**(1), 59–73.

Berman, R. E., and Butler, A. S. (2006). Preterm Birth: Causes, Consequences, and Prevention. National Academics, Washington, DC.

Bhutta, A. T., Cleves, M. A., Casey, P. H., Cradock, M. M., and Anand, K. (2002). Cognitive and behavioral outcomes of school-aged children who were born preterm: A meta-analysis. *JAMA* **288**(6), 728–737.

Blair, R. J. R. (2001). Neurocognitive models of aggression, the antisocial personality disorders, and psychopathy. *J. Neurol. Neurosurg. Psychiatry* **71**(6), 727.

Blair, R. J. R. (2010). Psychopathy, frustration, and reactive aggression: The role of ventromedial prefrontal cortex. *Br. J. Psychol.* **101**(Pt 3), 383–399.

Blair, R. J. R., Colledge, E., Murray, L., and Mitchell, D. G. (2001). A selective impairment in the processing of sad and fearful expressions in children with psychopathic tendencies. *J. Abnorm. Child Psychol.* **29**(6), 491–498.

Brake, W. G., Sullivan, R. M., and Gratton, A. (2000). Perinatal distress leads to lateralized medial prefrontal cortical dopamine hypofunction in adult rats. *J. Neurosci.* **20**(14), 5538–5543.

Brendgen, M., Dionne, G., Girard, A., Boivin, M., Vitaro, F., and Perusse, D. (2005). Examining genetic and environmental effects on social aggression: A study of 6-year-old twins. *Child Dev.* **76,** 930–946.

Brennan, P. A., and Mednick, S. A. (1997). Medical histories of antisocial individuals. *In* "Handbook of Antisocial Behavior" (D. M. Stoff, J. Breiling, and J. D. Maser, eds.), pp. 269–279. Wiley, New York.

Brennan, P. A., Mednick, S. A., and Raine, A. (1997). *In* "Biosocial Interactions and Violence: A Focus on Perinatal Factors" (A. Raine, ed.), pp. 163–174.

Brennan, P. A., Grekin, E. R., Mortensen, E. L., and Mednick, S. A. (2002). Relationship of maternal smoking during pregnancy with criminal arrest and hospitalization for substance abuse in male and female adult offspring. *Am. J. Psychiatry* **159**(1), 48–54.

Brennan, P. A., Hall, J., Bor, W., Najman, J. M., and Williams, G. (2003). Integrating biological and social processes in relation to early-onset persistent aggression in boys and girls. *Dev. Psychol.* **39** (2), 309–323.

Brennan, P. A., Hammen, C., Sylvers, P., Bor, W., Najman, J., Lind, P., Montgomery, G., and Smith, A. K. (2011). Interactions between the COMT Val108/158Met polymorphism and maternal prenatal smoking predict aggressive behavior outcomes. *Biol. Psychol.* **87,** 99–105.

Broidy, L. M., Nagin, D. S., Tremblay, R. E., Brame, B., Dodge, K., Fergusson, D., Horwood, J., Loeber, R., Laird, R., Lynam, D., Moffitt, T., Bates, J. E., *et al.* (2003). Developmental trajectories of childhood disruptive behaviors and adolescent delinquency: A six-site, cross-national study. *Dev. Psychol.* **39**(2), 222–245.

Brookes, K. J., Mill, J., Guindalini, C., Curran, S., Xu, X., Knight, J., Chen, C., Huang, Y., Sethna, V., Taylor, E., Chen, W., Breen, G., *et al.* (2006). A common haplotype of the dopamine transporter gene associated with attention-deficit/hyperactivity disorder and interacting with maternal use of alcohol during pregnancy. *Arch. Gen. Psychiatry* **63,** 74–81.

Bruin, J. E., Gerstein, H. C., and Holloway, A. C. (2010). Long-term consequences of fetal and neonatal nicotine exposure: A critical review. *Toxicol. Sci.* **116**(2), 364–374.

Budhani, S., and Blair, R. J. R. (2005). Response reversal and children with psychopathic tendencies. *J. Child Psychol. Psychiatry* **46**(9), 972–981.

Bush, G., Luu, P., and Posner, M. I. (2000). Cognitive and emotional influences in anterior cingulate cortex. *Trends Cogn. Sci.* **4**(6), 215–222.

Butter, C. M., and Snyder, D. R. (1972). Alterations in aversive and aggressive behaviors following orbital frontal lesions in rhesus monkeys. *Acta Neurobiol. Exp.* **32,** 525–565.

Cannon, M., Huttunen, M. O., Tanskanen, A. J., Arseneault, L., Jones, P. B., and Murray, R. M. (2002). Perinatal and childhood risk factors for later criminality and violence in schizophrenia: Longitudinal, population-based study. *Br. J. Psychiatry* **180**(6), 496–501.

Cases, O., Seif, I., Grimsby, J., Gaspar, P., Chen, K., Pournin, S., Müller, U., Aguet, M., Babinet, C., Shih, J. C., and De Maeyer, E. (1995). Aggressive behavior and altered amounts of brain serotonin and norepinephrine in mice lacking MAOA. *Science* **268,** 1763–1766.

Caspi, A., McClay, J., Moffitt, T. E., Mill, J., Martin, J., Craig, I. W., Taylor, A., and Poulton, R. (2002). Role of genotype in the cycle of violence in maltreated children. *Science* **297,** 851–854.

Caspi, A., Langley, K., Milne, B., Moffitt, T. E., O'Donovan, M., Owen, M. J., Polo Tomas, M., Poulton, R., Rutter, M., Taylor, A., Williams, B., and Thapar, A. (2008). A replicated molecular genetic basis for subtyping antisocial behavior in children with attention-deficit/hyperactivity disorder. *Arch. Gen. Psychiatry* **65,** 203–210.

Chen, Y.-C. I., Choi, J.-K., Andersen, S. L., Rosen, B. R., and Jenkins, B. G. (2005). Mapping dopamine D2/D3 receptor function using pharmacological magnetic resonance imaging. *Psychopharmacology* **180**(4), 705–715.

Chiavegatto, S., Dawson, V. L., Mamounas, L. A., Koliatsos, V. E., Dawson, T. M., and Nelson, R. J. (2001). Brain serotonin dysfunction accounts for aggression in male mice lacking neuronal nitric oxide synthase. *Proc. Natl. Acad. Sci. USA* **98**(3), 1277.

Cima, M., Smeets, T., and Jelicic, M. (2008). Self-reported trauma, cortisol levels, and aggression in psychopathic and non-psychopathic prison inmates. *Biol. Psychiatry* **78**(1), 75–86.

Clarke, A., and Schneider, M. (1993). Prenatal stress has long-term effects on behavioral responses to stress in juvenile rhesus monkeys. *Dev. Psychobiol.* **26**(5), 293–304.

Clarke, A., Wittwer, D., Abbott, D., and Schneider, M. (1994). Long-term effects of prenatal stress on HPA axis activity in juvenile rhesus monkeys. *Dev. Psychobiol.* **27**(5), 257–269.

Clarke, A., Soto, A., Bergholz, T., and Schneider, M. L. (1996). Maternal gestational stress alters adaptive and social behavior in adolescent rhesus monkey offspring. *Infant Behav. Dev.* **19**(4), 451–461.

Cloninger, C. R. (1987). A systematic method for clinical description and classification of personality variants: A proposal. *Arch. Gen. Psychiatry* **44**(6), 573.

Coccaro, E. F., and Kavoussi, R. J. (1997). Fluoxetine and impulsive aggressive behavior in personality-disordered subjects. *Arch. Gen. Psychiatry* **54**(12), 1081.

Coccaro, E. F., and Siever, L. J. (2002). Pathophysiology and treatment of aggression. In "Neuropsychopharmacology: The Fifth Generation of Progress" (K. Davis, D. Charney, J. Coyle, and C. Nemeroff, eds.), pp. 1709–1723. Lippincott Williams & Wilkins, Philadephia.

Coccaro, E. F., McCloskey, M. S., Fitzgerald, D. A., and Phan, K. L. (2007). Amygdala and orbitofrontal reactivity to social threat in individuals with impulsive aggression. *Biol. Psychiatry* **62**(2), 168–178.

Cohen, M. A., Piquero, A. R., and Jennings, W. G. (2010). Studying the costs of crime across offender trajectories. *Criminol. Public Policy* **9**(2), 279–305.

Corkin, S. (1979). Hidden-figures-test performance: Lasting effects of unilateral penetrating head injury and transient effects of bilateral cingulotomy. *Neuropsychologia* **17**(6), 585–605.

Côté, S., Vaillancourt, T., Leblanc, J. C., Nagin, D. S., and Tremblay, R. E. (2006). The development of physical aggression from toddlerhood to pre-adolescence: A nation wide longitudinal study of Canadian children. *J. Abnorm. Child Psychol.* **34**(1), 68–82.

Crick, N. R., and Dodge, K. A. (1996). Social information-processing mechanisms in reactive and proactive aggression. *Child Dev.* **67**(3), 993–1002.

Dabbs, J. M., and Morris, R. (1990). Testosterone, social class, and antisocial behavior in a sample of 4,462 men. *Psychol. Sci.* **1**(3), 209.

Dabbs, J. M., Frady, R. L., Carr, T. S., and Besch, N. F. (1987). Saliva testosterone and criminal violence in young adult prison inmates. *Psychosom. Med.* **49**(2), 174.

Davidson, R. J. (2000). Dysfunction in the neural circuitry of emotion regulation—A possible prelude to violence. *Science* **289**(5479), 591–594.

de Almeida, R. M. M., Ferrari, P. F., Parmigiani, S., and Miczek, K. A. (2005). Escalated aggressive behavior: Dopamine, serotonin and GABA. *Eur. J. Pharmacol.* **526**(1–3), 51–64.

Decety, J., Michalska, K. J., Akitsuki, Y., and Lahey, B. B. (2009). Atypical empathic responses in adolescents with aggressive conduct disorder: A functional MRI investigation. *Biol. Psychol.* **80**(2), 203–211.

Dennis, N. A., Need, A. C., Labar, K. S., Waters-Metenier, S., Cirulli, E. T., Kragel, J., Goldstein, D. B., and Cabeza, R. (2010). COMT Val108/158 Met genotype affects neural but not cognitive processing in healthy individuals. *Cereb. Cortex* **20**, 672–683.

Donchin, E., and Coles, M. G. H. (1988). Is the p300 component a manifestation of context updating? *Behav. Brain Sci.* **11**, 355–372.

D'Onofrio, B. M., Singh, A. L., Iliadou, A., Lambe, M., Hultman, C. M., Grann, M., Neiderhiser, J. M., Långström, N., and Lichtenstein, P. (2010). Familial confounding of the association between maternal smoking during pregnancy and offspring criminality: A population-based study in Sweden. *Arch. Gen. Psychiatry* **67,** 529–538.

Dougherty, D. D., Bonab, A. A., Ottowitz, W. E., Livni, E., Alpert, N. M., Rauch, S. L., *et al.* (2006). Decreased striatal D1 binding as measured using PET and [11C]SCH 23,390 in patients with major depression with anger attacks. *Depress Anxiety* **23**(3), 175–177.

Dow-Edwards, D. L. (1989). Long-term neurochemical and neurobehavioral consequences of cocaine use during pregnancy. *Ann. N. Y. Acad. Sci.* **562,** 280–289.

Dow-Edwards, D., Freed, L. A., and Milhorat, T. H. (1988). Stimulation of brain metabolism by perinatal cocaine exposure. *Brain Res.* **470,** 137–141.

Driscoll, C. D., Streissguth, A. P., and Riley, E. P. (1990). Prenatal alcohol exposure: Comparability of effects in humans and animal models. *Neurotoxicol. Teratol.* **12**(3), 231–237.

Dwyer, J. B., McQuown, S. C., and Leslie, F. M. (2009). The dynamic effects of nicotine on the developing brain. *Pharmacol. Ther.* **122**(2), 125–139.

Ehrenkranz, J., Bliss, E., and Sheard, M. H. (1974). Plasma testosterone: Correlation with aggressive behavior and social dominance in man. *Psychosom. Med.* **36,** 469–475.

Eisenberger, N. I., Way, B. M., Taylor, S. E., Welch, W. T., and Lieberman, M. D. (2007). Understanding genetic risk for aggression: Clues from the brain's response to social exclusion. *Biol. Psychiatry* **61,** 1100–1108.

Ekblad, M., Korkeila, J., Parkkola, R., Lapinleimu, H., Haataja, L., and Lehtonen, L. PIPARI Study Group (2010). Maternal smoking during pregnancy and regional brain volumes in preterm infants. *J. Pediatr.* **156**(2), 185–190.

Elliott, R., and Deakin, B. (2005). Role of the orbitofrontal cortex in reinforcement processing and inhibitory control: Evidence from functional magnetic resonance imaging studies in healthy human subjects. *Int. Rev. Neurobiol.* **65,** 89–116.

El-Sheikh, M., Erath, S. A., Buckhalt, J. A., Granger, D. A., and Mize, J. (2008). Cortisol and children's adjustment: The moderating role of sympathetic nervous system activity. *J. Abnorm. Child Psychol.* **36,** 601–611.

Emery, N. J., Capitanio, J. P., Mason, W. A., Machado, C. J., Mendoza, S. P., and Amaral, D. G. (2001). The effects of bilateral lesions of the amygdala on dyadic social interactions in rhesus monkeys (*Macaca mulatta*). *Behav. Neurosci.* **115**(3), 515–544.

Ferrari, P. F., Van Erp, A. M. M., Tornatzky, W., and Miczek, K. A. (2003). Accumbal dopamine and serotonin in anticipation of the next aggressive episode in rats. *Eur. J. Neurosci.* **17**(2), 371–378.

Finger, E. C., Marsh, A. A., Mitchell, D. G., Reid, M. E., Sims, C., Budhani, S., *et al.* (2008). Abnormal ventromedial prefrontal cortex function in children with psychopathic traits during reversal learning. *Arch. Gen. Psychiatry* **65**(5), 586.

Fitzgerald, P. (1999). Long-acting antipsychotic medication, restraint and treatment in the management of acute psychosis. *Australas. Psychiatry* **33**(5), 660–666.

Frankle, W. G., Lombardo, I., New, A. S., Goodman, M., Talbot, P. S., Huang, Y., *et al.* (2005). Brain serotonin transporter distribution in subjects with impulsive aggressivity: A positron emission study with [11C] McN 5652. *Am. J. Psychiatry* **162**(5), 915.

Gerra, G., Zaimovic, A., Avanzini, P., Chittolini, B., Giucastro, G., Caccavari, R., *et al.* (1997). Neurotransmitter-neuroendocrine responses to experimentally induced aggression in humans: Influence of personality variable. *Psychiatry Res.* **66**(1), 33–43.

Gerstle, J. E., Mathias, C. W., and Stanford, M. S. (1998). Auditory P300 and self-reported impulsive aggression. *Prog. Neuropsychopharmacol. Biol. Psychiatry* **22**(4), 575–583.

Goeders, N. E., and Smith, J. E. (1983). Cortical dopaminergic involvement in cocaine reinforcement. *Science* **221,** 773–775.

Greenberg, N., and Crews, D. (1983). Physiological ethology of aggression in amphibians and reptiles. In "Hormones and Aggressive Behavior" (B. Svare, ed.), pp. 469–506. Plenum, New York.

Gutteling, B. M., de Weerth, C., Willemsen-Swinkels, S. H., Huizink, A. C., Mulder, E. J., Visser, G. H., et al. (2005). The effects of prenatal stress on temperament and problem behavior of 27-month-old toddlers. Eur. Child Adolesc. Psychiatry 14(1), 41–51.

Hawes, D. J., Brennan, J., and Dadds, M. R. (2009). Cortisol, callous-unemotional traits, and pathways to antisocial behavior. Curr. Opin. Psychiatry 22(4), 357–362.

Heimburger, R. F., Whitlock, C. C., and Kalsbeck, J. E. (1966). Stereotaxic amygdalotomy for epilepsy with aggressive behavior. JAMA 198(7), 741–745.

Henry, B., and Moffitt, T. E. (1997). Neuropsychological and neuroimaging studies of juvenile delinquency and adult criminal behavior. In "Handbook of Antisocial Behavior" (D. M. Stoff, J. Breiling, and J. D. Maser, eds.), pp. 280–288. Wiley, New York.

Hicks, B., Krueger, R., Iacono, W., McGrue, M., and Patrick, C. (2004). Family transmission and heritability of externalizing disorders. Arch. Gen. Psychiatry 61, 922–928.

Higley, J. D., Mehlman, P., Taub, D., Higley, S., Suomi, S. J., Linnoila, M., et al. (1992). Cerebrospinal fluid monoamine and adrenal correlates of aggression in free-ranging rhesus monkeys. Arch. Gen. Psychiatry 49(6), 436.

Hubbard, J. A., Smithmyer, C. M., Ramsden, S. R., Parker, E. H., Flanagan, K. D., Dearing, K. F., et al. (2002). Observational, physiological, and self-report measures of children's anger: Relations to reactive versus proactive aggression. Child Dev. 73(4), 1101–1118.

Huebner, T., Vloet, T. D., Marx, I., Konrad, K., Fink, G. R., Herpertz, S. C., et al. (2008). Morphometric brain abnormalities in boys with conduct disorder. J. Am. Acad. Child Adolesc. Psychiatry 47(5), 540–547.

Huijbregts, S. C., Seguin, J. R., Zoccolillo, M., Boivin, M., and Tremblay, R. E. (2008). Maternal prenatal smoking, parental antisocial behavior, and early childhood physical aggression. Dev. Psychopathol. 20(2), 437–453.

Iacono, W. G., Carlson, S. R., Malone, S. M., and McGue, M. (2002). P3 event-related potential amplitude and risk for disinhibitory disorders in adolescent boys. Arch. Gen. Psychiatry 59, 750–757.

Insel, T. R., and Winslow, J. T. (1998). Serotonin and neuropeptides in affiliative behaviors. Biol. Psychiatry 44, 207–219.

Johns, J. M., Means, M. J., Bass, E., Means, L. W., Zimmerman, L. I., and McMillen, B. A. (1994). Prenatal exposure to cocaine: Effects on aggression in Sprague–Dawley rats. Dev. Psychobiol. 27 (4), 227–239.

Joseph, R. (1999). Environmental influences on neural plasticity, the limbic system, emotional development and attachment: A review. Child Psychiatry Hum. Dev. 29(3), 189–208.

Kahn, R. S., Khoury, J., Nichols, W. C., and Lanphear, B. P. (2003). Role of dopamine transporter genotype and maternal prenatal smoking in childhood hyperactive-impulsive, inattentive, and oppositional behaviors. J. Pediatr. 143, 104–110.

Kappeler, L., and Meaney, M. J. (2010). Epigenetics and parental effects. Bioessays 32, 818–827.

Kelly, S. J., Day, N., and Streissguth, A. P. (2000). Effects of prenatal alcohol exposure on social behavior in humans and other species. Neurotoxicol. Teratol. 22, 143–149.

Kelly, S. J., Goodlett, C. R., and Hannigan, J. H. (2009). Animal models of fetal alcohol spectrum disorders: Impact of the social environment. Dev. Disabil. Res. Rev. 15(3), 200–208.

Kennard, M. A. (1954). Effect of bilateral ablation of cingulate area on behaviour of cats. J. Neurophysiol. 18, 159–169.

Kim-Cohen, J., Caspi, A., Taylor, A., Williams, B., Newcombe, R., Craig, I. W., and Moffitt, T. E. (2006). MAOA, maltreatment, and gene–environment interaction predicting children's mental health: New evidence and a meta-analysis. Mol. Psychiatry 11, 903–913.

Kirsch, P. (2005). Oxytocin Modulates Neural Circuitry for Social Cognition and Fear in Humans. *J. Neurosci.* **25**(49), 11489–11493.

Kruk, M. (1992). Ethology and pharmacology of hypothalamic aggression in the rat. *Neurosci. Biobehav. Rev.* **15**, 527–538.

Lachman, H. M., Papolos, D. F., Saito, T., Yu, Y. M., Szumlanski, C. L., and Weinshilboum, R. M. (1996). Human catechol-O-methyltransferase pharmacogenetics: Description of a functional polymorphism and its potential application to neuropsychiatric disorders. *Pharmacogenetics* **6**, 243–250.

Langley, K., Turic, D., Rice, F., Holmans, P., van den Bree, M. B. M., Craddock, N., Kent, L., Owen, M. J., O'Donovan, M. C., and Thapar, A. (2008). Testing for gene-environment interaction effects in attention deficit hyperactivity disorder and associated antisocial behavior. *Am. J. Med. Genet. Part B* **147**, 49–53.

Lauder, J. M. (1988). Neurotransmitters as morphogens. *Prog. Brain Res.* **74**, 365–376.

Lauder, J. M., and Bloom, F. E. (1974). Ontogeny of monoamine neurons in the locus coeruleus, raphe, nuclei, and substantia nigra of the rat. *J. Comp. Neurol.* **155**, 469–481.

Lidov, H. G. W., and Molliver, M. E. (1982a). An immunohistochemical study of serotonin neuron development in the rat: Ascending pathways and terminal fields. *Brain Res. Bull.* **8**, 389–430.

Lidov, H. G. W., and Molliver, M. E. (1982b). Immunohistochemical study of the development of serotonergic neurons in the rat CNS. *Brain Res. Bull.* **9**, 559–604.

Lindblad, F., and Hjern, A. (2010). ADHD after fetal exposure to maternal smoking. *Nicotine Tob. Res.* **12**, 408–415.

Lipp, H. P., and Hunsperger, R. W. (1978). Threat, attack and flight elicited by electrical stimulation of the ventromedial hypothalamus of the marmoset monkey callithrix jacchus. *Brain Behav. Evol.* **15**(4), 260–293.

Liu, J. (2004). Prenatal and perinatal complications as predispositions to externalizing behavior. *J. Prenat. Perinat. Psychol. Health* **18**(4), 301–311.

Liu, J., Raine, A., Wuerker, A., Venables, P. H., and Mednick, S. (2009). The association of birth complications and externalizing behavior in early adolescents: Direct and mediating effects. *J. Res. Adolesc.* **19**(1), 93–111.

Lloyd, S. A., and Dixson, A. F. (1988). Effects of hypothalamic lesions upon the sexual and social behaviour of the male common marmoset (*Callithrix jacchus*). *Brain Res.* **463**(2), 317–329.

Loeber, R., and Pardini, D. (2008). Neurobiology and the development of violence: Common assumptions and controversies. *Philos. Trans. R. Soc. Lond. B Biol. Sci.* **363**, 2491–2503.

Loeber, R., Burke, J. D., Lahey, B. B., Winters, A., and Zera, M. (2000). Oppositional defiant and conduct disorder: A review of the past 10 years, part I. *J. Am. Acad. Child Adolesc. Psychiatry* **39**(12), 1468–1484.

Lopez-Duran, N. L., Hajal, N. J., Olson, S. L., Felt, B., and Vazquez, D. M. (2009). Hypothalamic pituitary adrenal axis functioning in reactive and proactive aggression in children. *J. Abnorm. Child Psychol.* **37**(2), 169–182.

Lorber, M. F. (2004). Psychophysiology of aggression, psychopathy, and conduct problems: A meta-analysis. *Psychol. Bull.* **130**(4), 531–552.

Machado, C. J., and Bachevalier, J. (2006). The impact of selective amygdala, orbital frontal cortex, or hippocampal formation lesions on established social relationships in rhesus monkeys (Macaca mulatta). *Behav. Neurosci.* **120**(4), 761–786.

Mann, J. J., McBride, P. A., Brown, R. P., Linnoila, M., Leon, A. C., DeMeo, M., *et al.* (1992). Relationship between central and peripheral serotonin indexes in depressed and suicidal psychiatric inpatients. *Arch. Gen. Psychiatry* **49**(6), 442–446.

Manuck, S. B., Flory, J. D., Ferrell, R. E., Mann, J. J., and Muldoon, M. F. (2000). A regulatory polymorphism of the monoamine oxidase-A gene may be associated with variability in aggression, impulsivity, and central nervous system serotonergic responsivity. *Psychiatry Res.* **95**, 9–23.

Marino, M. D., Bourdelat-Parks, B. N., Cameron Liles, L., and Weinshenker, D. (2005). Genetic reduction of noradrenergic function alters social memory and reduces aggression in mice. *Behav. Brain Res.* **161**(2), 197–203.

Marsh, A. A., Finger, E. C., Mitchell, D. G. V., Reid, M. E., Sims, C., Kosson, D. S., *et al.* (2008). Reduced amygdala response to fearful expressions in children and adolescents with callous-unemotional traits and disruptive behavior disorders. *Am. J. Psychiatry* **165**(6), 712–720.

Marsh, A. A., Yu, H. H., Pine, D. S., and Blair, R. J. R. (2010). Oxytocin improves specific recognition of positive facial expressions. *Psychopharmacology* **209**(3), 225–232.

Mattson, M. P. (1988). Neurotransmitters in the regulation of neuronal cytoarchitecture. *Brain Res. Rev.* **13**, 179–212.

Maughan, B., Taylor, A., Caspi, A., and Moffitt, T. E. (2004). Prenatal smoking and early childhood conduct problems: Testing genetic and environmental explanations of the association. *Arch. Gen. Psychiatry* **61**, 836–843.

Mayes, L. C. (1994). Neurobiology of prenatal cocaine exposure effect on developing monoamine systems. *Infant Ment. Health J.* **15**(2), 121–133.

McBurnett, K., Lahey, B. B., Rathouz, P. J., and Loeber, R. (2000). Low salivary cortisol and persistent aggression in boys referred for disruptive behavior. *Arch. Gen. Psychiatry* **57**(1), 38–43.

McCormick, M. C. (1985). The contribution of low birth weight to infant mortality and childhood morbidity. *N. Engl. J. Med.* **312**(2), 82–90.

McDougle, C. J., Holmes, J. P., Carlson, D. C., Pelton, G. H., Cohen, D. J., and Price, L. H. (1998). A double-blind, placebo-controlled study of risperidone in adults with autistic disorder and other pervasive developmental disorders. *Arch. Gen. Psychiatry* **55**(7), 633.

Mednick, S. A., and Kandel, E. (1988). Genetic and perinatal factors in violence. *In* "Biological Contributions to Crime Causation" (S. A. Mednick and T. E. Moffitt, eds.), pp. 121–131. Martinus Nijhoff Publishing, Dordrecht, Netherlands.

Miczek, K. A., Fish, E. W., De Bold, J. F., and De Almeida, R. M. (2002). Social and neural determinants of aggressive behavior: Pharmacotherapeutic targets at serotonin, dopamine and ?-aminobutyric acid systems. *Psychopharmacology* **163**(3–4), 434–458.

Miller, N. S., Gold, M. S., and Mahler, J. C. (1991). Violent behaviors associated with cocaine use: Possible pharmacological mechanisms. *Int. J. Addict.* **26**(10), 1077–1088.

Mirochnick, M., Meyer, J., Cole, J., Herren, T., and Zuckerman, B. (1991). Circulating catecholamine concentrations in cocaine-exposed neonates: A pilot study. *Pediatrics* **88**, 481–485.

Moffitt, T. E. (1993). Adolescence-limited and life-course-persistent antisocial behavior: A developmental taxonomy. *Psychol. Rev.* **100**(4), 674–701.

Moffitt, T. E., and Caspi, A. (2001). Childhood predictors differentiate life-course persistent and adolescence-limited antisocial pathways among males and females. *Dev. Psychopathol.* **13**, 355–375.

Monoghan, E., and Glickman, S. (1992). Hormones and aggressive behavior. *In* "Behavioral Endocrinology" (J. Becker, S. Breedlove, and D. Crews, eds.). MIT Press, Cambridge, MA.

Monuteaux, M. C., Blacker, D., Biederman, J., Fitzmaurice, G., and Buka, S. L. (2006). Maternal smoking during pregnancy and offspring overt and covert conduct problems: A longitudinal study. *J. Child Psychol. Psychiatry* **47**, 883–890.

Monuteaux, M. C., Biederman, J., Doyle, A. E., Mick, E., and Faraone, S. V. (2009). Genetic risk for conduct disorder symptom subtypes in an ADHD sample: Specificity to aggressive symptoms. *J. Am. Acad. Child Adolesc. Psychiatry* **48**, 757–764.

Mulder, E., de Medina, P., Huizink, A., Van den Bergh, B., Buitelaar, J., and Visser, G. (2002). Prenatal maternal stress: Effects on pregnancy and the (unborn) child. *Early Hum. Dev.* **70**(1–2), 3–14.

Nagin, D., and Tremblay, R. E. (1999). Trajectories of boys' physical aggression, opposition, and hyperactivity on the path to physically violent and nonviolent juvenile delinquency. *Child Dev.* **70**, 1181–1196.

Nelson, R. J., and Trainor, B. C. (2007). Neural mechanisms of aggression. *Nat. Rev. Neurosci.* **8**(7), 536–546.

New, A. S., Buchsbaum, M. S., Hazlett, E. A., Goodman, M., Koenigsberg, H. W., Lo, J., *et al.* (2004). Fluoxetine increases relative metabolic rate in prefrontal cortex in impulsive aggression. *Psychopharmacology* **176**(3–4), 451–458.

Nilsson, K. W., Sjöberg, R. L., Wargelius, H. L., Leppert, J., Lindström, L., and Oreland, L. (2007). The monoamine oxidase A (MAO-A) gene, family function and maltreatment as predictors of destructive behaviour during male adolescent alcohol consumption. *Addiction* **102**, 389–398.

Olweus, D. (1986). Aggression and hormones: Behavioral relationship with testosterone and adrenaline. In "Development of Antisocial and Prosocial Behavior" (D. Olweus, J. Block, and M. Radke-Yarrow, eds.), pp. 51–72. Academic press, New York.

Olweus, D., Mattsson, A., Schalling, D., and Low, H. (1988). Circulating testosterone levels and aggression in adolescent males: A causal analysis. *Psychosom. Med.* **50**, 261–272.

Parsey, R. V., Oquendo, M. A., Simpson, N. R., Ogden, R. T., Van Heertum, R., Arango, V., *et al.* (2002). Effects of sex, age, and aggressive traits in man on brain serotonin 5-HT1A receptor binding potential measured by PET using [C-11] WAY-100635. *Brain Res.* **954**(2), 173–182.

Patrick, C. J. (2008). Psychophysiological correlates of aggression and violence: An integrative review. *Philos. Trans. R. Soc. Lond. B Biol. Sci.* **363**(1503), 2543–2555.

Patterson, G. R., DeBaryshe, B. D., and Ramsey, E. (1989). A developmental perspective on antisocial behavior. *Am. Psychol.* **44**, 329–335.

Piquero, A., and Tibbetts, S. (1999). The impact of pre/perinatal disturbances and disadvantaged familial environment in predicting criminal offending. *Stud. Crime Crime Prev.* **8**(1), 52–70.

Poland, R. E., Lutchmansingh, P., Au, D., Edelstein, M., Lydecker, S., Hsieh, C., and McCracken, J. T. (1994). Exposure to threshold doses of nicotine in utero: I. Neuroendocrine response to restraint stress in adult male offspring. *Life Sci.* **55**(20), 1567–1575.

Popma, A., Doreleijers, T. A., Jansen, L. M., Van Goozen, S. H., Van Engeland, H., and Vermeiren, R. (2007a). The diurnal cortisol cycle in delinquent male adolescents and normal controls. *Neuropsychopharmacology* **32**(7), 1622–1628.

Popma, A., Vermeiren, R., Geluk, C. A. M. L., Rinne, T., van den Brink, W., Knol, D. L., *et al.* (2007b). Cortisol moderates the relationship between testosterone and aggression in delinquent male adolescents. *Biol. Psychiatry* **61**(3), 405–411.

Potegal, M., and Davidson, R. J. (2003). Temper tantrums in young children: 1. Behavioral composition. *J. Dev. Behav. Pediatr.* **24**, 140–147.

Radtke, K. M., Ruf, M., Gunter, H. M., Dohrmann, K., Schauer, M., Meyer, A., and Elbert, T. (2011). Transgenerational impact of intimate partner violence on methylation in the promoter of the glucocorticoid receptor. *Translational Psychiatry* **1**, e21.

Ragnauth, A. K., Devidze, N., Moy, V., Finley, K., Goodwillie, A., Kow, L. M., *et al.* (2005). Female oxytocin gene-knockout mice, in a semi-natural environment, display exaggerated aggressive behavior. *Genes Brain Behav.* **4**, 229–239.

Raine, A. (1993). The Psychopathology of Crime: Criminal Behavior as a Clinical Disorder. Academic, San Diego.

Raine, A. (2002a). Biosocial studies of antisocial and violent behavior in children and adults: A review. *J. Abnorm. Child Psychol.* **30**(4), 311–326.

Raine, A. (2002b). The role of prefrontal deficits, low autonomic arousal and early health factors in the development of antisocial and aggressive behavior in children. *J. Child Psychol. Psychiatry* **43**(4), 417–434.

Raine, A., Venables, P., and Williams, M. (1990). Relationships between central and autonomic measures of arousal at age 15 years and criminality at age 24 years. *Arch. Gen. Psychiatry* **47**(11), 1003.

Raine, A., Brennan, P., and Mednick, S. A. (1994). Birth complications combined with early maternal rejection at age 1 predispose to violent crime at age 18 years. *Arch. Gen. Psychiatry* **51**, 984–988.

Retz, W., Rosler, M., Supprian, T., Retz-Junginger, P., and Thome, J. (2003). Dopamine D3 receptor gene polymorphism and violent behavior: Relation to impulsiveness and ADHD-related psychopathology. *J. Neural Transm.* **110**(5), 561–572.

Rice, F., Harold, G. T., Boivin, J., Hay, D. F., van den Bree, M., and Thapar, A. (2009). Disentangling prenatal and inherited influences in humans with an experimental design. *Proc. Natl. Acad. Sci. USA* **106**(7), 2464–2467.

Robinson, B. W. (1967). Vocalization evoked from forebrain in *Macaca mulatta. Physiol. Behav.* **2**(4), 345–346.

Rolls, E. T., Hornak, J., Wade, D., and McGrath, J. (1994). Emotion-related learning in patients with social and emotional changes associated with frontal lobe damage. *J. Neurol. Neurosurg. Psychiatry* **57**(12), 1518–1524.

Rose, R. M., Holladay, J. W., and Bernstein, I. S. (1971). Plasma testosterone, dominance rank and aggressive behavior in male rhesus monkeys. *Nature* **231**, 366–368.

Ross, G., Lipper, E. G., and Auld, P. A. (1990). Social competence and behavior problems in premature children at school age. *Pediatrics* **86**, 391–397.

Rowe, R., Maughan, B., Worthman, C. M., Costello, E. J., and Angold, A. (2004). Testosterone, antisocial behavior, and social dominance in boys: Pubertal development and biosocial interaction. *Biol. Psychiatry* **55**, 546–552.

Sabol, S. Z., Hu, S., and Hamer, D. (1998). A functional polymorphism in the monamine oxidase A gene promoter. *Hum. Genet.* **103**, 273–279.

Sampson, R. J., and Laub, J. H. (2003). Life-course desisters—Trajectories of crime among delinquent boys followed to Age 70. *Criminology* **41**, 301–339.

Scarpa, A., and Raine, A. (1997). Psychophysiology of anger and violent behavior. *Psychiatr. Clin. North Am.* **20**(2), 375–394.

Scerbo, A. S., and Kolko, D. J. (1994). Salivary testosterone and cortisol in disruptive children: Relationship to aggressive, hyperactive, and internalizing behaviors. *J. Am. Acad. Child Adolesc. Psychiatry* **33**(8), 1174–1184.

Schneider, M. L. (1992a). Delayed object permanence development in prenatally stressed monkey infants (Macaca mulatta). *Occup. Ther. J. Res.* **12**, 96–110.

Schneider, M. L. (1992b). Prenatal stress exposure alters postnatal behavioral expression under conditions of novelty challenge in rhesus monkey infants. *Dev. Psychobiol.* **25**, 529–540.

Sengupta, S. M., Grizenko, N., Schmitz, N., Schwartz, G., Ben Amor, L., Bellingham, J., de Guzman, R., Polotskaia, A., Ter Stepanian, M., Thakur, G., and Joober, R. (2006). COMT Val108/158Met gene variant, birth weight, and conduct disorder in children with ADHD. *J. Am. Acad. Child Adolesc. Psychiatry* **45**, 1363–1369.

Shepard, G. M. (1988). Neurobiology. 2nd edn., Oxford University Press, New York.

Siegel, A., and Victoroff, J. (2009). Understanding human aggression: New insights from neuroscience. *Int. J. Law Psychiatry* **32**(4), 209–215.

Siever, L. J. (2008). Neurobiology of aggression and violence. *Am. J. Psychiatry* **165**(4), 429.

Slotkin, T. A., MacKillop, E. A., Rudder, C. L., Ryde, I. T., Tate, C. A., and Seidler, F. J. (2007). Permanent, sex-selective effects of prenatal or adolescent nicotine exposure, separately or sequentially, in rat brain regions: Indices of cholinergic and serotonergic synaptic function, cell signaling, and neural cell number and size at 6 months of age. *Neuropsychopharmacology* **32**, 1082–1097.

Sobrian, S. K., Vaughn, V. T., Ashe, W. K., Markovic, B., Djuric, V., and Jankovic, B. D. (1997). Gestational exposure to loud noise alters the development and post-atal responsiveness of humoral and cellular components of the immune system in offspring. *Environ. Res.* **73**, 227–241.

Soloff, P. H., Meltzer, C. C., Becker, C., Greer, P. J., Kelly, T. M., and Constantine, D. (2003). Impulsivity and prefrontal hypometabolism in borderline personality disorder. *Psychiatry Res.* **123** (3), 153–163.

Sood, B., Delaney-Black, V., Covington, C., Nordstrom-Klee, B., Ager, J., Templin, T., *et al.* (2001). Prenatal alcohol exposure and childhood behavior at age 6 to 7 years: I. Dose-response effect. *Pediatrics* **108**(2), 34–44.

Spear, L. P., Campbell, J., Snyder, K., Silveri, M., and Katovic, N. (1998). Animal behavior models. Increased sensitivity to stressors and other environmental experiences after prenatal cocaine exposure. *Ann. N. Y. Acad. Sci.* **846**, 76–88.

Stadler, C., Sterzer, P., Schmeck, K., Krebs, A., Kleinschmidt, A., and Poustka, F. (2007). Reduced anterior cingulate activation in aggressive children and adolescents during affective stimulation: Association with temperament traits. *J. Psychiatr. Res.* **41**(5), 410–417.

Sterzer, P., Stadler, C., Krebs, A., Kleinschmidt, A., and Poustka, F. (2005). Abnormal neural responses to emotional visual stimuli in adolescents with conduct disorder. *Biol. Psychiatry* **57**(1), 7–15.

Susman, E. J. (2006). Psychobiology of persistent antisocial behavior: Stress, early vulnerabilities and the attenuation hypothesis. *Neurosci. Biobehav. Rev.* **30**(3), 376–389.

Susman, E. J., Inoff-Germain, G., Nottelmann, E. D., Loriaux, D. L., Cutler, G. B., and Chrousos, G. P. (1987). Hormones, emotional dispositions, and aggressive attributes in young adolescents. *Child Dev.* **58**(4), 1114–1134.

Swanson, L. W. (2000). Cerebral hemisphere regulation of motivated behavior. *Brain Res.* **886**(1–2), 113–164.

Takahashi, L. K., Baker, E. W., and Kalin, N. H. (1990). Ontogeny of behavioral and hormonal responses to stress in prenatally stressed male rat pups. *Physiol. Behav.* **47**(2), 357–364.

Thapar, A., Fowler, T., Rice, F., Scourfield, J., van den Bree, M., Thomas, H., Harold, G., and Hay, D. (2003). Maternal smoking during pregnancy and attention deficit hyperactivity disorder symptoms in offspring. *Am. J. Psychiatry* **160**, 1985–1989.

Thapar, A., Langley, K., Fowler, T., Rice, F., Turic, D., Whittinger, N., Aggleton, J., Van den Bree, M., Owen, M., and O'Donovan, M. (2005). Catechol O-methyltransferase gene variant and birth weight predict early-onset antisocial behavior in children with attention-deficit/hyperactivity disorder. *Arch. Gen. Psychiatry* **62**, 1275–1278.

Toro, R., Leonard, G., Lerner, J. V., Lerner, R. M., Perron, M., Pike, G. B., Richer, L., Veillette, S., Pausova, Z., and Paus, T. (2008). Prenatal exposure to maternal cigarette smoking and the adolescent cerebral cortex. *Neuropsychopharmacology* **33**(5), 1019–1027.

Tow, P. M., and Whitty, C. W. (1953). Personality changes after operations on the cingulate gyrus in man. *J. Neurol. Neurosurg. Psychiatry* **16**, 186–193.

Tremblay, R. E. (2008). Development of physical aggression from early childhood to adulthood. *In* "Encyclopedia on Early Childhood Development" (R. E. Tremblay, R. G. Barr, R. De V. Peters, and M. Boivin, eds.), Rev. edn., pp. 1–6. Centre of Excellence for Early Childhood Development, Montreal, Quebec, http://www.child-encyclopedia.com/documents/TremblayANGxp_rev.pdf.

Tremblay, R. E., Nagin, D. S., Séguin, J. R., Zoccolillo, M., Zelazo, P. D., Boivin, M., Pérusse, D., and Japel, C. (2004). Physical aggression during early childhood: Trajectories and predictors. *Pediatrics* **114**(1), e43–e50.

Tunbridge, E. M., Harrison, P. J., and Weinberger, D. R. (2006). Catechol-o-methyltransferase, cognition, and psychosis: Val158Met and beyond. *Biol. Psychiatry* **60**, 141–151.

Van Bokhoven, I., Van Goozen, S. H. M., van Engeland, H., Schaal, B., Arseneault, L., Séguin, J. R., *et al.* (2005). Salivary cortisol and aggression in a population-based longitudinal study of adolescent males. *J. Neural Transm.* **112**(8), 1083–1096.

van Bokhoven, I., van Goozen, S. H. M., van Engeland, H., Schaal, B., Arseneault, L., Séguin, J. R., *et al.* (2006). Salivary testosterone and aggression, delinquency, and social dominance in a population-based longitudinal study of adolescent males. *Horm. Behav.* **50**(1), 118–125.

van Goozen, S., Matthys, W., Cohen-Kettenis, P., Buitelaar, J., and van Engeland, H. (2000). Hypothalamic-pituitary-adrenal axis and autonomic nervous system activity in disruptive children and matched controls. *J. Am. Acad. Child Adolesc. Psychiatry* **39**(11), 1438–1445.

Veenema, A. H., Torner, L., Blume, A., Beiderbeck, D. I., and Neumann, I. D. (2007). Low inborn anxiety correlates with high intermale aggression: Link to ACTH response and neuronal activation of the hypothalamic paraventricular nucleus. *Horm. Behav.* **51**(1), 11–19.

Veit, R., Flor, H., Erb, M., Hermann, C., Lotze, M., Grodd, W., *et al.* (2002). Brain circuits involved in emotional learning in antisocial behavior and social phobia in humans. *Neurosci. Lett.* **328**(3), 233–236.

Virkkunen, M., Rawlings, R., Tokola, R., Poland, R. E., Guidotti, A., Nemeroff, C., *et al.* (1994). CSF biochemistries, glucose metabolism, and diurnal activity rhythms in alcoholic, violent offenders, fire setters, and healthy volunteers. *Arch. Gen. Psychiatry* **51**(1), 20–27.

Volavka, J. (1999). The neurobiology of violence: An update. *J. Neuropsychiatry Clin. Neurosci.* **11**(3), 307.

Volkow, N. D., Tancredi, L. R., Grant, C., Gillespie, H., Valentine, A., Mullani, N., *et al.* (1995). Brain glucose metabolism in violent psychiatric patients: A preliminary study. *Psychiatry Res.* **61**(4), 243–253.

Vom Saal, F. (1983). Models of early hormonal effects on intrasex aggression in mice. In "Hormones and Aggressive Behavior" (B. Svare, ed.), pp. 197–222. Plenum, New York.

Wakschlag, L. S., Pickett, K. E., Cook, E., Jr., Benowitz, N. L., and Leventhal, B. L. (2002). Maternal smoking during pregnancy and severe antisocial behavior in offspring: A review. *Am. J. Public Health* **92**(6), 966–974.

Wakschlag, L. S., Kistner, E. O., Pine, D. S., Biesecker, G., Pickett, K. E., Skol, A. D., Dukic, V., Blair, R. J., Leventhal, B. L., Cox, N. J., Burns, J. L., Kasza, K. E., *et al.* (2010). Interaction of prenatal exposure to cigarettes and MAOA genotype in pathways to youth antisocial behavior. *Mol. Psychiatry* **15**(9), 928–937.

Wallace, J. A., and Lauder, J. M. (1983). Development of the serotonergic system in the rat embryo: An immunocytochemical study. *Brain Res. Bull.* **10**, 459–479.

Ward, I. L., and Weisz, J. (2011). Differential effects of maternal stress on circulating levels of corticosterone, progesterone, and testosterone in male and female rat fetuses and their mothers. *Endocrinology* **114**(5), 1635–1644.

Weaver, K., Campbell, R., Mermelstein, R., and Wakschlag, L. (2007). Pregnancy smoking in context: The influence of multiple levels of stress. *Nicotine Tob. Res.* **10**(6), 1065–1073.

Weinberg, J., Taylor, A. N., and Gianoulakis, C. (1996). Fetal ethanol exposure: Hypothalamic-pituitary-adrenal and beta-endorphin responses to repeated stress. *Alcohol Clin. Exp. Res.* **20**, 122–131.

White, J. L., Moffitt, T. E., Caspi, A., Bartusch, D. J., Needles, D. J., and Stouthamer-Loeber, M. (1994). Measuring impulsivity and examining its relationship to delinquency. *J. Abnorm. Psychol.* **103**(2), 192–205.

Wingfield, J. C., and Hahn, T. P. (1994). Testosterone and sedentary behaviour in sedentary and migratory sparrows. *Anim. Behav.* **47**(1), 77–89.

Young, L. J., Lim, M. M., Gingrich, B., and Insel, T. R. (2001). Cellular mechanisms of social attachment. *Horm. Behav.* **40**(2), 133–138.

Neurocriminology

**Benjamin R. Nordstrom,* Yu Gao,† Andrea L. Glenn,‡
Melissa Peskin,§ Anna S. Rudo-Hutt,§ Robert A. Schug,¶
Yaling Yang,‖ and Adrian Raine*,§,#**

*Department of Psychiatry, University of Pennsylvania, Philadelphia, USA
†Department of Psychology, Brooklyn College, New York, USA
‡Department of Child and Adolescent Psychiatry, Institute of Mental Health,
Singapore, Singapore
§Department of Psychology, University of Pennsylvania, Philadelphia, USA
¶Department of Criminal Justice, California State University, Long Branch,
USA
‖Laboratory of Neuro Imaging, University of California, Los Angeles, USA
#Department of Criminology, University of Pennsylvania, Philadelphia, USA

0065-2660/11 $35.00
DOI: 10.1016/B978-0-12-380858-5.00006-X

ABSTRACT

In the past several decades there has been an explosion of research into the biological correlates to antisocial behavior. This chapter reviews the state of current research on the topic, including a review of the genetics, neuroimaging, neuropsychological, and electrophysiological studies in delinquent and antisocial populations. Special attention is paid to the biopsychosocial model and gene–environment interactions in producing antisocial behavior.   © 2011, Elsevier Inc.

I. INTRODUCTION

In 1977, George Engel wrote an essay in Science to advocate for a new model in medicine that would serve as a corrective to the biomedical reductionism that he noted in the field and in psychiatry in particular (Engel, 1977). The model he suggested was called the biopsychosocial model and, in understanding the disease in question, took into account the biological aspect of the individual, their psychological state, and the social context in which they exist. This model has since become the dominant paradigm in psychiatric treatment.

In recent years, a tremendous amount of research has been done to elucidate the biological correlates and causes of antisocial behavior. This work has been conducted in an environment that has been, at times, hostile to this kind of research, as the dominant paradigm in criminology research has focused on social theories of crime. What we hope to accomplish in this chapter is to present the evidence for a biopsychosocial model of crime.

We will present the data that argues that there is an inherited propensity for criminal behavior. The behavioral phenotype of those who criminally offend is demonstrably and obviously different from those who do not; we will show that their biological phenotypes are also different. We will marshal the data that suggest that the various brain areas that perform cognitive processes relevant to criminal offending are structurally and functionally different in antisocial people compared to others. We will discuss how these brain differences are also evident in techniques that elucidate the mind–body connection. We will also discuss how various social, or environmental, events can have multiplicative interactions with the biological risk factors to produce criminal offending.

That there is not one standard diagnosis to identify the behavioral phenotype of interest to criminologists is a limitation of any review of this literature. Some research teams studying children use the diagnosis of oppositional defiant disorder or conduct disorder. Others use a broader category of disruptive behavior, attention deficit hyperactivity disorder. Researchers interested in adults may use the diagnosis of antisocial personality disorder to identify participants, while others might use the more stringent diagnosis of psychopathy. Other teams use self-reports of violence or aggression, a history of arrests, or various scores on personality inventories to identify the population of interest. Our stance on this is that although these differences make it difficult to directly compare the results of studies using different identifying criteria, all add the potential to better understand the biological correlates of problematic behaviors.

II. PSYCHODYNAMIC THEORIES

For the first part of the twentieth century, psychoanalytic models of crime and/or criminality (Holmes and Holmes, 1998; Wittels, 1937), cases of murder (Abrahamsen, 1973; Arieti and Schreiber, 1981; Bromberg, 1951; Cassity, 1941; Evseef and Wisniewski, 1972; Karpman, 1951a,b; Lehrman, 1939; Morrison, 1979; Revitch and Schlesinger, 1981, 1989; Wertham, 1949, 1950; Wittels, 1937), and even homicide wound patterns (DeRiver, 1951) appeared in the psychiatric literature. A common feature of psychoanalytic criminological theory centers on unconscious processes (i.e., drives, instincts, and motivations, and the defense mechanisms used to control them which operate outside of a person's conscious awareness) which are maladaptive and lead to antisocial and criminal behavior (Alexander and Staub, 1931).

A number of psychodynamic theorists have posited that early problems of attachment to parents (especially mothers) can predispose individuals to unstable personality structure and later criminal offending (Bowlby, 1944, 1969, 1973, 1980). Some theorists have posited that early experiences with rejecting mothers can lead children to mentally internalize fragments of this "bad mother," which can then be externalized onto later female victims (Liebert, 1972). More recent authors have posited that such malformed attachments might be at work in cases of serial homicide (Whitman and Akutagawa, 2004).

Although some of these notions of psychodynamic theories may appear quaint compared to the astonishing technological achievements used in the studies described later, it is worth noting that the psychodynamic theorists may be using a different language to describe phenomena other researchers frame in more exacting biological terms. For example, we will later turn to discussions of how biology and environment can interact in ways that increase the likelihood of criminal offending. We will see that some of this data involves birth complications, and in what might be a partial affirmation of attachment theory, maternal rejection.

III. NEUROIMAGING

Using neuroanatomy as a tool to study criminal propensity is an idea that dates back to the early eighteenth century when the German physician Franz Joseph Gall developed phrenology. Phrenology purported to analyze the shape of cranial bones to make scientific inferences as to both the size and function of underlying brain areas. Later technological advances replaced the pseudoscience of phrenology, allowing for the scientific study of how brain structure and function relate to antisocial behavior. A comprehensive review of the neuroanatomic literature as it pertains to antisocial behavior is available (Yang et al., 2008).

The two main technological advances that allowed images of the brain itself to be generated are computerized axial tomography (CAT) scans and magnetic resonance imaging (MRI) scans. CAT scans are produced using a series of X-rays taken along the axis of the body. The X-rays pass unevenly through tissues of different densities, allowing for distinctions between fluid, bone, and brain tissue to be made. A computer then assembles these slices into a sequence of cross-sectional images. MRI scans are created by using powerful magnetic fields to orient all the hydrogen atoms (primarily found in water molecules) in the brain in the same direction. A radiofrequency electromagnetic field is introduced which then produces a signal that is detected by the MRI scanner's receiver. These signals are then assembled into high-resolution images that can distinguish the gray matter from the white matter of the brain. MRI scans don't use radiation and produce more detailed pictures than do CAT scans, but they also take much longer to obtain and are much more expensive.

A. Structural imaging studies

Structural neuroimaging studies the size of brain regions of interest (ROI). One early study found that when CAT scans of the brains of sexual sadists were studied, about 50% of them had abnormal brain structures, especially in the temporal lobes (Langevin et al., 1988). Other researchers found that nearly 50% of 19 murder suspects studied had atrophic brains on CAT scan (Blake et al., 1995). Later studies using MRI scans found brain atrophy as well, especially in the frontotemporal region (Aigner et al., 2000; Sakuta and Fukushima, 1998).

Another qualitative structural imaging study found that 6 of 10 of the violent psychiatric inpatients they studied had atrophic temporal regions (Chesterman et al., 1994).

A number of studies in structural imaging have used larger samples and reported their findings in quantitative terms. Raine and colleagues viewed 21 individuals with antisocial personality disorder and compared them to a matched group of substance users and normal controls (Raine et al., 2000, 2010). This work reported an 11% reduction in the gray matter of the prefrontal cortices of the antisocial group. A second study by this group found reduced prefrontal cortical gray matter volumes in unsuccessful psychopaths (i.e., psychopaths who had been criminally convicted at least once), compared to successful psychopaths (i.e., psychopaths who had never been convicted of a crime), and normal controls (Yang et al., 2005). In addition, Yang et al. revealed reduced cortical gray matter thickness in the frontal and temporal regions in psychopaths when compared to normal controls (Yang et al., 2009). Other groups have found that, compared to normal controls, subjects with antisocial personality disorder have smaller temporal lobes (Dolan et al., 2002; Laakso et al., 2002), as well as reductions in their dorsolateral, medial frontal, and orbitofrontal cortices (Laakso et al., 2002).

One team of researchers demonstrated smaller gray matter volumes in the orbitofrontal and temporal lobes of children with conduct disorder compared to normal controls (Huebner et al., 2008). Reductions in gray matter concentration have also been observed in the frontal and temporal lobes of criminal psychopaths compared to normal controls (Muller et al., 2008). Along these same lines, another research group found insignificant prefrontal lobe volume reductions, but significant temporal lobe volume reductions, in conduct disordered children (Kruesi et al., 2004).

Buried deep in the temporal lobe is the amygdala, which is associated with fear conditioning, and the hippocampus, a structure associated with learning and memory. Laakso and colleagues found, in a group of violent offenders with alcoholism and antisocial personality disorder, that smaller posterior hippocampus measures matched higher psychopathy rating scores (Laakso et al., 2000, 2001). Other researchers report that adolescents with conduct disorder demonstrate reduced gray matter volumes in the insula and amygdala compared to normal controls (Sterzer et al., 2007).

A relatively new technique called diffusion tensor imaging (DTI) allows images to be taken of the structural integrity of the white matter tracts connecting various parts of the brain. One DTI study showed evidence of abnormal white matter tract structure in the frontotemporal regions of adolescents with disruptive behavior compared to normal controls (Li et al., 2005). A second study showed similar evidence of abnormal white matter tracts connecting the amygdalas and orbitofrontal cortices of criminal psychopaths when compared to normal controls (Craig *et al.*, 2008). Other studies looking at abnormalities in connectivity have focused on white matter structures. Raine *et al.* (2003a,b) found that compared to a normal comparison group, psychopathic, antisocial subjects had a longer, thinner corpus callosum with overall increased volume. They also found a correlation between psychopathy scores and larger callosal volumes (Raine *et al.*, 2003a).

B. Functional imaging studies

Not only does current technology allow us to study the structure and connectivity of brain regions, it also allows us to image the functioning of brain areas as well. One form of functional neuroimaging is photon emission tomography (PET). This technique relies on injecting subjects with radioactively labeled substance such as glucose. Images of their brains can then be obtained. Areas of higher radioactive signal have more glucose metabolism and are thought to be more active (Yang *et al.*, 2008). A second form of functional neuroimaging is single photon emission tomography (SPECT). This form of imaging also involves the injection of a radioactive tracer. The camera detects the amount of radiation coming from different parts of the brain. These differences are due to differences in regional cerebral blood flow (rCBF) and are thought to reflect different levels of activity in various parts of the brain (Yang *et al.*, 2008) Functional magnetic resonance imaging (fMRI) studies measure changes in blood oxygen in ROI in the brain before and after cognitive tasks are undertaken. These blood oxygen level dependent (BOLD) signals are used as a proxy for how active a region of the brain is. By comparing subjects of interest with matched controls, the patterns of activation or inactivation in the brain can be studied to learn how the functioning of various brain regions relates to the condition at hand (Yang *et al.*, 2008).

One early PET scan study showed that, compared to controls, antisocial subjects demonstrated reduced glucose metabolism in the prefrontal and temporal areas of their brains (Volkow *et al.*, 1995). A SPECT study of aggressive psychiatric patients also found reduced rCBF in the prefrontal cortex, as well as increased blood flow to the left temporal and the anterior medial frontal cortices (Amen *et al.*, 1996).

Other PET studies have investigated how glucose metabolism responds to difficult cognitive tests, such as a continuous performance task (CPT). One group found that the number of impulsive–aggressive acts perpetrated by subjects with personality disorders, including antisocial personality disorder, was negatively correlated to glucose metabolism in the orbitofrontal, anterior medial frontal, and left anterior frontal cortices (Goyer *et al.*, 1994). Raine *et al.* (1994a,b) found that after a CPT, a sample of murderers demonstrated reduced glucose metabolism in the anterior medial prefrontal, orbitofrontal, and superior frontal cortices compared to a normal comparison group (Raine *et al.*, 1994b). A follow up study with a larger sample but a similar methodology found the same pattern of reduced glucose metabolism in the anterior frontal cortices, and in the amygdalas and hippocampi as well (Raine *et al.*, 1997a).

The amygdala is a structure in the brain that plays a significant role in emotion processing. This makes it an important structure in associative learning, in which individuals assign an affective valence to the consequences of their actions. These associations can be positive, such as learning to feel good after helping someone, or negative, such as learning to feel guilty or bad after harming someone. It has been theorized that associating harmful actions with the distress of others could thus discourage antisocial behavior (Blair, 2006a,b).

A study using PET technology looked at a sample of normal controls, a sample of schizophrenic patients with a history of repeated violent offending and a sample of schizophrenic patients with a history of nonrepetitive violent offending (Wong *et al.*, 1997). This team found that, compared to the normal controls, the patient samples had reduced glucose metabolism in the anterior inferior temporal lobes. This reduction was bilateral in the nonrepetitively violent group, but isolated to the left side in the repetitively violent group. Later, a research group using SPECT found that, compared to normal controls, antisocial populations have reduced rCBF to the frontal cortex and temporal cortex, and that psychopathy scores are negatively correlated with the degree of rCBF reduction to these areas (Soderstrom *et al.*, 2000, 2002).

One fMRI study looked at patterns of brain activation in 13 adolescent aggressive conduct disordered males and 14 matched controls as they looked at neutral pictures and pictures with a strong negative affective valence. It was found that when the conduct disordered youth viewed the distressing pictures they had significantly reduced activity to their left amygdalas compared to the control subjects (Sterzer *et al.*, 2005). Similar findings have been described in adult populations (Kiehl *et al.*, 2004; Muller *et al.*, 2003).

Another group used a similar methodology to study the reaction of a sample of 36 children and adolescents as they viewed photographs of neutral, angry, or fearful faces. 12 of the participants had callous–unemotional traits and oppositional defiant disorder or conduct disorder, 12 had attention deficit hyperactivity disorder and 12 were comparison subjects. When compared to the other two groups, the group with callous–unemotional traits demonstrated significantly reduced amygdala activation on viewing the fearful (but not the angry or neutral) faces (Marsh *et al.*, 2008). In addition, in a functional connectivity analysis, the callous–unemotional children showed reduced connectivity between the ventromedial prefrontal cortex and the amygdala. Further, the degree of reduction in this connectivity was negatively correlated with the score on the scale that measured the degree of callous–unemotional traits. This is particularly interesting as the ventromedial prefrontal cortex has been implicated in processing punishment and reward (Rolls, 2000), affective theories of the mind (Shamay-Tsoory *et al.*, 2005), response inhibition (Aron *et al.*, 2004; Vollm *et al.*, 2006), and emotional regulation (Ochsner *et al.*, 2005).

IV. NEUROPSYCHOLOGICAL TESTING

Neuropsychological tests provide another method for testing the capabilities and functioning of various brain areas. One of the most consistent findings in the neuropsychological aspects of criminality is that antisocial populations have lower verbal IQs compared to nonantisocial groups (Brennan *et al.*, 2003; Déry *et al.*, 1999; Raine, 1993; Teichner and Golden, 2000). Researchers have found that verbal deficits on testing at age 13 predict delinquency at age 18 (Moffitt *et al.*, 1994). A number of authors have found evidence that such neuropsychological deficits show interactive effects when they are present in children with social risk factors (Aguilar *et al.*, 2000; Brennan *et al.*, 2003; Raine, 2002a,b).

Executive functioning is another neuropsychological function of interest in criminology (Moffitt, 1990, 1993). Executive functioning refers to the group of cognitive processes that produce goal-directed, flexible, and strategically effective behavior (Lezak *et al.*, 2004; Luria, 1996; Spreen and Strauss, 1998). Executive dysfunction involves impairments in impulse control, self-regulation, abstract reasoning, concept formation, sustained attention, planning, organization, problem solving, and cognitive flexibility (Raine, 2002a,b). A meta-analysis of 39 studies incorporating data from 4589 individuals studied the relationship between executive dysfunction and antisocial behavior(Morgan and Lilienfeld, 2000). These authors found significant effect sizes ($d = 0.86$ for juvenile delinquency and $d = 0.46$ for conduct disorder) for the association between antisocial behavior and executive dysfunction.

Another neuropsychological test that has been studied in antisocial populations tests selective attention, or the ability to attend to one or more stimuli while ignoring others. The dichotic listening test is used to probe selective attention by having subjects wear headphones and then sending different auditory stimuli to each ear, while instructing them to respond to only 1 ton and ignore others. Both adult (Hare and Jutai, 1988) and juvenile (Raine *et al.*, 1990a) populations with psychopathic traits have been shown to have abnormalities on this test when verbal stimuli are used. These researchers have hypothesized that this reduced lateralization of linguistic processes might indicate that people with psychopathic traits have a reduced use of language to regulate their behavior.

Other neuropsychological tests have focused on how antisocial populations respond to affectively charged stimuli. Loney *et al.* (2003) found that juveniles with callous–unemotional traits showed slower reaction times after being presented with emotionally negative words, while those with impulsive traits showed faster reaction times to such stimuli. Adult psychopaths have been found to have deficits in passive-avoidance learning tasks (Newman and Kosson, 1986) and adolescent psychopaths have been shown to demonstrate hyperresponsivity to rewards (Scerbo *et al.*, 1990). Taken together, these data suggest that psychopathic individuals will be less sensitive to punishment and more sensitive to the possibility of rewards as a consequence to their behavior. Also, given the executive functioning literature, they may be less able to plan, act in a rationally self-interested fashion, control their impulses and respond flexibly to the various problems encountered in everyday life.

V. PSYCHOPHYSIOLOGICAL EVIDENCE

The autonomic underarousal and hyporesponsivity noted in various electrophysiological studies have given rise to fearlessness theory. This theory posits that the low level of arousal noted in the somewhat stressful testing situations can be taken as evidence of a lack of normal fear (Raine, 1993, 1997). An alternative to fearlessness theory is the stimulation-seeking theory, which presumes that the observed hypoarousal is experienced by affected individuals as unpleasant, and is compensated for using risk-taking/thrill-seeking behaviors. Supporting this hypothesis is the observation that 3-year-old children who show high levels of sensation seeking and lower levels of fearlessness demonstrate increased levels of aggression at age 11 (Raine *et al.*, 1998).

It is likely that stimulation-seeking and fearlessness explain some part of the low resting heart rate shown in antisocial youth, but a causal link between the low resting heart rate and criminal behavior is more elusive (Raine, 2002a,b). A third theory, the prefrontal deficit theory, argues that the low arousal seen arises from abnormalities in the prefrontal cortical–subcortical circuits involved with arousal and stress response (Raine, 2002a,b).

A number of psychophysiological studies have also elucidated biological correlates of criminal behavior. These studies have typically focused on heart rate, skin conductance and electrocortical measurements. In-depth descriptions of the methodologies used in psychophysiological research are available (Cacioppo *et al.*, 2007).

A. Electrocortical measures

1. Electroencephalogram (EEG)

The electrical activity in the cerebral cortex can be measured by a noninvasive test, the EEG (Hugdahl, 2001). In an EEG, the subject has electrodes placed in specific points over the scalp. These electrodes detect the brain's electrical impulses, which are then recorded and analyzed by a computer. The frequency and amplitude of the resultant signals are then interpreted.

Increasing frequency is associated with increasing arousal, and lower frequency is associated with lower arousal (Hugdahl, 2001). Slower EEG activity in children and adolescents is associated with later criminal behavior (Mednick et al., 1981; Petersen et al., 1982). Raine and colleagues demonstrated that, compared to their peers with higher arousal, 15-year-old boys with lower arousal as measured by resting EEG were more likely to become criminals at age 24 (Raine et al., 1990b). Children with externalizing and antisocial behaviors have been noted to demonstrate abnormal patterns of EEG asymmetry in their frontal lobes (Ishikawa and Raine, 2002; Santesso et al., 2006).

It has been noted that dominant EEG frequencies increase with age (Dustman et al., 1999). The EEG abnormalities noted with respect to criminal behavior have been hypothesized to be due to cortical immaturity (Volavka, 1987). It has been suggested that abnormal frontal EEG asymmetry might belie language and analytic reasoning deficits, thus impairing emotion regulation (Santesso et al., 2006).

2. Event-related potentials (ERPs)

A stimulus perceived by the brain will cause a change in the brain's electrical activity. An ERP is a measure of the magnitude of that change after the presentation of specific stimuli. The change, or deflection, may be positive or negative in direction, and occurs within milliseconds of the onset of the stimulus. Typically an ERP is measured several times, and the average of all the trials is taken (Hugdahl, 2001). The P300 is a waveform that typically occurs approximately 300 ms after the presentation of a stimulus. Early onset of drug abuse and criminal behavior has been shown to be related to smaller P300 amplitudes (Iacono and McGue, 2006). Other studies have demonstrated that greater negative amplitude at 100 ms and faster latency at 300 ms at age 15 are predict criminal behavior at age 24 (Raine et al., 1990b). A meta-analysis of studies of ERP in antisocial populations found that, in general, antisocial individuals have smaller P300 amplitudes and longer latencies (Gao and Raine, 2009).

3. Low resting heart rate

Low resting heart rate is the best-replicated biological correlate of antisocial behavior in juvenile samples (Ortiz and Raine, 2004). In a meta-analytic review of 29 samples, the average effect size was 0.56. This effect was demonstrated in both genders and irrespective of measurement technique (Raine, 1996). This relationship is not artifactual, as confounding variables such as height, weight, body composition, muscle tone, poor school performance, low IQ, hyperactivity, low attention, drug and alcohol use, participation in sports and exercise, social class, and family size and composition have all been ruled out (Farrington, 1997; Raine et al., 1990b, 1997b; Wadsworth, 1976).

The finding that low resting heart rate predicts later crime has been replicated in the United States, Germany, England, Canada, Mauritius, and New Zealand (Farrington, 1997; Mezzacappa et al., 1997; Moffitt and Caspi, 2001; Raine et al., 1997b; Rogeness et al., 1990; Schmeck and Poustra, 1993). In longitudinal studies, low resting heart rate has been shown to accurately identify individuals who are at risk for later developing antisocial behavior. This finding is specific for antisocial behavior (Rogeness et al., 1990) and has not been shown in other psychiatric syndromes.

In the Cambridge Study in Delinquent Development, a series of six regression analyses were used to identify the best independent risk factors of violence (Farrington, 1997). Only two risk factors, low resting heart rate and poor concentration, were found, independently of all other risk factors, to predict violence. This same study found evidence of an interaction between low resting heart rate and several environmental risk factors (e.g., coming from a large family, having a teenaged mother, being of low socioeconomic status) in producing violent behavior. Lastly, it has been shown that having a high resting heart rate is negatively correlated with later violent behavior (i.e., a high resting heart rate is a protective factor against developing crime development) (Raine et al., 1995).

4. Skin conductance

The ease with which the skin can conduct electrical impulses is a function of sympathetic nervous system activity. Increased sweating leads to improved electrical conductance along the surface of the skin. In times of stress, sympathetic nervous system activity increases, and skin conductance will also increase. A classically conditioned fear response (as measured by an increase in skin conductance) can be produced by pairing a stressful stimulus, such as a noxious sound, with a neutral stimulus, such as a light turning on. Studying skin conductance under different paradigms can thus provide insight into the functioning of the sympathetic nervous system.

Low skin conductance has been shown to be associated with conduct problems (Lorber, 2004). Boys with conduct disorder have been shown to have reduced fluctuations in skin conductance and impairments in conditioned fear responses (Fairchild et al., 2008; Herpertz et al., 2005). Longitudinal studies have demonstrated that reduced skin conductance arousal at age 15 has been associated with criminal offending at age 24 (Raine et al., 1995) and that low skin conductance at age 11 predicts institutionalization at age 13 (Kruesi et al., 1992).

Impaired fear conditioning as measured by skin conductance at age 3 has been shown to predict aggression at age 8 and criminal behavior at age 23 (Gao *et al.*, 2010a,b).

Low sympathetic reactivity has been shown in psychopathy-prone adolescents and in children with conduct disorder and callous–unemotional traits (Anastassiou-Hadjichara and Warden, 2008; Kimonis *et al.*, 2006; Loney *et al.*, 2003). At age 3, having an abnormal skin conductance response to unpleasant stimuli is a risk factor for displaying psychopathy in adulthood (Glenn *et al.*, 2007).

VI. GENETICS

As described in more detail in a separate chapter in this volume, a growing body of evidence has shown that there is a strong genetic contribution to juvenile delinquency (Popma and Raine, 2006). Although a number of genes have been shown to have an association with antisocial behavior, no one gene seems to "explain" criminal behavior (Goldman and Ducci, 2007). Investigating the potential genetic basis for complex behaviors is inherently complicated as they are likely to involve multiple genes, in contrast to conditions where there is a single-gene effect, as in classic Mendelian genetics (Uhl and Grow, 2004). Studies report heritability estimates that range widely, although the majority of investigators find heritability estimates that fall between 40% and 60% (Arsenault *et al.*, 2003; Beaver *et al.*, 2009; Jaffee *et al.*, 2004, 2005; Lyons *et al.*, 1995; Moffitt, 2005; Rhee and Waldman, 2002; Slutske *et al.*, 1997).

A. Twin studies

One way to investigate a genetic component to a behavior is by comparing the frequency with which the disease occurs in different kinds of siblings. Monozygotic (also called identical) twins arise from a single fertilized ovum, meaning they have exactly the same genetic material. Dizygotic (also called fraternal) twins arise from two separate fertilized ova. Like any siblings, they share 50% of the same genes. A twin pair demonstrates concordance when both individuals demonstrate the condition in question, while twin pairs with only one affected individual are said to show discordance. The heritability of a disease can be estimated by comparing the rates of concordance and discordance in both monozygotic and dizygotic twins (Jorde *et al.*, 1995).

One study that investigated the genetic contribution to childhood antisocial and aggressive behavior investigated 605 families of 9- to 10-year-old twins and triplets (Baker *et al.*, 2007). In this economically and ethnically diverse sample, such behavior was strongly heritable. Another study analyzed self-report measures of aggression in 182 monozygotic and 118 dizygotic twins (Coccaro *et al.*, 1997a). The investigators found significant heritability for three out of the four forms of aggression studied. Although twin studies provide the opportunity study individuals with identical genetic make-ups or identical pre-natal histories, other methodologies have also sought to gain understanding into the relative contribution of genes and parenting on later problematic behavior.

B. Adoption studies

Adoption studies provide another mechanism for studying the genetic versus the environmental contributions to antisocial behavior. In such studies, the char-acteristics of a child's biological and adoptive parents are considered relative to the child's own behavior. One early such study (Bohman, 1978) found evidence for a genetic predisposition to alcohol, but not to criminality, while another study from that same year (Cadoret, 1978) found evidence for heritability of antisocial behavior. A study of 862 Swedish male adoptees found that genetic influences were the most significant contributor to later criminal behavior (Cloninger *et al.*, 1982). Another large sample of adoptees in Denmark similarly found strong evidence for a genetic propensity for criminal behavior (Gabrielli and Mednick, 1984).

C. Molecular genetics

Although it was previously noted complex behavioral syndromes don't follow ordinary Mendelian patterns of inheritance, there is a notable exception to this. One group of researchers identified a family that demonstrated X-linked inheri-tance of borderline intellectual functioning and "abnormal behavior." (Brunner *et al.*, 1993) The behaviors exhibited by affected males included aggression, rape, exhibitionism, and arson. The researchers found that all had inherited a defi-ciency in the gene that coded for monoamine oxidase A (MAO-A).

D. ACE model

In behavioral genetic research, the heritability (i.e., the portion of the pheno-typic variance explained by genetic factors) is represented by the letter "A." The letter "C" is used to represent the family-wide, common, or shared environment. This includes influences that siblings would share, such as parenting styles or neighborhood characteristics. The letter "E" is used to represent environmental conditions uniquely encountered by an individual, such as getting a head injury. These are also called nonshared environmental influences.

One such study investigated the genetic basis for psychopathy (Larsson *et al*., 2006). The researchers found that "A" accounted for 63% of the variance, "C" accounted for 0%, and "E" accounted for 37% of the variance.

One meta-analytic study that used the ACE model found that, in children, genes ("A") and shared environment ("C") were equally important in explaining aggressive behavior (Miles and Carey, 1997). The researchers also found that heritability was slightly higher for males than for females, and that in adulthood the role of heritability increased while the role of shared environment fell to inconsequential levels.

Another meta-analytic study of over 100 behavioral genetic studies showed that 40–50% of the variance of antisocial behavior is due to heritability, 30% is due to the nonshared environmental influences, and 15–20% is due to shared environmental influences (Rhee and Waldman, 2002).

It has been demonstrated that the influence of genes on criminal behavior varies over the life course (Goldman and Ducci, 2007). The majority of reports find that heritability estimates for antisocial behavior are lower, and shared environmental effects on antisocial behavior are higher, in childhood than in adolescence (Jacobson *et al*., 2002; Lyons *et al*., 1995; Miles and Carey, 1997). It further seems that some genes affect the propensity for criminal involvement in adolescence, while others exert their effects in adulthood (Goldman and Ducci, 2007).

The effects of genetics are also moderated by the type of criminal offending being considered. Heritability estimates for aggressive offending are higher than those for nonaggressive offending, such as rule breaking and theft (Eley *et al*., 2003). The opposite appears to be true for nonaggressive offending, which may be influenced more by shared environmental factors, such as family criminality, family poverty, and poor parenting, although research suggests that genetic influences also affect several of these risk factors (Moffitt, 2005).

E. Gene–environment interaction

Other studies have focused on how a person's genetic endowment interacts with the environment in which the person lives. In the Swedish adoption study described above (Cloninger *et al*., 1982), the researchers found that if a person had both a biological parent and an adoptive parent who were criminals, then the person's likelihood of criminal behavior was greater than the sum of the individual risks. In other words, there was a multiplicative effect of having a biological predisposition to crime and then being raised in a criminogenic environment.

Another large study of gene–environment interaction identified people who carried a genotype that conferred a low expression of MAO-A activity (Caspi *et al*., 2002). The researchers looked at the people with high versus low

MAO-A activity and also whether or not the individual had been abused as a child. They found evidence of a strong interaction between low MAO-A activity and childhood maltreatment in the likelihood of developing conduct disorder.

VII. NONGENETIC RISK FACTORS

There are many different types of the kinds of environmental risk factors captured by the ACE model. Researchers have identified a number of intriguing risk factors, some of which could be shared by siblings, some of which are less likely to be, which have been associated with later problematic behavior. These risk factors can be broken into those that arise during pregnancy (prenatal), those that arise during birth (perinatal) and those that arise in childhood (postnatal).

A. Prenatal

1. Minor physical anomalies (MPAs)

MPAs are subtle physical defects such as having a curved fifth finger, a single palmar crease, low seated ears, or a furrowed tongue, are thought to arise from abnormalities in fetal development. These are thought to serve as biomarkers for abnormalities in neural development as well. MPAs may have a genetic basis, but they might also be due to anoxia, bleeding, or infection (Guy *et al.*, 1983). Early studies showed an increase in the prevalence of MPAs in school-aged boys exhibiting behavioral problems (Halverson and Victor, 1976). MPAs have also been shown to be correlated to aggressive behaviors in children as young as 3 years old (Waldrop *et al.*, 1978). It has also been shown that MPAs identified at age 14 predict violence at age 17 (Arsenault *et al.*, 2000).

Mednick and Kandel studied MPAs in a sample of 129 12-year-old boys (Mednick and Kandel, 1988). They found MPAs were related to violent offending as assessed 9 years later when subjects were 21 years old. Interestingly, when subjects were divided into those from unstable (i.e., non-intact) homes versus those from stable homes, a biosocial interaction was observed. MPAs only predicted violence in those individuals raised in unstable home environments.

Similarly, a study of 72 male offspring of psychiatrically ill parents found that those with both MPAs and family adversity had especially high rates of adult violent offending (Brennan *et al.*, 1997). Another study showed that the presence of MPAs significantly interacted with environmental risk factors (e.g., poverty, marital conflict) to predict conduct problems in adolescence (Pine *et al.*, 1997).

2. Tobacco

There is a significant body of evidence that demonstrates that maternal smoking during pregnancy predisposes children to developing antisocial behavior (Wakschlag *et al.*, 2002). Maternal prenatal smoking predicts externalizing behaviors in childhood and criminal behavior in adolescence (Fergusson *et al.*, 1993, 1998; Orlebeke *et al.*, 1997; Rantakallio *et al.*, 1992b; Wakschlag *et al.*, 1997). Researchers have elucidated a clear dose-dependent relationship between smoking and later criminal behavior (Brennan *et al.*, 1999; Maughan *et al.*, 2001, 2004).

Although the mechanism by which smoking produces these effects is unknown, basic science research has shown that the byproducts of smoking may affect the brain's dopaminergic and noradrenergic systems (Muneoka *et al.*, 1997) and glucose metabolism (Eckstein *et al.*, 1997). Smoking may affect various brain structures, for example, the basal ganglia, cerebral, and cerebellar cortices—implicated in the deficits observed in violent offenders (Olds, 1997; Raine, 2002a,b).

3. Alcohol

There is also a great deal of evidence that prenatal exposure to alcohol predisposes individuals to antisocial behavior (Fast *et al.*, 1999; Olson *et al.*, 1997; Streissguth *et al.*, 1996). Although Fetal Alcohol Syndrome (FAS) does not arise in all children exposed to alcohol *in utero*, evidence shows that children who do not display the full FAS syndrome can have some of the functional deficits characteristic of the syndrome (Schonfeld *et al.*, 2005). Children who do not meet diagnostic criteria for FAS, yet were exposed to high levels of alcohol *in utero*, are at increased risk of antisocial behavior (Roebuck *et al.*, 1999).

B. Perinatal risk factors

Obstetrical complications are untoward events that occur at the time of delivery and include such things as maternal preeclampsia, premature birth, low birth weight, use of forceps in delivery, transfer to a neonatal intensive care unit, anoxia, and low Apgar scores. Maternal complications have been shown to have deleterious effects on neonatal brain function (Liu, 2004; Liu and Wuerker, 2005). Newborns who suffer obstetrical complications are more likely to exhibit externalizing behaviors at age 11 than those without complications (Liu *et al.*, 2009). Obstetrical complication was found to mediate the relationship between low IQ and externalizing behaviors.

It has also been demonstrated that obstetrical complications interact with other environmental factors to predict later antisocial behavior. Raine *et al.* (1994a,b) investigated a cohort 4269 Danish men. The investigators found that birth complication significantly interacted with severe maternal rejection (e.g., efforts to abort the pregnancy, reporting the pregnancy as unwanted, or attempting to give up custody of the baby) to predict violent crime in adolescence (Raine *et al.*, 1994a). This study has since been replicated in the United States, Sweden, Finland, and Canada, and has repeatedly shown that birth complications interact with a number of psychosocial risk factors to produce antisocial behavior (Arsenault *et al.*, 2002; Hodgins *et al.*, 2001; Kemppainen *et al.*, 2001; Tibbetts and Piquero, 1999).

C. Postnatal

Poor nutrition has been investigated as a risk factor for criminal behavior for some time. An association between aggressive behavior and vitamin and mineral deficiency has been described (Breakey, 1997; Werbach, 1995). The exact mechanism by which malnutrition affects later antisocial behavior is not well understood, it has been hypothesized that proteins or minerals may either regulate neurotransmitters and hormones, or ameliorate neurotoxins (Coccaro *et al.*, 1997b; Liu and Raine, 2006).

Although most studies have focused on nutrition in the postnatal period, one study investigated the role of malnutrition in the prenatal period in producing antisocial behavior (Neugebauer *et al.*, 1999). This group studied the offspring of women who were pregnant during the German food blockade of Dutch cities in World War II. The blockade produced near starvation and severe food shortages. The researchers found that the male offspring of women who were in the first and second trimesters (but not the third trimester) of pregnancy during this time had two and a half times the rate of antisocial personality disorder than did the offspring of women who were not affected by food shortages.

Another study of prenatal nutrition studied a sample of 11,875 pregnant women. Those women who ate less seafood (i.e., less than 340 g a week), which is rich in omega-3 fatty acids, had offspring that demonstrated significantly lower scores on a number of neurodevelopmental outcomes, including prosocial behavior, than the offspring of mothers who ate more seafood (Hibbeln *et al.*, 2007).

Studies have also shown that deficiency in nutrients such as proteins, zinc, iron, and docosahexaenoic acid (a component of omega-3 fatty acid) can lead to impaired brain functioning and a predisposition to antisocial behavior in childhood and adolescence (Arnold *et al.*, 2000; Breakey, 1997; Fishbein, 2001; Lister *et al.*, 2005; Liu and Raine, 2006; Rosen *et al.*, 1985).

Longitudinal studies have shown that malnutrition in infancy is associated with aggressive behavior and attentional deficits in childhood (Galler and Ramsey, 1989; Galler *et al.*, 1983a,b). Liu *et al.* conducted a prospective longitudinal study to investigate how early malnutrition can predispose to behavior problems later in life (Liu and Raine, 2006). The researchers found that, compared to controls, children with protein, iron, or zinc deficiencies at age 3 had significantly more aggressive and hyperactive behavior at age 8, more antisocial behavior at age 11, and more excessive motor activity and conduct disorder at age 17. Significantly, this team also found a dose-dependent relationship between the extent of malnutrition and the extent of later behavior problems.

1. Traumatic brain injury (TBI)

Another risk factor that has been studied in relation to antisocial behavior is TBI. One group of investigators found that half of the juvenile delinquents in their sample had a history of TBI, and a third of the delinquents with TBI were thought by their parents to have neuropsychological sequelae from their injuries (Hux *et al.*, 1998). Another study, which used more severe criteria in the definition of TBI than the previous study, found that 27.7% of the delinquents in their sample had a history of TBI (Carswell *et al.*, 2004). A number of large, longitudinal studies of have repeatedly shown an increased incidence of delinquent behavior among youth with a history of TBI (Asarnow *et al.*, 1991; Bloom *et al.*, 2001; Butler *et al.*, 1997; McAllister, 1992; Rantakallio *et al.*, 1992a; Rimel *et al.*, 1981; Rivera *et al.*, 1994).

VIII. THE LIMITATIONS AND POTENTIAL OF NEUROCRIMINOLOGY

The field of neurocriminology has struggled to free itself from associations to earlier efforts to incorporate biology into the field of criminology. The reductionism of Lombroso's biological positivism, the pseudoscience of phrenology, and the appalling racism of social Darwinists have all cast long shadows that have affected how contemporary efforts have been received by sociologically oriented criminologists.

There is a danger that the kind of neurocriminological data could be used in a sensationalistic or superficial manner to implicate or exculpate individual offenders. Although the current state of imaging and other forms of biological research have not advanced to the point where an individual's data could be confidently compared against a reliable database of normal controls, the possibility exists that such databases could be created and validated (Yang *et al.*, 2008). Until such time, however, it is important to note that studies of the sort reviewed in this chapter cannot be taken to imply that any one biological factor causes criminal behavior. Rather, the presence of these factors only increases the probability that problematic behavior will be present in people with a given biological risk factor.

We have now reviewed a number of studies that describe such increased probabilities of biological risk factors in criminal behavior. The sociological roots have crime have also been widely studied. This chapter has also reviewed a number of examples of how biological and environmental forces can interact to produce problematic behavior. We can see that criminal behavior can be investigated and explained at many different levels of abstraction.

The psychiatrist and philosopher Kenneth Kendler illustrates this phenomenon with a hypothetical case of a pharmacologist running a randomized controlled trial of a medication for a psychiatric condition (Kendler, 2005). Although it is undoubtedly true that the medication is a molecule, and molecules are made up of atoms and atoms are made up of subatomic particles, it does not necessarily make sense to consult with a particle physicist in conducting the study. Thus, some levels of abstraction may be more or less efficient in explaining the phenomenon in question. However, each time a new level is identified, new possibilities for intervention arise as well. In uncovering biological leads relevant to crime the potential for new strategies for crime prevention are created as well.

IX. MODIFIABLE RISK FACTOR INTERVENTIONS

Not all risk factors for criminal behavior (e.g., male gender, having a biological parent with a history of criminal behavior) are modifiable. There are a number of risk factors (e.g., smoking, nutrition), however, that can be modified. Successful interventions have been developed to reduce prenatal alcohol exposure (Chang et al., 1999, 2005). Interventions have also been designed to reduce smoking in pregnancy, but these have been notably less effective than the interventions for alcohol use (Ershoff et al., 2004). Of note, women who persist in smoking throughout pregnancy are more likely than those who quit to have a personal history of conduct disorder (Kodl and Wakschlag, 2004).

Other studies have sought to correct nutritional deficits. One randomized, double blind, placebo-controlled study was performed in a sample of 486 public schoolchildren to see if a daily multivitamin and mineral supplement could reduce antisocial behavior (Schoenthaler and Bier, 2000). The researchers found that, compared to the placebo group, the treatment group had a 47% reduction in antisocial behavior after 4 months. Previously, this team had investigated the effect of vitamin and mineral supplementation in a group of juvenile delinquents confined to a correctional setting. The results of this randomized, double blind, placebo-controlled trial showed that, compared to the placebo group, the treatment group had significantly less violent and nonviolent antisocial behaviors (Schoenthaler et al., 1997). Another randomized, double blind, placebo-controlled trial of omega-3 fatty acid supplementation was done in a sample of 50 children. Compared to the placebo condition, the intervention group had a 42.7% reduction in conduct disorder problems (Stevens *et al.*, 2003). A study using omega-3 fatty acid supplementation in ADHD failed to reveal a benefit (Hirayama *et al.*, 2004).

Other interventions address more than one risk factor at a time. For example, one highly successful intervention for prevention of later criminal and antisocial behavior involves home nursing visits for pregnant and new mothers. Parenting, health, and nutritional guidance are provided in the sessions (Olds *et al.*, 1998). Other authors have also shown that prenatal education on nutrition, health, and parenting can lead to reductions in juvenile delinquency at age 15 (Lally *et al.*, 1988).

Another multidimensional intervention was tested in a randomized, controlled fashion (Raine *et al.*, 2003b). In this study, an intervention consisting of physical exercise and nutritional and educational enrichment was tested on a sample of 3–5 year olds. The study found that the intervention significantly reduced antisocial behavior at age 17 and criminal behavior at age 23. The intervention was found to be especially effective for the subgroup of children who displayed signs of malnutrition at age 3, suggesting the nutritional aspect of the treatment was particularly beneficial. The intervention was shown to produce lasting psychophysiological changes at age 11, including increased skin conductance, more orienting, and more arousal on EEG (Raine *et al.*, 2001, 2003b). These changes might then protect against the development of criminal offending (Raine *et al.*, 1995, 1996).

X. CONCLUSION

Human beings are biological creatures. Whatever the truest essence of our souls may be, our subjective mental lives are mediated by and expressed through a system that is undeniably biological. This biological self exists in a specific social reality, which, in turn, shapes and alters the biological self in ways that will find some biological expression. What this chapter has sought to do is clarify how these biological aspects of the self can be used to understand, identify and, hopefully, predict individuals who criminally offend. Understanding these processes is the first step in then being able to modify risk factors or target at-risk individuals for services designed to attenuate their criminal propensity.

References

Abrahamsen, D. (1973). The Murdering Mind. Harper & Row, New York.

Aguilar, B., Sroufe, A., *et al.* (2000). Distinguishing the early-onset/persistent and adolescent-onset antisocial behavior types: From birth to 6 years. *Dev. Psychopathol.* **12**, 109–132.

Aigner, M. R., Eher, R., *et al.* (2000). Brain abnormalities and violent behavior. *J. Psychol. Human Sex.* **11**, 57–64.

Alexander, F., and Staub, H. (1931). The Criminal, the Judge and the Public: A Psychological Analysis. The Macmillan Company, New York.

Amen, D. G., Stubblefield, M., et al. (1996). Brain SPECT findings and aggressiveness. Ann. Clin. Psychiatry 8, 129–137.

Anastassiou-Hadjichara, X., and Warden, D. (2008). Physiologically-indexed and self-perceived affective empathy in conduct-disordered children high and low on callous-unemotional traits. Child Psychiatry Hum. Dev. 39, 503–517.

Arieti, S., and Schreiber, F. R. (1981). Multiple murders of a schizophrenic patient: A psychodynamic interpretation. J. Am. Acad. Psychoanal. 9(4), 501–524.

Arnold, L. E., Pinkham, S. M., et al. (2000). Does zinc moderate essential fatty acid and amphetamine treatment of attention-deficit/hyperactivity disorder? J. Child Adolesc. Psychopharmacol. 10, 111–117.

Aron, A. R., Robbins, T. W., et al. (2004). Inhibition and the right infrerior frontal cortex. Trends Cogn. Sci. 8(4), 170–177.

Arsenault, L., Tremblay, R. E., et al. (2000). Minor physical anomalies and family adversity as risk factors for violent delinquency in adolescence. Am. J. Psychiatry 157, 917–923.

Arsenault, L., Tremblay, R. E., et al. (2002). Obstetrical complications and violent delinquency: Testing two developmental pathways. Child Dev. 73, 496–508.

Arsenault, L., Moffitt, T. E., et al. (2003). Strong genetic effects on cross-situational antisocial behaviour among 5-year-old children according to mothers, teachers, examiners-observers, and twins' self-reports. J. Child Psychol. Psychiatry 44(6), 832–848.

Asarnow, R., Satz, P., et al. (1991). Behavior problems and adaptive functioning in children with mild and severe closed head injury. J. Pediatr. Psychol. 16, 543–555.

Baker, L. A., Jacobson, K. C., et al. (2007). Genetic and environmental bases of childhood antisocial behavior: A multi-informant twin study. J. Abnorm. Psychol. 116(2), 219–235.

Beaver, K. M., DeLisi, M., et al. (2009). Gene environment interplay and delinquent involvement: Evidence of direct and indirect, and interactive effects. J. Adolesc. Res. 24, 147–168.

Blair, R. J. R. (2006a). The emergence of psychopathy: Implications for the neuropsychological approach to developmental disorders. Cognition 101, 414–442.

Blair, R. J. R. (2006b). Subcortical brain systems in psychopathy. In "Handbook of Psychopathy" (C. J. Patrick, ed.), pp. 296–312.

Blake, P. Y., Pincus, J. H., et al. (1995). Neurologic abnormalities in murderers. Neurology 45, 1641–1647.

Bloom, D. R., Levin, H. S., et al. (2001). Lifetime and novel psychiatric disorders after pediatric brain injury. J. Am. Acad. Child Adolesc. Psychiatry 40, 572–579.

Bohman, M. (1978). Some genetic aspects of alcoholism and criminality. A population of adoptees. Arch. Gen. Psychiatry 35(3), 269–276.

Bowlby, J. (1944). Forty-four thieves: Their character and home life. Int. J. Psychoanal. 25, 19–52.

Bowlby, J. (1969). Attachment and Loss: Volume 1 Attachment. Basic Books, New York.

Bowlby, J. (1973). Attachment and Loss. Vol. 2. Separation: Anxiety and Anger. Basic Books, New York.

Bowlby, J. (1980). Attachment and Loss: Vol. 3. Loss: Sadness and Depression. Basic Books, New York.

Breakey, J. (1997). The role of diet and behaviour in childhood. J. Paediatr. Child Health 33, 190–194.

Brennan, P. A., Mednick, S. A., et al. (1997). Biosocial interactions and violence: A focus on perinatal factors. In "Biosocial Bases of Violence" (A. Raine, P. Brennan, D. P. Farrington, and S. A. Mednick, eds.). Plenum, New York.

Brennan, P. A., Grekin, E. R., et al. (1999). Maternal smoking during pregnancy and adult male criminal outcomes. Arch. Gen. Psychiatry 56, 215–219.

Brennan, P. A., Hall, J., *et al.* (2003). Integrating biological and social processes in relation to early-onset persistent aggression in boys and girls. *Dev. Psychopathol.* **39**(2), 309–323.

Bromberg, W. (1951). A psychological study of murder. *Int. J. Psychoanal.* **32**, 117–127.

Brunner, H. G., Nelen, M., *et al.* (1993). Abnormal behavior associated with a point mutation in the structural gene for monoamine oxidase A. *Science* **262**(5133), 578–580.

Butler, G., Chadwick, O., *et al.* (1997). A typology of psychosocial functioning in pediatric closed-head injury. *Child Neuropsychol.* **3**, 98–133.

Cacioppo, J. T., and Tassinary, L. G. *et al.* (eds.) (2007). *In* "Handbook of Psychophysiology". Cambridge University Press, New York.

Cadoret, R. J. (1978). Psychopathology in adopted-away offspring of biologic parents with antisocial behavior. *Arch. Gen. Psychiatry* **35**(2), 176–184.

Carswell, K., Maughan, B., *et al.* (2004). The psychosocial needs of young offenders and adolescents from an inner city area. *J. Adolesc.* **27**, 415–428.

Caspi, A., McClay, J., *et al.* (2002). Role of genotype in the cycle of violence in maltreated children. *Science* **297**(5582), 851–854.

Cassity, J. H. (1941). Personality study of 200 murderers. *J. Crim. Psychopathol.* **2**, 296–304.

Chang, G., Wilkins-Haug, L., *et al.* (1999). Brief intervention for alcohol use in pregnancy: A randomized trial. *Addiction* **94**(10), 1499–1508.

Chang, G., McNamara, T. K., *et al.* (2005). Brief intervention for prenatal alcohol use: A randomized trial. *Obstet. Gynecol.* **10**(5 Pt. 1), 991–998.

Chesterman, L., Taylor, P., *et al.* (1994). Multiple measures of cerebral state in dangerous mentally disordered inpatients. *Crim. Behav. Ment. Health* **4**, 228–239.

Cloninger, C. R., Sigvardsson, S., *et al.* (1982). Predisposition to petty criminality in Swedish adoptees. II. Cross-fostering analysis of gene-environment interaction. *Arch. Gen. Psychiatry* **39**(11), 1242–1247.

Coccaro, E. F., Bergeman, C. S., *et al.* (1997a). Heritability of aggression and irritability: A twin study of the Buss-Durkee aggression scales in adult male subjects. *Biol. Psychiatry* **41**(3), 273–284.

Coccaro, E. F., Kavoussi, R. J., *et al.* (1997b). Serotonin function and anti-aggressive response to fluoxetine: A pilot study. *Biol. Psychiatry* **42**, 546–552.

Craig, M. C., Catani, M., *et al.* (2008). Altered connections on the road to psychopathy. *Mol. Psychiatry* **14**(10), 946–953.

DeRiver, J. P. (1951). The Sexual Criminal: A Psychoanalytic Study. Charles C. Thomas, Springfield, IL.

Déry, M., Toupin, J., *et al.* (1999). Neuropsychological characteristics of adolescents with conduct disorder: Association with attention-deficit-hyperactivity and aggression. *J. Abnorm. Child Psychol.* **27**(3), 225–236.

Dolan, M., Deakin, J. F. W., *et al.* (2002). Quantitative frontal and temporal structural MRI studies in personality-disordered offenders and control subjects. *Psychiatry Res.* **116**, 133–149.

Dustman, R. E., Shearer, D. E., *et al.* (1999). Life-span changes in EEG spectral amplitude, amplitude variability and mean frequency. *Clin. Neurophysiol.* **110**, 1399–1409.

Eckstein, L. W., Shibley, I. J., *et al.* (1997). Changes in brain glucose levels and glucose transporter protein isoforms in alcohol- or nicotine- treated chick embryos. *Brain Res. Dev. Brain Res.* **15**, 383–402.

Eley, T. C., Lichtenstein, P., *et al.* (2003). A longitudinal behavioral genetic analysis of the etiology of aggressive and nonaggressive antisocial behavior. *Dev. Psychopathol.* **15**, 383–402.

Engel, G. L. (1977). The need for a new medical model: A challenge for biomedicine. *Science* **196**(4286), 129–136.

Ershoff, D. H., Ashford, T. H., *et al.* (2004). Helping pregnant women quit smoking: An overview. *Nicotine Tob. Res.* **6**(2), S101–S105.

Evseef, G. S., and Wisniewski, E. M. (1972). A psychiatric study of a violent mass murderer. *J. Forensic Sci.* **17**(3), 371–376.

Fairchild, G., Van Goozen, S. H. M., *et al.* (2008). Fear conditioning and affective modulation of the startle reflex in male adolescents with early-onset or adolescence-onset conduct disorder and health control subjects. *Biol. Psychiatry* **63,** 279–285.

Farrington, D. P. (1997). The relationship between low resting heart rate and violence. *In* "Biosocial Bases of Violence" (A. Raine, P. A. Brennan, D. P. Farrington, and S. A. Mednick, eds.), pp. 89–106. Prenum Press, New York.

Fast, D. K., Conry, J., *et al.* (1999). Identifying Fetal Alcohol Syndrome among youth in the criminal justice system. *J. Dev. Behav. Pediatr.* **20,** 370–372.

Fergusson, D. M., Horwood, L. J., *et al.* (1993). Maternal smoking before and after pregnancy. *Pediatrics* **92,** 815–822.

Fergusson, D. M., Woodward, L. J., *et al.* (1998). Maternal smoking during pregnancy and psychiatric adjustment in late adolescence. *Arch. Gen. Psychiatry* **55,** 721–727.

Fishbein, D. (2001). Biobehavioral Perspectives in Criminology. Wadsworth/Thomson Learning, Belmont, CA.

Gabrielli, W. F., and Mednick, S. A. (1984). Urban environment, genetics, and crime. *Criminology* **22**(4), 645–652.

Galler, J. R., and Ramsey, F. (1989). A follow-up study of the influence of early malnutrition on development. *J. Am. Acad. Child Adolesc. Psychiatry* **26,** 23–27.

Galler, J. R., Ramsey, F., *et al.* (1983a). The influence of early malnutrition on subsequent behavioral development. II. Classroom behavior. *J. Am. Acad. Child Adolesc. Psychiatry* **22,** 16–22.

Galler, J. R., Ramsey, F., *et al.* (1983b). The influence of early malnutrition on subsequent behavioural development. I. Degree of impairment of intellectual performance. *J. Am. Acad. Child Adolesc. Psychiatry* **22,** 8–15.

Gao, Y., and Raine, A. (2009). P3 event-related potential impairments in antisocial and psychopathic individuals: A meta-analysis. *Biol. Psychiatry* **83,** 199–210.

Gao, Y., Raine, A., *et al.* (2010a). Poor childhood fear conditioning predisposes to adult crime. *Am. J. Psychiatry* **167,** 56–60.

Gao, Y., Raine, A., *et al.* (2010b). Reduced electrodermal fear conditioning from ages 3 to 8 years is associated with aggressive behavior at age 8 years. *J. Child Psychol. Psychiatry* **51**(5), 550–558.

Glenn, A. L., Raine, A., *et al.* (2007). Early temperamental and psychophysiological precursors of adult psychopathic personality. *J. Abnorm. Psychol.* **116,** 508–518.

Goldman, D., and Ducci, F. (2007). The genetics of psychopathic disorders. *In* "International Handbook of Psychopathic Disorders and the Law" (A. R. Felthous and H. Sa, eds.), Vol. 1, pp. 149–169. John Wiley and Sons, Ltd., West Sussex, England.

Goyer, P. F., Andreason, P. J., *et al.* (1994). Positron-emission tomography and personality disorders. *Neuropsychopharmacology* **10,** 21–28.

Guy, J. D., Majorski, L. V., *et al.* (1983). The incidence of minor physical anomalies in adult male schizophrenics. *Schizophr. Bull.* **9,** 571–582.

Halverson, C. F., and Victor, J. B. (1976). Minor physical anomalies and problem behavior in elementary schoolchildren. *Child Dev.* **47,** 281–285.

Hare, R. D., and Jutai, J. W. (1988). Psychopathy and cerebral asymmetry in semantic processing. *Pers. Indiv. Differ.* **9,** 329–337.

Herpertz, S. C., Mueller, B., *et al.* (2005). Emotional responses in boys with conduct disorder. *Am. J. Psychiatry* **162,** 1100–1107.

Hibbeln, J. R., Davis, J. M., *et al.* (2007). Maternal seafood consumption in pregacy and neurodevelopmental outcomes in childhood (ALSPAC study): An observational cohort study. *Lancet* **369**(9561), 578–585.

Hirayama, S., Hamazaki, T., *et al.* (2004). Effect of docosahexaenoic acid-containing food administration on symptoms of attention-deficit/hyperactivity disorder- a placebo-controlled double-blind study. *Eur. J. Clin. Nutr.* **58,** 467–473.

Hodgins, S., Kratzer, L., *et al.* (2001). Obstetric complications, parenting, and risk of criminal behavior. *Arch. Gen. Psychiatry* **58,** 746–752.

Holmes, R. M., and Holmes, S. T. (1998). Serial Murder. Thousand Oaks, CA, Sage.

Huebner, T., Vloet, T. D., *et al.* (2008). Morphometric brain abnormalities in boys with conduct disorder. *J. Am. Acad. Child Adolesc. Psychiatry* **47,** 540–547.

Hugdahl, K. (2001). Psychophysiology: The Mind-Body Perspective. Harvard University Press, Cambridge, MA.

Hux, K., Bond, V., *et al.* (1998). Parental report of occurrences and consequences of traumatic brain injury among delinquent and non-delinquent youth. *Brain Inj.* **12,** 667–681.

Iacono, W. G., and McGue, M. (2006). Association between P3 event-related brain potential amplitude and adolescent problem behavior. *Psychophysiology* **43,** 465–469.

Ishikawa, S. S., and Raine, A. (2002). Psychophysiological correlates of antisocial behavior: A central control hypothesis. *In* "The Neurobiology of Criminal Behavior" (J. Glicksohn, ed.), pp. 187–229. Kluwer, Boston, MA.

Jacobson, K. C., Prescott, C. A., *et al.* (2002). Sex differences in the genetic and environmental influences on the development of antisocial behavior. *Dev. Psychopathol.* **14,** 395–416.

Jaffee, S. R., Caspi, A., *et al.* (2004). Physical maltreatment victim to antisocial child: Evidence of an environmentally-mediated process. *J. Abnorm. Psychol.* **113,** 44–55.

Jaffee, S. R., Caspi, A., *et al.* (2005). Nature x nurture: Genetic vulnerabilities interact with physical maltreatment to promote conduct problems. *Dev. Psychopathol.* **17,** 67–84.

Jorde, L. B., Carey, J. C., *et al.* (1995). Medical Genetics. Mosby, New York.

Karpman, B. (1951a). A psychoanalytic study of a case of murder. *Psychoanal. Rev.* **38**(2), 139–157.

Karpman, B. (1951b). A psychoanalytic study of a case of murder. *Psychoanal. Rev.* **38**(2), 245–270.

Kemppainen, L., Jokelainen, J., *et al.* (2001). The one-child family and violent criminality: A 31-year follow-up study of the Northern Finland 1966 birth cohort. *Am. J. Psychiatry* **158,** 960–962.

Kendler, K. S. (2005). Toward a philosophical structure for psychiatry. *Am. J. Psychiatry* **162,** 433–440.

Kiehl, K., Smith, A. M., *et al.* (2004). Temporal lobe abnormalities in semantic processing by criminal psychopaths as revealed by functional magnetic resonance imaging. *Psychiatry Res.* **130,** 27–42.

Kimonis, E. R., Frick, P. J., *et al.* (2006). Psychopathy, aggression, and the processing of emotional stimuli in non-referred girls and boys. *Behav. Sci. Law* **24,** 21–37.

Kodl, M. M., and Wakschlag, L. S. (2004). Does a childhood history of externalizing problems predict smoking during pregnancy? *Addict. Behav.* **29**(2), 273–279.

Kruesi, M. J. P., Hibbs, E. D., *et al.* (1992). A 2-year prospective follow-up study of children and adolescents with disruptive behavior disorders. Prediction by cerebrospinal fluid 5-hydroxyindoleacetic acid, homovanillic acid, and autonomic measures? *Am. J. Psychiatry* **49**(6), 429–435.

Kruesi, M. J. P., Casanova, M. V., *et al.* (2004). Reduced temporal lobe volume in early-onset conduct disorder. *Psychiatry Res.* **132,** 1–11.

Laakso, M. P., Vaurio, O., *et al.* (2000). A volumetric MRI study of the hippocampus in type 1 and 2 alcoholism. *Behav. Brain Res.* **109,** 117–186.

Laakso, M. P., Vaurio, O., *et al.* (2001). Psychopathy and the posterior hippocampus. *Behav. Brain Res.* **118,** 187–193.

Laakso, M. P., Gunning-Dixon, F., *et al.* (2002). Prefrontal volume in habitually violent subjects with antisocial personality disorder and type 2 alcoholism. *Psychiatry Res.* **114,** 95–102.

Lally, J. R., Mangione, P. L., *et al.* (1988). Long-range impact of an early intervention with low income children and their families. *In* "Parent Education as Early Childhood Intervention" (D. R. Powell, ed.), pp. 79–104. Ablex, Norwood, NJ.

Langevin, R., Ben-Aron, M., *et al.* (1988). The sex killer. *Ann. Sex Res.* **1,** 263–301.

Larsson, H., Andershed, H., *et al.* (2006). A genetic factor explains most of the variation in the psychopathic personality. *J. Abnorm. Psychol.* **115**(2), 2211–2230.

Lehrman, P. R. (1939). Some unconscious determinants in homicide. *Psychoanal. Q.* **13**(4), 606–621.

Lezak, M. D., Howieson, D. B., *et al.* (2004). Neuropsychological Assessment. 4th edn., Oxford University Press, New York.

Li, T. Q., Mathews, V. P., *et al.* (2005). Adolescents with disruptive behavior disorder investigated using an optimized MR diffusion tensor imaging protocol. *Ann. N. Y. Acad. Sci.* **1064,** 184–192.

Liebert, J. A. (1972). Contributions of psychiatric consultation in the investigation of serial murder. *Int. J. Offender Ther. Comp. Criminol.* **29,** 187–200.

Lister, J. P., Blatt, G. J., *et al.* (2005). Effect of prenatal protein malnutrition on numbers of neurons in the principal cell layers of the adult rat hippocampal formation. *Hippocampus* **15,** 393–403.

Liu, J. (2004). Childhood externalizing behavior: Theory and implications. *J. Child Adolesc. Psychiatr. Nurs.* **17,** 93–103.

Liu, J., and Raine, A. (2006). The effect of childhood malnutrition on externalizing behaviors. *Curr. Opin. Pediatr.* **18,** 565–570.

Liu, J., and Wuerker, A. (2005). Biosocial bases of aggressive and violent behavior- implications for nursing studies. *Int. J. Nurs. Stud.* **42,** 229–241.

Liu, J., Raine, A., *et al.* (2009). The association of birth complications and externalizing behavior in early adolescents. *J. Res. Adolesc.* **19,** 93–111.

Loney, B. R., Frick, P. J., *et al.* (2003). Callous-unemotional traits, impulsivity and emotional processing in adolescents with antisocial behavior problems. *J. Clin. Child Adolesc. Psychol.* **32,** 66–80.

Lorber, M. F. (2004). Psychophysiology of aggression, psychopathy, and conduct problems: A meta-analysis. *Psychol. Bull.* **130,** 531–552.

Luria, A. (1996). Higher Cortical Functions in Man. Basic Books, New York.

Lyons, M. J., True, W. R., *et al.* (1995). Differential heritability of adult and juvenile traits. *Arch. Gen. Psychiatry* **52,** 906–915.

Marsh, A. A., Finger, E. C., *et al.* (2008). Reduced amygdala response to fearful expressions in children and adolescents with callous-unemotional traits and disruptive behavior disorders. *Am. J. Psychiatry* **165,** 712–720.

Maughan, B., Taylor, C., *et al.* (2001). Pregnancy smoking and childhood conduct problems: A causal association? *J. Child Psychol. Psychiatry* **42,** 1021–1028.

Maughan, B., Taylor, A., *et al.* (2004). Prenatal smoking and early childhood conduct problems. *Arch. Gen. Psychiatry* **61,** 836–843.

McAllister, T. (1992). Neuropsychiatric sequelae of head injuries. *Psychiatr. Clin. North Am.* **15,** 661–665.

Mednick, S. A., and Kandel, E. S. (1988). Congential determinants of violence. *Bull. Am. Acad. Psychiatry Law* **16,** 101–109.

Mednick, S. A., Volavka, J., *et al.* (1981). EEG as a predictor of antisocial behavior. *Criminology* **19,** 219–229.

Mezzacappa, E., Tremblay, R. E., *et al.* (1997). Anxiety, antisocial behavior and heart rate regulation in adolescent males. *J. Child Psychol. Psychiatry* **38,** 457–468.

Miles, D. R., and Carey, G. (1997). Genetic and environmental architecture of human aggression. *J. Pers. Soc. Psychol.* **72,** 207–217.

Moffitt, T. E. (1990). Juvenile delinquency and attention-deficit disorder: Developmental trajectories from age three to fifteen. *Child Dev.* **61,** 893–910.

Moffitt, T. E. (1993). Adolescence-limited and life-course-persistent antisocial behavior: A developmental taxonomy. *Psychol. Rev.* **100**(2), 674–701.

Moffitt, T. E. (2005). The new look of behavioral genetics in developmental psychopathology: Gene-environment interplay in antisocial behaviors. *Psychol. Bull.* 533–554.

Moffitt, T. E., and Caspi, A. (2001). Childhood predictors differentiate life-course persistent and adolescence-limited antisocial pathways among males and females. *Dev. Psychopathol.* **13,** 355–375.

Moffitt, T. E., Lynam, D. R., *et al.* (1994). Neuropsychological tests predicting persistent male delinquency. *Criminology* **32**(2), 277–300.

Morgan, A. B., and Lilienfeld, S. O. (2000). A meta-analytic review of the relationship between antisocial behavior and neuropsychological measures of executive function. *Clin. Psychol. Rev.* **20,** 113–136.

Morrison, H. L. (1979). Psychiatric observations and interpretations of bite mark evidence in multiple murders. *J. Forensic Sci.* **24**(2), 492–502.

Muller, J. L., Sommer, M., *et al.* (2003). Abnormalities in emotion processing within cortical and subcortical regions in criminal psychopaths: Evidence from a functional magnetic imaging study using pictures with emotional content. *Psychiatry Res.* **54,** 152–162.

Muller, J. L., Ganssbauer, S., *et al.* (2008). Gray matter changes in right superior temporal gyrus in criminal psychopaths. Evidence from voxel-based morphometry. *Psychiatry Res.* **163,** 213–222.

Muneoka, K., Ogawa, T., *et al.* (1997). Prenatal nicotine exposure affects the development of the central serotonergic system as well as the dopaminergic system in rat offspring: Involvement of route of drug administrations. *Brain Res. Dev. Brain Res.* **102**(1), 117–126.

Neugebauer, R., Hoek, H. W., *et al.* (1999). Prenatal exposure to wartime famine and development of antisocial personality disorder in early adulthood. *J. Am. Med. Assoc.* **4,** 479–481.

Newman, J. P., and Kosson, D. S. (1986). Passive avoidance learning in psychopathic and non-psychopathic offenders. *J. Abnorm. Psychol.* **95,** 252–256.

Ochsner, K. N., Beer, J. S., *et al.* (2005). The neural correlates of direct and reflected self-knowledge. *Neuroimage* **28**(4), 797–814.

Olds, D. (1997). Tobacco exposure and impaired development: A review of the evidence. *Ment. Retard. Dev. Disabil. Res. Rev.* **3,** 257–269.

Olds, D., Henderson, C. R. J., *et al.* (1998). Long-term effects of nurse home visitation on children's criminal and antisocial behavior: 15-year follow-up of a randomized controlled trial. *J. Am. Med. Assoc.* **280,** 1238–1244.

Olson, H. C., Streissguth, A. P., *et al.* (1997). Association of prenatal alcohol exposure with behavioral and learning problems in early adolescence. *J. Am. Acad. Child Adolesc. Psychiatry* **36,** 1187–1194.

Orlebeke, J. F., Knol, D. L., *et al.* (1997). Increase in child behavior problems resulting from maternal smoking during pregnancy. *Arch. Environ. Health* **52,** 317–321.

Ortiz, J., and Raine, A. (2004). Heart rate level and antisocial behavior in children and adolescents: A meta analysis. *J. Am. Acad. Child Adolesc. Psychiatry* **43,** 154–162.

Petersen, K. G. I., Matousek, M., *et al.* (1982). EEG antecedents of thievery. *Acta Psychiatr. Scand.* **65,** 331–338.

Pine, D. S., Shaffer, D., *et al.* (1997). Minor physical anomalies: Modifiers of environmental risks for psychiatric impairment? *J. Am. Acad. Child Adolesc. Psychiatry* **36,** 395–403.

Popma, A., and Raine, A. (2006). Will future forensic assessment be neurobiologic? *Child Adolesc. Psychiatr. Clin. N. Am.* **36,** 395–444.

Raine, A. (1993). The Psychopathology of Crime: Criminal Behavior as a Clinical Disorder. Academic Press, San Diego, CA.

Raine, A. (1996). Autonomic nervous system activity and violence. *In* "Neurobiological Approaches to Clinical Aggression Research" (D. M. Stoff and R. B. Cairns, eds.), pp. 145–168. Lawrence Erlbaum, Mahwah, NJ.

Raine, A. (1997). Psychophysiology and antisocial behavior: A biosocial perspective an a prefrontal dysfunction hypothesis. *In* "Handbook of Antisocial Behavior" (D. M. Stoff, J. Breiling, and J. D. Maser, eds.), pp. 289–304. Wiley, New York.

Raine, A. (2002a). Annotation: The role of prefrontal deficits, low autonomic arousal and early health factors in the development of antisocial and aggressive behavior in children. *J. Child Psychol. Psychiatry* **43,** 417–434.

Raine, A. (2002b). Biosocial studies of antisocial and violent behavior in children and adults: A review. *J. Abnorm. Child Psychol.* **304**(4), 311–326.

Raine, A., O'Brien, M., *et al.* (1990a). Reduced lateralization in verbal dichotic listening in adolescent psychopaths. *J. Abnorm. Psychol.* **99,** 272–277.

Raine, A., Venables, P. H., *et al.* (1990b). Relationships between CNS and ANS measures of arousal at age 15 years as protective factors against criminal behavior at age 29 years. *Am. J. Psychiatry* **152,** 1595–1600.

Raine, A., Brennan, P., *et al.* (1994a). Birth complications combined with early maternal rejection at age 1 year predispose to violent crime at age 18 years. *Arch. Gen. Psychiatry* **51,** 984–988.

Raine, A., Buchsbaum, M., *et al.* (1994b). Selective reductions in prefrontal glucose metabolism in murderers. *Biol. Psychiatry* **36,** 365–373.

Raine, A., Venables, P. H., *et al.* (1995). High autonomic arousal and electrodermal orienting at age 15 years as protective factors against criminal behavior at age 29 years. *Am. J. Psychiatry* **152,** 1595–1600.

Raine, A., Venables, P. H., *et al.* (1996). Better autonomic conditioning and faster electrodermal half-recovery time at age 15 years as possible protective factors against crime at age 29 years. *Dev. Psychol.* **32,** 624–630.

Raine, A., Buchsbaum, M., *et al.* (1997a). Brain abnormalities in murderess indicated by positron emission tomography. *Biol. Psychiatry* **42,** 495–508.

Raine, A., Venables, P. H., *et al.* (1997b). Low resting heart rate age 3 years predisposes to aggression at age 11 years: Evidence from the Mauritius Child Health Project. *J. Am. Acad. Child Adolesc. Psychiatry* **36,** 1457–1464.

Raine, A., Reynolds, C., *et al.* (1998). Fearlessness, stimulation-seeking, and large body size at age 3 years as early predispositions to childhood aggression at age 11 years. *Arch. Gen. Psychiatry* **55,** 745–751.

Raine, A., Lencz, T., *et al.* (2000). Reduced prefrontal gray matter volume and reduced autonomic activity in antisocial personality disorder. *Arch. Gen. Psychiatry* **57,** 119–127.

Raine, A., Venables, P. H., *et al.* (2001). Early educational and health enrichment at age 3–5 years is associated with increased autonomic and central nervous system arousal and orienting at age 11 years: Evidene from the Mauritius Child Health Project. *Psychophysiology* **38,** 254–266.

Raine, A., Lencz, T., *et al.* (2003a). Corpus callosum abnormalities in psychopathic individuals. *Arch. Gen. Psychiatry* **160,** 1627–1635.

Raine, A., Mellingen, K., *et al.* (2003b). Effects of environmental enrichment at ages 3–5 years on schizotypal personality and antisocial behavior at ages 17 and 23 years. *Am. J. Psychiatry* **160,** 1627–1635.

Raine, A., Lee, L., *et al.* (2010). Neurodevelopmental marker for limbic maldevelopment in antisocial personality disorder and psychopathy. *Br. J. Psychiatry* **197,** 186–192.

Rantakallio, P., Koiranen, M., *et al.* (1992a). Association of prenatal events, epilepsy, and central nervous system trauma with juvenile delinquency. *Arch. Dis. Child.* **67,** 1459–1461.

Rantakallio, P., Laara, E., *et al.* (1992b). Maternal smoking during pregnancy and delinquency of the offspring: An association without causation? *Int. J. Epidemiol.* **21,** 1106–1113.

Revitch, E., and Schlesinger, L. B. (1981). The Psychopathology of Homicide. Charles C. Thomas, Springfield, IL.

Revitch, E., and Schlesinger, L. B. (1989). Sex Murder and Sex Aggression: Phenomenology, Psychopathology, Psychodynamics and Prognosis. Charles C. Thomas, Springfield, IL.

Rhee, S. H., and Waldman, I. D. (2002). Genetic and environmental influences on antisocial behavior: A meta-analysis of twin and adoption studies. *Psychol. Bull.* **128**, 490–529.

Rimel, R., Giordani, B., *et al.* (1981). Disability caused by minor head injury. *Neurosurgery* **9**, 1459–1461.

Rivera, J., Jaffee, K., *et al.* (1994). Family functioning and children's academic performance and behavior problems in the year following brain injury. *Arch. Phys. Med. Rehabil.* **75**, 369–379.

Roebuck, T. M., Mattson, S. N., *et al.* (1999). Behavioral and psychosocial profiles of alcohol-exposed children. *Alcohol. Clin. Exp. Res.* **23**, 1070–1076.

Rogeness, G. A., Cepeda, C., *et al.* (1990). Differences in heart rate and blood pressure in children with conduct disorder, major depression and separation anxiety. *Psychiatry Res.* **33**, 199–206.

Rolls, E. T. (2000). The orbitofrontal cortex and reward. *Cereb. Cortex* **10**, 284–294.

Rosen, G. M., Deinard, A. S., *et al.* (1985). Iron deficiency among incarcerated juvenile delinquents. *J. Adolesc. Health Care* **6**, 419–423.

Sakuta, A., and Fukushima, A. (1998). A study on abnormal findings pertaining to the brain in criminals. *Int. Med. J.* **5**, 283–292.

Santesso, D. L., Reker, D. L., *et al.* (2006). Frontal electroencephalogram activation asymmetry, emotional intelligence, and externalizing behaviors in 10-year-old children. *Child Psychiatry Hum. Dev.* **36**, 311–328.

Scerbo, A., Raine, A., *et al.* (1990). Reward dominance and passive avoidance learning in adolescent psychopaths. *J. Abnorm. Child Psychol.* **18**, 451–463.

Schmeck, K., and Poustra, F. (1993). Psychophysiologische Reacktionsmuster und psychische auffalligkeiten im kindesalter. *In* "Bioloogische Psychiatrie Der Gegenwart" (P. Baumann, ed.). Springer-Verlag, Wien.

Schoenthaler, S. J., and Bier, I. D. (2000). The effect of vitamin-mineral supplementation on juvenile delinquency among American schoolchildren: A randomized double blind placebo-controlled trial. *J. Altern. Complement. Med.* **6**, 19–29.

Schoenthaler, S. J., Amos, S. P., *et al.* (1997). The effect of randomized vitamin-mineral supplementation on violent and non-violent antisocial behavior among incarcerated juveniles. *J. Nutr. Environ. Med.* **7**, 343–352.

Schonfeld, A. M., Mattson, S. N., *et al.* (2005). Moral maturity and delinquency after prenatal alcohol exposure. *J. Stud. Alcohol* **6**, 19–29.

Shamay-Tsoory, S. G., Tomer, R., *et al.* (2005). Impaired 'affective theory of mind' is associated with right ventromedial prefrontal damage. *Cogn. Behav. Neurol.* **18**(1), 55–67.

Slutske, W. S., Heath, A. C., *et al.* (1997). Modeling genetic and environmental influences in the etiology of conduct disorder: A study of 2,682 adult twin pairs. *J. Abnorm. Psychol.* **106**, 266–279.

Soderstrom, H., Tullberg, M., *et al.* (2000). Reduced regional cerebral blood flow in non-psychotic violent offenders. *Psychiatry Res.* **98**, 29–41.

Soderstrom, H., Hultin, L., *et al.* (2002). Reduced frontotemporal perfusion in psychopathic personality. *Psychiatry Res.* **114**, 81–94.

Spreen, O., and Strauss, E. (1998). A Compendium of Neuropsychological Tests. 2nd edn., Oxford University Press, New York.

Sterzer, P., Stadler, C., *et al.* (2005). Abnormal neural responses to emotional visual stimuli in adolescents with conduct disorder. *Biol. Psychiatry* **57**, 7–15.

Sterzer, P., Stadler, C., *et al.* (2007). A structural neural deficit in adolescents with conduct disorder and its association with lack of empathy. *Neuroimage* **37**, 335–342.

Stevens, L., Zhang, W., *et al.* (2003). EFA supplementation in children with inattention, hyperactivity, and other disruptive behaviors. *Lipids* **38**, 1007–1021.

Streissguth, A. P., Barr, H. M., *et al.* (1996). Understanding the Occurrence of Secondary Disabilities in Clients with Fetal Alcohol Syndrome (FAS) and Fetal Alcohol Effects (FAE). Centers for Disease Control and Prevention, Washington, D.C.

Teichner, G., and Golden, C. J. (2000). The relationship of neuropsychological impairment to conduct disorder in adolescence: A conceptual review. *Aggress. Violent Behav.* 5(6), 509–528.

Tibbetts, S. G., and Piquero, A. R. (1999). The influence of gender, low birth weight, and disadvantaged environment in predicting early onset of offending: A test of Moffitt's interactional hypothesis. *Criminology* 37(4), 843–878.

Uhl, G. R., and Grow, R. W. (2004). The burden of complex genetics in brain disorders. *Arch. Gen. Psychiatry* 61, 223–229.

Volavka, J. (1987). Electroencephalogram among criminals. In "The Causes of Crime: New Biological Approaches" (S. A. Mednick, T. E. Moffitt, and S. Stack, eds.), pp. 137–145. Cambridge University Press, Cambridge.

Volkow, N. D., Tancredi, L. R., *et al.* (1995). Brain glucose metabolism in violent psychiatric patients: A preliminary study. *Psychiatry Res.* 61, 243–253.

Vollm, B., Richardson, P., *et al.* (2006). Serotonergic modulation of neuronal responses to behavioural inhibition and reinforcing stimuli: An fMRI study in healthy volunteers. *Eur. J. Neurosci.* 23, 552–560.

Wadsworth, M. E. J. (1976). Delinquency, pulse rate and early emotional deprivation. *Br. J. Criminol.* 16, 245–256.

Wakschlag, L. S., Lahey, B. B., *et al.* (1997). Maternal smoking during pregnancy and the risk of conduct disorder in boys. *Arch. Gen. Psychiatry* 54, 670–676.

Wakschlag, L. S., Pickett, K. E., *et al.* (2002). Maternal smoking during pregnancy and severe antisocial behavior in offspring: A review. *Am. J. Public Health* 92, 966–974.

Waldrop, M. F., Bell, R. Q., *et al.* (1978). Newborn minor physical anomalies predict short attention span, peer aggression, and impulsivity at age 3. *Science* 199, 563–564.

Werbach, M. (1995). Nutritional influences on aggressive behavior. *J. Orthomol. Med.* 7, 45–51.

Wertham, F. (1949). The Show of Violence. Doubleday & Company, Garden City, NY.

Wertham, F. (1950). Dark Legend. Doubleday & Company, Garden City, NY.

Whitman, T. A., and Akutagawa, D. (2004). Riddles in serial murder: A synthesis. *Aggress. Violent Behav.* 9, 693–703.

Wittels, F. (1937). The criminal psychopath in the psychoanalytic system. *Psychoanal. Rev.* 24(1), 276–291.

Wong, M. T., Fenwick, P. B., *et al.* (1997). Positron emission tomography in male violent offenders with schizophrenia. *Psychiatry Res.* 68, 111–123.

Yang, Y., Raine, A., *et al.* (2005). Volume reduction in prefrontal gray matter in unsuccessful criminal psychopaths. *Biol. Psychiatry* 57, 1103–1108.

Yang, Y., Glenn, A. L., *et al.* (2008). Brain abnormalities in antisocial individuals: Implications for the law. *Behav. Sci. Law* 26, 65–83.

Yang, Y., Raine, A., *et al.* (2009). Localization of deformities within the amygdala in individuals with psychopathy. *Arch. Gen. Psychiatry* 66, 986–994.

Index

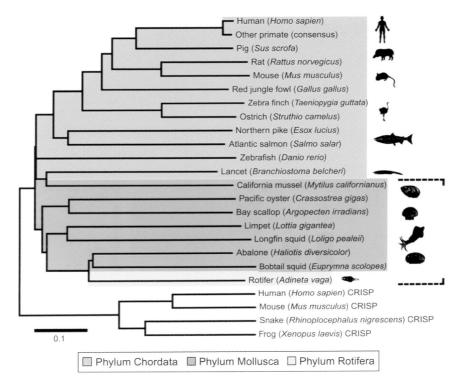

Chapter 3, Figure 3.6. (See Page 43 of this volume).

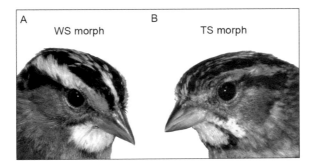

Chapter 5, Figure 5.5. (See Page 104 of this volume).